THE LAW, ECONOMICS AND POLITICS OF INTERNATIONAL STANDARDISATION

In an era of increased reliance on private regulatory bodies and globalised economic activity, standardisation is the field in which politics, technical expertise and strategic behaviour meet and interact. International standard-setting bodies exemplify the rise of transnational governance and the challenges that it brings in relating to institutional choice, legitimacy and procedural and substantive fairness or transparency. This book takes a more empirically based approach, focusing on the mechanics of international standard-setting. It constitutes a multidisciplinary inquiry into the foundations of international standard-setting, an empirically under-researched yet important area of international informal law-making. The contributors expertly examine the peculiarities of international standardisation in selected issue areas and legal orders and shed light on the attributes of international standard-setters, allowing comparisons among standard-setting bodies with a view to identifying best practices and improving our understanding of standardisation processes.

PANAGIOTIS DELIMATSIS is Professor of European and International Trade Law and Director of the Tilburg Law and Economics Center (TILEC) at Tilburg University, the Netherlands. He is a renowned expert on trade and regulatory issues relating to services industries, technical barriers to trade, financial regulation and energy. He has published extensively in top European and international refereed journals.

CAMBRIDGE INTERNATIONAL TRADE AND ECONOMIC LAW

As the processes of regionalisation and globalisation have intensified, there have been accompanying increases in the regulations of international trade and economic law at the levels of international, regional and national laws.

The subject matter of this series is international economic law. Its core is the regulation of international trade, investment and cognate areas such as intellectual property and competition policy. The series publishes books on related regulatory areas, in particular human rights, labour, environment and culture, as well as sustainable development. These areas are vertically linked at the international, regional and national level, and the series extends to the implementation of these rules at these different levels. The series also includes works on governance, dealing with the structure and operation of related international organisations in the field of international economic law, and the way they interact with other subjects of international and national law.

Books in the series:

The Law, Economics and Politics of International Standardisation
Panagiotis Delimatsis

The WTO and International Investment Law: Converging Systems
Jürgen Kurtz

Export Restrictions on Critical Minerals and Metals: Testing the Adequacy of WTO Disciplines
Ilaria Espa

Optimal Regulation and the Law of International Trade: The Interface between Societal Values and WTO Law
Boris Rigod

The Social Foundations of World Trade: Norms, Community and Constitution
Sungjoon Cho

Public Participation and Legitimacy in the WTO
Yves Bonzon

The Challenge of Safeguards in the WTO
Fernando Piérola

General Interests of Host States in International Investment Law
Edited by Giorgio Sacerdoti, Pia Acconci, Mara Valenti and Anna De Luca

The Law of Development Cooperation: A Comparative Analysis of the World Bank, the EU and Germany
Philipp Dann

WTO Disciplines on Subsidies and Countervailing Measures: Balancing Policy Space and Legal Constraints
Dominic Coppens

Domestic Judicial Review of Trade Remedies: Experiences of the Most Active WTO Members
Müslüm Yilmaz

International Organizations in WTO Dispute Settlement: How Much Institutional Sensitivity?
Marina Foltea

Public Services and International Trade Liberalization: Human Rights and Gender Implications
Barnali Choudhury

The Law and Politics of WTO Waivers: Stability and Flexibility in Public International Law
Isabel Feichtner

African Regional Trade Agreements as Legal Regimes
James Thuo Gathii

Liberalizing International Trade after Doha
David Gantz

Processes and Production Methods (PPMs) in WTO Law: Interfacing Trade and Social Goals
Christiane R. Conrad

Non-Discrimination in International Trade in Services: 'Likeness' in WTO/GATS
Nicolas Diebold

The Law, Economics and Politics of Retaliation in WTO Dispute Settlement
Edited by Chad P. Bown and Joost Pauwelyn

The Multilateralization of International Investment Law
Stephan W. Schill

Trade Policy Flexibility and Enforcement in the WTO: A Law and Economics Analysis
Simon A. B. Schropp

THE LAW, ECONOMICS AND POLITICS OF INTERNATIONAL STANDARDISATION

Edited by

PANAGIOTIS DELIMATSIS

CAMBRIDGE
UNIVERSITY PRESS

CAMBRIDGE
UNIVERSITY PRESS

University Printing House, Cambridge CB2 8BS, United Kingdom

Cambridge University Press is part of the University of Cambridge.

It furthers the University's mission by disseminating knowledge in the pursuit of education, learning and research at the highest international levels of excellence.

www.cambridge.org
Information on this title: www.cambridge.org/9781107128330

First published 2015

A catalogue record for this publication is available from the British Library

Library of Congress Cataloguing in Publication data
The law, economics and politics of international standardisation / edited by Panagiotis Delimatsis.
pages cm. – (Cambridge international trade and economic law)
This edited volume is a collection of the papers first presented at an international, multidisciplinary conference on international standardisation that took place in Tilburg, the Netherlands, on 7–8 November 2013, organized by the Tilburg Law and Economics Center (TILEC) – ECIP Acknowledgments.
Includes index.
ISBN 978-1-107-12833-0 (hardback)
1. Standardization – Law and legislation – Congresses. I. Delimatsis, Panagiotis, editor.
II. Tilburg Law and Economics Center, sponsoring body.
K3840.A6L39 2015
389'.6 – dc23 2015021706

ISBN 978-1-107-12833-0 Hardback
ISBN 978-1-107-57194-5 Paperback

CONTENTS

FIGURES

TABLES

CONTRIBUTORS

ALESSANDRA ARCURI, Associate Professor of International Economic Law at Erasmus School of Law, Erasmus University Rotterdam, the Netherlands

STEPHANIE BIJLMAKERS, PhD Candidate and Research Fellow at the Leuven Centre for Global Governance Studies, University of Leuven, Belgium

KNUT BLIND, Professor of Innovation Economics at Technische Universität Berlin and Fraunhofer Institute of Open Communication Systems FOKUS, Germany, and endowed Chair on Standardisation at Rotterdam School of Management, Erasmus University Rotterdam, the Netherlands

OLIVIER CADOT, Professor of Economics and Director of the Institute of Applied Economics at the University of Lausanne, Switzerland

GEERT VAN CALSTER, Professor of Law at University of Leuven, Belgium; King's College, London, United Kingdom; Monash University, Australia; and practising lawyer at the Brussels Bar

PANAGIOTIS DELIMATSIS, Professor of European and International Trade Law and Director of the Tilburg Law and Economics Center (TILEC) at Tilburg University, the Netherlands

ANGELOS DIMOPOULOS, Senior Lecturer of Law at Queen Mary University of London, United Kingdom

ANNE-CÉLIA DISDIER, Researcher at the Paris School of Economics (INRA), France

LIONEL FONTAGNÉ, Professor of Economics at the Paris School of Economics (University of Paris 1, Panthéon-Sorbonne) and CEPII, France

NEIL GANDAL, Professor of Economics at the Berglas School of Economics, Tel Aviv University, Israel, and Research Fellow at the Centre for Economic Policy Research (CEPR)

ELENI KOSTA, Associate Professor of Technology Regulation at the Tilburg Institute of Law, Technology and Society (TILT) at Tilburg University, the Netherlands

PIERRE LAROUCHE, Professor of Competition Law at Tilburg University, the Netherlands, and Visiting Professor at the College of Europe, Bruges

HANS LINDAHL, Professor of Legal Philosophy at Tilburg University, the Netherlands, and Research Fellow of the Stellenbosch Institute of Advanced Study, South Africa

JUAN A. MARCHETTI, Counsellor at the Trade in Services Division of the World Trade Organization (WTO), Geneva, Switzerland

AXEL MARX, Deputy Director of the Leuven Centre for Global Governance Studies at University of Leuven, Belgium

WALTER MATTLI, Professor of International Political Economy, Department of Politics and International Relations and Fellow of St John's College, University of Oxford, United Kingdom

PETROS C. MAVROIDIS, Edwin B. Parker Professor of Foreign and Comparative Law at Columbia Law School, New York, United States of America and Professor of Law at the University of Neuchâtel, Switzerland (on leave at European University Institute, Florence, Italy)

GEERTRUI VAN OVERWALLE, Professor of Intellectual Property Law at the University of Leuven, Belgium, and Visiting Professor at Tilburg University, the Netherlands

AWILO OCHIENG PERNET, Chair of the Codex Alimentarius Commission, Rome/Geneva, and Advisor at the Federal Food Safety and Veterinary Office (FSVO), Switzerland

PIERRE RÉGIBEAU, Vice President at Charles River Associates and Visiting Research Fellow at Imperial College London, United Kingdom

HARM SCHEPEL, Professor of Economic Law at the University of Kent, United Kingdom

JACK SEDDON, PhD Candidate in the Department of Politics and International Relations and Member of St John's College, University of Oxford, United Kingdom

KEES STUURMAN, Professor of Law at the Tilburg Institute for Law, Technology and Society (TILT), Tilburg University, the Netherlands

HENK J. DE VRIES, Associate Professor of Standardisation at Rotterdam School of Management, Erasmus University Rotterdam, the Netherlands, and President of the European Academy for Standardisation (EURAS)

JAN WOUTERS, Professor of International Law and the Law of International Organisations and Jean Monnet Chair at University of Leuven, Belgium

ACKNOWLEDGEMENTS

This edited volume is a collection of the papers first presented at an international, multidisciplinary conference on international standardisation that took place in Tilburg, the Netherlands, on 7–8 November 2013. The conference was organised by the Tilburg Law and Economics Center (TILEC), a Center of Excellence at Tilburg University active in the interdisciplinary study of economic activity, on the occasion of its tenth anniversary. In line with TILEC tradition, a mix of experts, academics, policy officials, practitioners and regulators joined to assist in asking the right questions and addressing the issues identified as thoroughly as possible. The main objective of the conference was to improve our knowledge about and understanding of standard-setting processes in various legal orders and organisations, whether public, private or hybrid. Of particular interest for the organisers was the identification of any common patterns in the mechanics of standards development in the areas under discussion; the conditions and circumstances that may change the dynamics in some issue areas but not in others; and the traits of the new generation of standard-setters and the standards they produce in such diverse areas as finance, sustainable manufacturing or the information and communication technology markets.

TILEC has been generating cutting-edge research in the areas of intellectual property, standardisation and innovation at the borders of law and economics since 2006, which became possible due to multiyear sponsorship agreements that TILEC signed with Qualcomm and Microsoft. The cooperation with Qualcomm in particular has allowed TILEC researchers to have a better grasp of industry's concerns in the standard-setting processes. The research programme associated with these agreements is always executed in line with the rules of academic independence of the Royal Academy of Arts and Sciences.

This volume would not have seen the light of the day without the dedication, patience and relentless efforts of its contributors who had to revise their papers several times to serve the coherence of this book.

For their commitment beyond the call of duty since the first day the idea of such a conference was brought forward, I am grateful to Ilse Streng, TILEC's administrative director, and Maartje van Genk, TILEC's event coordinator. For his support and collegiality, I also thank Cédric Argenton, TILEC director on the side of economics.

I thank the three anonymous referees for their insightful comments. I am grateful to Cambridge University Press and Kim Hughes for their professionalism and helpful guidance throughout the publication process. I feel honoured that Thomas Cottier and Lorand Bartels accepted this volume to include in the Cambridge International Trade and Economic Law series. This volume is dedicated to Despina, Melina and Lydia Elodie as a minor attempt to give something in return for their unconditional love, patience and encouragement in times of doubt.

Panagiotis Delimatsis
Tilburg, March 2015

ABBREVIATIONS

AB	Appellate Body
ACP	African, Caribbean, and Pacific States
ADB	Asian Development Bank
ADI	Acceptable Daily Intake
AFNOR	Association Française de Normalisation
AFTA	ASEAN Free Trade Area
AIP	Apparel Industry Partnership
ALADI	Association Latino-Américaine d'Intégration
ANSI	American National Standards Institute
APEC	Asia-Pacific Economic Cooperation
ASC	Accredited Standards Committee
ASEAN	Association of South-East Asian Nations
ASME	American Society of Mechanical Engineers
ATM	Automated teller machine
ATS	Alien Tort Statute
AVBWasserV	Verordnung über Allgemeine Bedingungen für die Versorgung mit Wasser
BCBS	Basel Committee on Banking Supervision
BEREC	Body of European Regulators for Electronic Communications
BGH	Bundesgerichtshof
BGHZ	Entscheidungen des Bundesgerichtshofs in Zivilsachen
BIS	Bank for International Settlements
BITs	Bilateral investment treaties
BSI	British Standards Institution
CAC	Codex Alimentarius Commission
CACM	Central American Common Market
CAFTA	Central America FTA
CCEXEC	Codex Executive Committee
CCFA	Codex Committee on Food Additives
CCGP	Codex Committee on General Principles
CCM	Cloud Controls Matrix

CCPs	Central Counterparties
CCRVDF	Codex Committee on Residues of Veterinary Drugs in Foods
CCSL	Cloud Certification Schemes List
CEBR	Centre for Economic and Business Research
CEN	Comité Européen de Normalisation
CEN/ISSS/CENELEC	Comité Européen de Normalisation Électrotechnique
CEO	Chief executive officer
CEPII	Centre d'Études Prospectives et d'Informations Internationales
CEPT	Common Effective Preferential Tariff
CESR	Committee of European Securities Regulators
CFTC	Commodity Futures Trading Commission
CGF	Consumer Goods Forum
CGFS	Committee on the Global Financial System
CICICFTC	Interim Compliant Identifier
CIE	Centre for International Economics
CISI	Chartered Institute for Securities and Investment
CJEU	Court of Justice of the European Union
CoC	Chain-of-custody
COMESA	Common Market for Eastern and Southern Africa
COMTRADE	United Nations Commodity Trade Statistics Database
COOL	Certain Country of Origin Labeling
COPOLCO	ISO Committee on Consumer Policy
CPSS	Committee on Payment and Settlement Systems
CSA	Cloud Security Alliance
CSD	Central securities depository
CSR	Corporate social responsibility
CTF	Codex Trust Fund
CUP	China UnionPay
CWA	CEN Workshop Agreement
DAPIX	Working Party on Information Exchange and Data Protection
DC	Developing countries
DFID	Department for International Development
DG	Directorate General of European Commission
DIIS	Danish Institute for International Studies
DIN	Deutsches Institut für Normung
DIS	Draft International Standard
DIVX	Digital Video Express
DMS	Database Management Software
DNT	Do not track
DRAM	Dynamic random-access memory

DS	Dispute settlement
DSU	Understanding on Rules and Procedures Governing the Settlement of Disputes
DTCC	Depository Trust and Clearing Corporation
DTI	UK Department of Trade and Industry
DVGW	Deutsche Vereinigung des Gas – und Wasserfaches
EAC	East African Community
EC	European Community
ECB	European Central Bank
ECJ	European Court of Justice
ECN	European Competition Network
ECOFIN	Economic and Financial Affairs Council
ECR	European Court Reports
EEA	European Economic Area
EEA	The European Environment Agency
EFSA	European Food and Safety Authority
EFTA	European Free Trade Association
EIOPA	European Insurance and Occupational Pensions Authority
EIPR	European Intellectual Property Review
EITI	Extractive Industries Transparency Initiative
ENISA	European Union Agency for Network and Information Security
EP	European Parliament
EPO	European Patent Office
EPS	Electronic payment services
ERGEG	European Regulators Group for Electricity and Gas
ESMA	European Securities and Markets Authority
ESME	European Securities Markets Expert Group
ESO	European Standardisation Organisation
ETNO	European Telecommunications Network Operators Association
ETRI	Electronics and Telecommunications Research Institute
ETSG	European Trade Study Group
ETSI	European Telecommunications Standards Institute
EU	European Union
EUI	European University Institute
EURAS	European Academy for Standardization
FAO	Food and Agriculture Organization
FATF	Financial Action Task Force on Money Laundering
FDA	US Food and Drug Administration
FDI	Foreign Direct Investment
FDIS	Final Draft International Standard

FEEDAP	EFSA Panel on Additives and Products or Substances used in Animal Feed
FESCO	Forum of European Securities Commissions
FIN-USE	Forum of User Experts in the Area of Financial Services
FIX	Financial Information Exchange
FLA	Fair Labour Association
FLO	Fairtrade Labelling Organization
FOKUS	Fraunhofer Institute for Open Communication Systems
FpML	Financial Products Mark-up Language
FRAND	Fair, Reasonable, and Non-Discriminatory
FSAP	Financial Sector Assessment Program
FSAP	Financial Services Action Plan
FSB	Financial Stability Board
FSC	Forest Stewardship Council
FSF	Financial Stability Forum
FSSAs	Financial Sector Stability Assessments
FTC	US Federal Trade Commission
FVO	Food and Veterinary Office
G20	Group of Twenty
GAP	Good Agricultural Practices
GATS	General Agreement on Trade in Services
GATT	General Agreement on Tariffs and Trade
GDP	Gross Domestic Product
GDPR	General Data Protection Regulation
GEF	Global Environmental Facility
GFSI	Global Food Safety Initiative
GHP	Good Hygienic Practices
GiZ	Deutsche Gesellschaft für Internationale Zusammenarbeit
GMOs	Genetically modified organisms
GMP	Good Manufacturing Practices
GPL	General public license
GRI	Global Reporting Initiative
HACCP	Hazard Analysis and Critical Control Point
HIPAA	Health Insurance Portability and Accountability Act
HS	Harmonized System
HTML	Hyper Text Markup Language
IAASB	International Auditing and Assurance Standards Board
IADI	International Association of Deposit Insurers
IAF	International Accreditation Forum
IAIS	International Association of Insurance Supervisors
IASB	International Accounting Standards Board
IASC	International Accounting Standards Committee

IATA	International Air Transport Association
ICSID	International Centre for Settlement of Investment Disputes
ICT	Information and Communication Technology
IEC	International Electrotechnical Commission
IEEE	Institute of Electrical and Electronics Engineers
IETF	Internet Engineering Task Force
IFAC	International Federation of Accountants
IFIs	International Finance Institutions
IFRS	International Financial Reporting Standards
IGO	Inter-Governmental Organisation
IIA	International Investment Agreement
IIED	International Institute for Environment and Development
IISD	International Institute for Sustainable Development
ILO	International Labour Organization
IMF	International Monetary Fund
IMO	International Maritime Organization
IMRL	International Maximum Residue Level
IMS	Identity Management Systems
IOP	Interoperability, Data Portability, and Reversibility
IOSCO	International Organisation of Securities Commissions
IP	Intellectual property
IPC	International Patent Classification
IPPC	International Plant Protection Convention
IPRs	Intellectual Property Rights
IPSE	Initiative for Privacy Standardisation in Europe
ISEA	Inter-Society for the Electronic Arts
ISEAL	International Social and Environmental Accreditation and Labeling
ISIN	International Securities Identification Number
ISO	International Organisation for Standardisation
ISO SC	ISO subcommittees
ISSB	International Standard-Setting Body
ISSS	Information Society Standardisation System
ISTPA	International Security, Trust, and Privacy Alliance
IT	Information Technology
ITC	International Trade Centre
iTIP	WTO Trade Intelligence Portal
ITU	International Telecommunication Union
IWGDPT	International Working Group on Data Protection in Telecommunications
JECFA	Joint WHO/FAO Expert Committee on Food Additives
JEDEC	Joint Electron Device Engineering Council

JEMRA	Joint FAO/WHO Expert Meetings on Microbiological Risk Assessment
JISC	The Japanese Industrial Standards Committee
JMPR	Joint FAO/WHO Meetings on Pesticides Residues
JORF	Journal Officiel de la République Française
JTC	Joint Technical Committee
KPCS	United Nations Kimberly Process Certification Scheme
LDCs	Least developed countries
LEI	Legal Entity Identifier
LIBE	Committee on Civil Liberties, Justice, and Home Affairs
MAST	Multi-Agency Support Team
MIGA	Multilateral Investment Guarantee Agency
ML	Maximum limits
MNC	Multinational company
MNE	Multinational enterprise
MoU	Memorandum of Understanding
MRL	Maximum Residue Level
MSC	Marine Stewardship Council
MSS	Management Systems Standardization
NAFTA	North American Free Trade Agreement
NAMA	Non-Agricultural Market Access
NASA	National Aeronautics and Space Administration
NASDAQ	National Association of Securities Dealers Automated Quotations
NBSIR	National Bureau of Standards Information Report
NCBA	US National Cattlemen's Beef Association
NGO	Non-Governmental Organization
NIS	Network and Information Security
NIST	National Institute of Standards and Technology
NP	New Work Item Proposal
NPEs	Non-Producing Entities
NRC	US National Research Council
NTBMM	NTB Monitoring Mechanism
NTBs	Non-Tariff Barriers
NTMs	Non-Tariff Measures
OASIS	Organization for the Advancement of Structured Information Standards
OCF	Open Certification Framework
ODCA	Open Data Center Alliance
OECD	Organisation for Economic Co-operation and Development
OIE	World Organization for Animal Health
OJ	Official Journal of the European Union

OTC	Over-the-counter
P3P	Platform for Privacy Preferences
PDCA	'Plan-Do-Check-Act'
PEFC	Program for the Endorsement of Forest Certification
PET	Privacy-enhancing technology
PIA	Privacy impact assessment
PIN	Personal identification number
PIOB	Public Interest Oversight Board
PMS	Privacy Management Systems
PPMs	Production and Process Methods
PSE	Paris School of Economics
PTA	Preferential Trade Agreement
QRs	Quantitative restrictions
R&D	Research and development
RAND	Reasonable and non-discriminatory
RASFF	Rapid Alert System for Food and Feed
RFID	Radio-frequency identification
RMB	Renminbi
ROC	Regulatory Oversight Committee
ROSCs	Reports on the Observance of Standards and Codes
RTA	Regional trade agreement
S&DT	Special and Differential Treatment
SADC	South African Development Community
SAFTA	South Asian Free Trade Area
SAGE	Strategic Advisory Group on the Environment
SAGISO	Strategic Advisory Group on Social Responsibility
SAI	Social Accountability International
SC	Supreme Court of the United States
SCORE	Stockholm Venter for Organizational Research
SCVPH	Scientific Committee on Veterinary Measures relating to Public Health
SD	Sustainable development
SDO	Standards developing organisation
SEC	US Securities and Exchange Commission
SEP	Standard-Essential Patent
SI	International System of Units
SLA	Service level agreements
SMEs	Small and medium-sized enterprises
SPS	Agreement on the Application of Sanitary and Phytosanitary Measures
SR	Social responsibility
SRI	Socially responsible investment

SSB	Standard-setting body
SSO	Standard-setting organisation
STDF	Standards and Trade Development Facility
STP	Straight-through processing
STS	Science, Technology, and Society
SWIFT	Society for Worldwide Interbank Financial Telecommunication
TBT	Agreement on Technical Barriers to Trade
TC	ISO Technical Committee
TFEU	Treaty on the Functioning of the European Union
TIA	Telecommunications Industry Association
TILEC	Tilburg Law and Economics Center
TMB	ISO Technical Management Board
TMCMSADC	Trade Monitoring and Compliance Mechanism
TNC	Transnational corporation
TPP	Trans-Pacific Partnership
TPRC	Telecommunications Policy Research Conference
TQM	Total quality management
TRs	Technical regulations
TTIP	Transatlantic Trade and Investment Partnership
UKFSA	UK Food Safety Act
UN	United Nations
UNCITRAL	United Nations Commission on International Trade Law
UNCTAD	United Nations Conference on Trade and Development
UNECE	UN Economic Commission or Europe
UNEP	United Nations Environment Programme
UNFSS	United Nations Forum on Sustainability Standards
UNIDO	United Nations Industrial Development Organization
USDA	US Department of Agriculture
VCLT	Vienna Convention on the Law of Treaties
VSS	Voluntary Sustainability Standards
VTDF	Voluntary Technology Dialogue Framework
W-CDMA	Wideband Code Division Multiple Access
W3C	World Wide Web Consortium
WB	World Bank
WEF	World Economic Forum
WG	ISO Working Group
WHA	World Health Assembly
WHO	World Health Organization
WR	ISO Working Report
WS/DPP	CEN/ISSS Data Protection and Privacy Workshop
WSC	World Standards Cooperation

WSSD	World Summit on Sustainable Development
WTO	World Trade Organization
WTOA	Agreement Establishing the World Trade Organization
WWF	World Wildlife Fund
XBRL	eXtensible, Business Reporting language
XML	Extensible Markup Language

Introduction: continuity and change in international standardisation

PANAGIOTIS DELIMATSIS

The objective of this volume is to explore the phenomenon of international standardisation by shedding light on standard-setting processes and the development of standards in several legal orders and subject matters. The contributors bring to light an area of international rule-making that has had an increasingly important impact on the production and distribution of, and trade in, goods and services globally. The contributions chosen for inclusion in this volume critically review standardisation processes and actors in various public, private and hybrid standard-setting bodies (SSBs)[1] and identify potential problematic instances in terms of openness, transparency or inclusiveness.

The idea for this volume came out of the recognition that although standards increase trade and facilitate the functioning of global supply chains, we still know little about international standardisation processes and their properties. An ever-increasing body of scholarship in economics and management touches on standard-setting processes, notably by focusing on the creation of particular standards, but legal scholarship has not paid due attention to the complex phenomena that lead to the creation of technical standards. Questions relating to institutional structures and decision-making processes are rather underexplored to date. Although theories on transnational, private-driven law abound today,[2] empirical

[1] In this volume, different terms are used to identify entities that set standards. In the economics literature, such entities are typically called 'standard-setting organisations'. For our purposes, we adopt the broader concept of 'body' rather than 'organisation'. However, the terms are used interchangeably.

[2] See also A. Peters, L. Koechlin, T. Förster and G.F. Zinkernagel (eds), *Non-State Actors as Standard Setters* (Cambridge University Press, 2009).

evidence is still scarce with respect to the mechanics of transnational standard-setting. This volume fills this gap and improves our knowledge about the peculiarities and drivers of standard-setting as a highly important area of transnational rule-making.

More specifically, this volume contributes to the international economic law literature by discussing the process of developing standards in selected issue areas; the institutional structures; the dynamics *within* and *among* SSBs; the interaction of standard-setters at the national, regional and international levels; the relationship between standard-*makers* (such as the International Organisation for Standardisation [ISO] and the Codex Alimentarius Commission) and standard-*takers* (such as the World Trade Organization [WTO]) and the effects on trade, development or innovation. The contributors were requested to reflect on and identify any common patterns in the mechanics of developing standards in the issue areas under discussion; to pinpoint particular concerns that change the dynamics (and ultimately the traits of the adopted standard) in some issue areas but not in others but also to analyse the characteristics of the new generation of standard-setters (in particular, private standard-setting in the financial, sustainable production and information and communication technology [ICT] markets).

For this volume, we have invited scholars from law, economics and management but also policy makers as well as officials of regulatory authorities and international organisations, including SSBs. Contributors were carefully selected to include both renowned experts in the field of standardisation, trade and innovation, and younger, enthusiastic scholars with a genuine willingness to delve into the peculiarities of international standard-setting, both conceptually and empirically.

The contributions included in this volume were presented at an international multidisciplinary conference on international standardisation held in Tilburg, the Netherlands, in November 2013. The event was organised on the occasion of the tenth anniversary of the Tilburg Law and Economics Center (TILEC), the largest interdisciplinary research centre of its kind in Europe, renowned for its high-quality interdisciplinary work on the governance of economic activity. Research on the mechanics of standard-setting covers a substantial part of TILEC's research programme on trade, finance and investment for the years 2012–2017.

A. Some introductory thoughts on international standard-setting

I. Standard-setting as alternative, informal (transnational) rule-making

Increased international standardisation exemplifies achievements in technological innovation, the expansion of global trade, the interest of producers in reducing compliance costs and ensuring interoperability; and this without compromising incentives for innovation and the increased attention paid to social and sustainability issues due to the demand of consumers for better and safer products.

The conventional theory of international standardisation would be as follows: standards, no matter how well-crafted, can still impede trade. This is mainly because standards reflect preferences and values of a given society that may – and usually do – diverge, thereby inflating compliance costs for companies.[3] If developed internationally, then substantial gains can be made through the diminution of such costs and by addressing network externalities, interoperability issues and information asymmetries.[4] Thus, and quite inevitably, the main locus of standardisation is to be found outside national borders.[5] With the emergence of global supply chains, the importance of international standards increases, suggesting that compatibility standards of high quality can yield substantial network effects that can make such standards self-enforcing.[6] However, the reduction of compliance costs may be only a long-term effect; in the short run, the effect of international standards may vary in that compliance costs will rise for some firms because the new standard used may be more sophisticated. At the same time, an international standard, the theory suggests, would bring about a diminution of consumer costs as information becomes more readily available and prices more readily comparable.[7]

[3] R. Staiger and A. Sykes, 'International Trade, National Treatment and Domestic Regulation', 40 *Journal of Legal Studies* (2011) 149.

[4] See WTO, *World Trade Report 2005.*

[5] T. Büthe and W. Mattli, 'Setting International Standards – Technological Rationality or Primacy of Power ?' 56 *World Politics* (2003) 1; K. Tamm Hallström, *Organizing International Standardization – ISO and the IASC in Quest of Authority* (Edward Elgar, 2004).

[6] World Economic Forum (WEF), *The Global Enabling Trade Report 2012 – Reducing Supply Chain Barriers*, 2012.

[7] See WTO, *World Trade Report 2012*, p. 136.

The globalisation of markets and law, together with the concomitant collapse of the rigid public–private divide,[8] are particularly apparent in and constantly transform standard-setting institutions.[9] This is a reflection of transformations taking place domestically and affecting legal ordering beyond borders. Modern states have resolutely conceded part of their powers to other actors that can act more efficiently and swiftly not only at the domestic level but increasingly at the transnational level as well; the world of standard-setting is a straightforward reflection of this shift.[10]

Contextually, international standardisation is part of the undisputed rise in transnational law-making as international SSBs become more private actor–driven. Such transnational regimes have often evolved in a vacuum, avoiding any frontal confrontation with state law. In the case of standard-setting at the international level, whereas there are private-driven SSBs, in practice, state-driven actors such as regulators or government-sponsored bodies may develop partnerships – be these formal or informal – with private actors to generate what can be termed 'informal law'.[11] Just to complicate the picture, international SSBs often compete with each other, leading to purely private constellations claiming authority through their standard-setting processes. It is not only the informal character of the actors; informality extends to the output: whereas domestic standardisation can encompass both binding and voluntary technical standards and specifications, international standardisation typically involves standards with which compliance is *voluntary*.[12]

Standardisation has escaped public scrutiny for a long time mainly because of its voluntary nature. Absent any backing by a state's coercive power, voluntary standardisation was indifferent for many. This was the case even if standards display public good characteristics in the sense that, once produced, one cannot totally limit the use of such standards.

Standardisation is emblematic of the increasing complexity in defining exactly the confines of 'law'. Standard-setters function within a continuum 'between state and private, between norm and regulation, between

[8] C. Cutler, *Private Power and Global Authority – Transnational Merchant Law and the Global Political Economy* (Cambridge University Press, 2003).

[9] W. Hesser and A. Inklaar, *An Introduction to Standards and Standardisation* (Beuth Verlag, 1998).

[10] N. Brunsson and B. Jacobsson (eds), *A World of Standards* (Oxford University Press 2000).

[11] Cf. J. Pauwelyn, R. Wessel and J. Wouters (eds), *Informal International Lawmaking* (Oxford University Press, 2012).

[12] See C. Murphy and J. Yates, *The International Organization for Standardization (ISO) – Global Governance through Voluntary Consensus* (Routledge, 2009).

precursor and holding action. As use of standards increases, their place along any axis of these continuums also changes.[13] Standardisation is a quasi-legal form of self-regulation and, depending on the circumstances and the legal context, it can be a form of co-regulation, or else a (hybrid) public–private partnership.[14] Some regional standardisation bodies such as the European Committee for Standardisation (CEN) fall under the latter category.[15]

Standardisation serves a complementary function to traditional command-and-control regulation.[16] Therefore, non-binding, or 'soft', norms such as standards can act as gap-fillers for 'harder' forms of law. This derives from the acknowledgement that, in an increasingly complex and diversified world with an inconceivable pace of technological progress, the traditional state does not have the resources or the savvy to regulate efficiently in all areas. Modern states concede part of their powers to other actors that can act more effectively and swiftly, mainly because of their expertise,[17] thereby allowing non-state voices to be heard and accordingly reshuffling its regulatory behaviour and supervisory role (for instance, by focusing to *ex post* control of a certain activity). Nowadays, 'cooperation' is the buzzword that embodies the philosophy of the modern state.[18]

The advantages of cooperation and 'soft' forms of regulation transform states into catalysts, coordinators ('orchestrators') and supporters of certain activities at the national or transnational level.[19] More often than

[13] N. Roht-Arriaza, '"Soft Law" in a "Hybrid" Organization: The International Organization for Standardisation' in D. Shelton (ed), *Commitment and Compliance – The Role of Non-Binding Norms in the International Legal System* (Oxford University Press, 2000), 263; and D. Vogel, 'Private Global Business Regulation', 11 *Annual Review of Political Science* (2008) 261.

[14] N. Brunsson and B. Jacobsson (eds), *A World of Standards* (Oxford University Press, 2000).

[15] F. Cafaggi and H. Muir Watt, *The Regulatory Function of European Private Law* (Edward Elgar, 2009).

[16] As Schepel puts it, 'regulators count on [private standards], markets cannot function without them'. See H. Schepel, 'Constituting Private Governance Regimes: Standards Bodies in American Law', in C. Joerges, I.-J. Sand and G. Teubner (eds), *Transnational Governance and Constitutionalism* (Hart, 2004), 164.

[17] L. Senden, *Soft Law in European Community Law* (Hart, 2004).

[18] J. Jackson, *Sovereignty, the WTO and Changing Fundamentals of International Law* (Cambridge University Press, 2006); also H. Spruyt, *The Sovereign State and Its Competitors* (Princeton University Press, 1994).

[19] K. Abbott and D. Snidal, 'The Governance Triangle: Regulatory Standards Institutions and the Shadow of the State' in W. Mattli and N. Woods (eds), *The Politics of Global Regulation* (Princeton University Press, 2009); *but* see Mattli and Seddon's contribution in this volume (Chapter 8).

not, states are rule-*takers* in the area of standardisation. An underlying assumption appears to be that the voluntary character of a standard (i.e. that compliance is voluntary) will ensure that the best available technology will be adopted in the standard. In other words, the industry as a self-maximising entity will necessarily converge toward a standard that is best for it and, by implication, for society overall. In certain instances, standards acquire legal status through reference in official legal instruments, which, again, points to a peculiar, uneven relationship between the state and non-state actors involved in all stages of standard-setting.[20] Again, this approach is consistent with the premises of technological rationality[21] – a sort of technocratic legitimacy – and is considered the result of low sovereignty costs for governments that such delegation of power entails.[22]

As high levels of technical expertise are warranted, states prefer to adopt a hands-off approach vis-à-vis more specialised private parties, which have the necessary savvy to produce – sometimes complex – technical standards. This approach seems to be negatively correlated with the complexity and fast-changing nature of a given sector. In other words, the more complex a given area of standardisation is, the more likely a hands-off approach will be chosen. As a result, standard-setting consortia in highly volatile and constantly changing areas such as ICT are mushrooming. However, a deferential stance may not always constitute good politics.[23] For instance, under certain circumstances, it may be worrisome if this type of soft law pre-empts hard forms of law, which may be justifiably more intrusive, seeking higher levels of protection.[24]

II. Standard-setting, standards battles and strategic behaviour

SSBs are important coordination devices that can lead, at least in theory, to substantial efficient outcomes. Indeed, because there are manifold

[20] See Regulation 1025/2012 of the European Parliament and the Council on European standardisation [2012] L 316/12.

[21] L. Cabral and T. Kretschmer, 'Standards battles and public policy' in S. Greenstein and V. Stango (eds), *Standards and Public Policy* (Cambridge University Press, 2007); also T. Loya and J. Boli, 'World Polity: Technical Rationality over Power' in J. Boli and G. Thomas (eds), *Constructing World Culture – International Nongovernmental Organizations since 1875* (Stanford University Press, 1999).

[22] See K. Abbott and D. Snidal, 'Hard and Soft Law in International Governance', 54:3 *International Organization* (2000) 421, at 441.

[23] A. Feenberg, *Questioning Technology* (Routledge, 1999).

[24] Roht-Arriaza (above n. 13), p. 272.

technological approaches, an SSB offers a forum where competitors and competing vested interests can resolve conflicts.[25] Given the importance of such standardisation fora, substantial financial resources and efforts can be invested in them.[26] The increase of standards-related patent disputes and the emergence of industry-sponsored consortia, but also actions against anticompetitive practices within SSBs such as patent hold-up or royalty stacking – all these are indicative of the significance of standardisation today.[27] As a result, public authorities have recently renewed their attention to standard-setting practices.[28]

In standard-setting, firms that have launched alternative technologies typically compete for inclusion in a given standard that may consolidate the best available technology.[29] In practice, competition not only among firms but also among standard-setting groups constitutes a typical feature of private standard-setting. Standard-setting groups compete in offering the most attractive institutional setting for the development of standards. However, in many cases, strategic behaviour and market power will still not prevail over technological strength, which may be due to the dynamic process that characterises standardisation.[30] What will many times determine the choice of forum (i.e. SSB) is whether for a given firm the possibility to dictate a standard carries more weight in its standards-related behaviour than the reputation of a given SSB – that is, whether reputation costs are lower than the benefits of dictating the standard in a second-best standard-setting scheme. This choice will essentially depend on the size of the market and the attractiveness of the technology.[31] Thus, SSBs are important not only as coordination devices but also as certification agents.

[25] See, generally, S. Greenstein and Stango (eds), *Standards and Public Policy* (Cambridge University Press, 2006).

[26] See also J. Farrell and T. Simcoe, 'Choosing the Rules for Consensus Standardisation', 43:2 *RAND Journal of Economics* (2012) 235.

[27] See M. Lemley, 'Ten Things to Do about Patent Holdup of Standards (and One Not To)', 48:1 *Boston College Law Review* (2007) 149. See also Larouche and van Overwalle's contribution to this volume (Chapter 15).

[28] US Federal Trade Commission, *The Evolving IP Marketplace: Aligning Patent Notice and Remedies with Competition*, 2011.

[29] D. Spulder, 'Innovation Economics: The Interplay among Technology Standards, Competitive Conduct, and Economic Performance', 9:4 *Journal of Competition Law and Economics* (2013) 777, at 784.

[30] See B.J. DeLacey, K. Herman, D. Kiron and J. Lerner, 'Strategic Behavior in Standard-Setting Organizations', Harvard NOM Working Paper, 2006.

[31] J. Lerner and J. Tirole, 'A Model of Forum Shopping', 96 *American Economic Review* (2006) 1091.

As insinuated earlier, although technological rationality and technical strength are important factors in standardisation activities, standardisation is nonetheless a highly politicised process.[32] Many times, (voluntary) standards are the precursors of domestic (mandatory) technical regulations. Even though standards are adopted mainly through soft-law processes by non-state actors, these actors aspire to capitalise on their success and see the initially non-binding norms they champion transformed into hard law to gain rents from first mover advantages through expedited enforcement.[33] Thus, because the stakes are high, strategic behaviour is sometimes observable. The more important standardisation becomes, the fiercer is the competition for increased influence in SSBs.[34] Empirical evidence, although limited and rather scattered, suggests that power politics and regulatory capture may be endemic in certain SSBs.[35] This pattern is reminiscent of the regulatory capture theory and the associated doctrine of special interest groups, extensively discussed in social sciences, which has traditionally been associated with the function of the *state* and *public* authorities.[36]

The beneficial effects of standardisation can be undermined if, for instance, SSBs cannot resist market power or do not have the mechanisms in place that would allow taking into account important societal values. In addition, standardisation in certain fora displays high levels of path dependency whereby a typically small group of parties dominates the standardisation process and advocates for a particular, well-established method of behaving, coordinating and decision-making. This can be problematic, as it may undermine the effectiveness and inclusiveness of standard-setting processes. It can also discourage new entrants from joining these fora. Thus, despite its voluntary nature and consensus-driven decision-making, standardisation may lead to exclusion of certain economic actors and create undesirable anticompetitive effects.

[32] T. Büthe and W. Mattli, *The New Global Rulers: The Privatization of Regulation in the World Economy* (Princeton University Press, 2011); J. Swinnen and T. Vandemoortele (2012), 'Trade and the Political Economy of Standards', 11:3 *World Trade Review* (2012) 390; P. Marquez, 'Standardisation and Capture: The Rise of Standardisation in International Industrial Regulation and Global Administrative Law', 7:3 *Global Jurist* (2007) Article 5.

[33] M. O'Connell, 'The Role of Soft Law in a Global Order' in D. Shelton (ed) (above n. 13), 100.

[34] S. Besen and J. Farrell, 'Choosing How to Compete: Strategies and Tactics in Standardisation', 8:2 *Journal of Economic Perspectives* (1994) 117.

[35] D. Drezner, *All Politics Is Global: Explaining International Regulatory Regimes* (Princeton University Press, 2007); also J. Braithwaite and P. Drahos, *Global Business Regulation* (Cambridge University Press, 2000).

[36] G. Grossman and E. Helpman, *Special Interest Politics* (MIT Press, 2001).

III. *The business model of SSBs and the WTO*

The current business model in various SSBs can in fact widen the divide between those parties that have achieved high levels of technical sophistication and those that try to catch up. Alternatively, it may lead to a fragmented landscape for standard-setting in which parties that are unhappy with particular processes in SSBs adopt 'exit strategies' and decide to create other, competing fora or consortia in an attempt to impose their own standards.[37] Both situations are suboptimal: they result in unduly higher costs for certain economic actors and affect future patterns in trade, development and innovation.

At the international level, whereas the principle of consensus is one of the pillars of international standardisation, this does not mean that inclusiveness and representativeness are ensured in the standard-setting processes used. Often, procedural and substantive guarantees are missing from the matrix of rules (both written and unwritten) within which SSBs function. Thus, a grey area of legality in international law is created that one cannot neglect, particularly when viewed through the lens of the objective of promoting inclusiveness in international bodies with a view to progressively shortening the development divide between developed and developing countries.

This became even clearer after the entry into force of the WTO agreement in the mid-1990s that required countries that until then had abstained from standardisation activities to use international standards when they exist, thereby transforming voluntary standards into de facto mandatory ones. In early WTO cases, international standards developed outside the WTO were used to defeat national regulatory autonomy, thereby transforming voluntary standards into quasi-mandatory ones. In this way, the WTO became a prominent 'promoter' of international standards, shying away from examining the very legitimacy of SSBs.

However, contemporary demands from emerging, developing economies, non-governmental organisations (NGOs) and civil society for more transparency and openness within global governance institutions[38] are also relevant for international, largely private-driven, standard-setting. Just to corroborate the importance of this concern for transnational governance and the topicality of the subject-matter of this edited volume, a recent WTO ruling (*US – Tuna II*) also raises questions as to

[37] Cf. A. Hirschman, *Exit, Voice, and Loyalty – Responses to Decline in Firms, Organizations, and States* (Harvard University Press, 1970).

[38] See, generally, A. Bianchi and A. Peters (eds), *Transparency in International Law* (Cambridge University Press, 2013).

the processes of SSBs and the ensuing properties of their standards.[39] By the same token, public regulators started investigating more closely the functioning of SSBs with a view to ensuring equality of opportunities among the actors involved, transparency and openness. Thus, where discussions on institutional reform in SSBs are expected to emerge, a proper understanding of standardisation processes and dynamics is necessary for such discussions to be meaningful. This is so all the more if one considers that, at this level of rule-making, procedural defects will most likely have a bearing on the very substance of the final product – that is, the standard adopted.

B. Standardisation processes and their impact on development, trade and innovation

This edited volume is a multidisciplinary inquiry into the foundations of international standard-setting, an empirically under-researched yet increasingly important area of international informal law-making. Standardisation has drawn little attention in academic literature, not only because of its voluntary nature but also because of the complex institutional structures that SSBs sometimes display. Typically, such regimes have their own preferred idiom, structural bias and ethos, trying to shield their activities from external scrutiny. In addition, and admittedly, they have been quite effective in the tasks assigned to them: technological progress in recent years has been unprecedented. Nevertheless, measuring the legitimacy of a given institution based on the outcomes of its policies (output legitimacy) is merely one way of assessing legitimacy because legitimacy more broadly has multiple facets. Furthermore, a more functionalist approach, if accepted, would rather downgrade the importance of the question of legitimacy.

Regardless of the viewpoint taken, one of the starting assumptions of this volume is that the current state of the literature is insufficient to draw safe conclusions as to the mechanics of SSBs. Empirical evidence is inconclusive as to whether foreclosure, exclusion, strategic behaviour and other unfair practices are present in standard-setting processes. Therefore, a more systematic analysis of the properties of standard-setting processes and detailed mapping of case studies will allow for an informed judgment regarding the appropriateness of standard-setting processes.

[39] For more details, see the contributions of Delimatsis (Chapter 5) and Schepel (Chapter 9) in this volume.

Contributions to this volume take a careful look at the peculiarities of international standardisation in selected issue areas without any pre-existing bias against SSBs as institutions per se. Most contributions delve into the internal structures of various SSBs and take real-life examples of standards to test empirically their hypotheses as to standardisation processes in specific SSBs such as the ISO, the Codex Alimentarius Commission or other transnational, private-driven SSBs. Through this exercise, we can identify instances of procedural and substantive deficiency but also instances that demonstrate the limits of inclusiveness, participation and openness to the detriment of efficient decision-making or technological progress.

While claiming to be anything but comprehensive, this edited volume represents a balanced account that brings to the fore the importance of institutional design for standardisation and allows for a better understanding by our readership of the role of private and state actors within SSBs and the increasing importance of standards in general. It further elucidates a significant area of international informal law-making, contributing to the discussion on the new role of the state in an ever-changing international landscape but also encouraging more empirical accounts of standard-setting, a highly important area of rule-making that enables and thus determines the fate and pace of transnational economic activity.

This edited volume is divided into five parts covering essential areas on which standard-setting may have an impact, such as trade, development and innovation.

In Part I, Henk de Vries, Hans Lindahl and Knut Blind set the scene and identify some of the most pressing questions related to standard-setting. More specifically, de Vries highlights the importance of analysing standardisation processes from various angles and the need for further research on how standard-setting works and evolves over time by advocating a more holistic approach on how we deal with the phenomenon of standardisation and its entry as a new discipline in academia.

Lindahl offers a fresh look at the analysis and normative content of standardisation processes by using the ISO as a telling example of authoritative collective action. He first shows that the ISO is a powerful yet nascent legal order that few have attempted to conceptualise. Second, Lindahl underscores the importance of the political element, along with the technical aspects of, standard-setting and the ensuing ramifications in terms of representation and deliberation.

The contribution by Blind concludes Part I. After revisiting the current literature on the economic effects of standardisation, Blind tackles the important, emerging but largely neglected issue of quality infrastructure – that is, all those enabling mechanisms that allow producers to improve their products and production processes. In this context, standardisation is just one of various stages in a comprehensive system of quality infrastructure. The other stages, such as conformity assessment or metrology, are not mutually exclusive, but *complementary* to standardisation. Thus, standardisation is just another part of a broader equation with multiple interdependent components.

Part II deals with the role of international standards in the global trade regime. Alessandra Arcuri examines the dynamics between the WTO and the standardisation activity within the Codex Alimentarius Commission (CAC) and critically reviews the evolving role of technocracy in the global regime for food safety standards. Crucially, she highlights that technocracy acquired a more prominent role due to the adoption of the SPS agreement and the importance that the latter gave to risk assessment and expert knowledge. On the occasion of the notorious standard-setting process within the CAC relating to ractopamine,[40] Arcuri challenges the deferential approach to technocracy and scientific ethos and welcomes an alleged shift toward a mix of science and democratic values when setting global food standards.

Panagiotis Delimatsis examines the stance of the Agreement on Technical Barriers to Trade (TBT) vis-à-vis international standards in the light of the relevant WTO jurisprudence to date. After an analysis of the current state of affairs, he criticises the deferential use of international standards in international trade adjudication. Delimatsis argues that several international standards are adopted through processes that do not appear to respect the fundamental tenets of transparency and due process, whereas representativeness and inclusiveness are not guaranteed in several ISSBs. Because good governance appears to be one of the *desiderata* in the area, Delimatsis presents a conceptual framework for reviewing international standards as relevant benchmarks to be used in WTO adjudication to identify disguised protectionism in domestic regulations.

Technical standard-setting in financial services is one of the most important yet underdeveloped areas in international trade regulation

[40] As Arcuri explains, ractopamine hydrochloride is a beta-agonist used in meat (mainly pork and cattle) to achieve fast growth and leanness of meat. The CAC adopted standards for maximum residue levels for ractopamine in July 2012 after a highly controversial process that culminated to a divisive voting. For more details, see Arcuri's chapter in this volume (Chapter 4, Section D).

literature. Juan Marchetti's chapter is a tentative scrutiny of this emerging area of standard-setting that raises awareness and improves our understanding of the stakes at issue, notably as far as the General Agreement on Trade in Services (GATS) is concerned. Indeed, although seemingly neutral at first blush, technical standards in financial services such as trading and settlement can constitute trade barriers because they modify the conditions of competition and affect the effective equality of opportunities in the relevant services market.

Petros Mavroidis builds on Marchetti's contribution to reflect on the need for a TBT analogue for the GATS. Mavroidis answers this question in the negative, based on the fact that, in his view, internationalisation of services standards has yet to occur. In Mavroidis's view, the current modest wording of Article VI of the GATS also reflects the penury of international developments in services standard-setting.

In Part III, the volume shifts gears to discuss the dynamics and political economy of international and regional standardisation. Walter Mattli and Jack Seddon discuss the political dimension of standard-setting and the relative power of standard-setters when acting in what they call 'governance partnerships'. One of the most powerful findings of their contribution is that important distributional effects of varying power endowments among partners in such constellations are to be observed, favouring those who write the rules ('penholders') to the detriment of those that have a rule-supporting role. Mattli and Seddon thus cast doubt on the alleged equality among partners in such collaborative arrangements.

Harm Schepel reflects on recent case law within the European Union relating to the normative power of domestic standard-setters. Schepel questions the usefulness of the concept of 'de facto mandatory standards' after a critical review of the *Fra.bo* judgment delivered by the Court of Justice of the European Union and the *Tuna II* ruling within the WTO.

Axel Marx and Jan Wouters map the various private sustainability regimes that increasingly dominate the markets of the developed – and increasingly the developing – world, identifying the drivers for this phenomenon. Absent any coordination mechanism, such regimes for voluntary sustainability standards (VSS) many times overlap, creating a fragmented landscape that hardly facilitates market access, notably for the less sophisticated producers. Marx and Wouters identify a credibility gap, increased costs to gain certification by multiple VSS systems and a potential race to the bottom due to the multiplicity of supply of VSS. They conclude by discussing two remedial mechanisms, mutual recognition and meta-regulation.

Concluding Part III, Olivier Cadot, Anne-Claire Disdier and Lionel Fontagné bring to the forefront the practical challenges that North–South harmonisation entails and the dilemmas that Southern (that is, developing country) partners face in their attempts to increase trade opportunities. Cadot, Disdier and Fontagné examine whether harmonisation of standards and technical regulations promotes trade expansion (both at the North–South and the South–South level). They find that there is a causal relationship between harmonisation of technical requirements in a North–South preferential trade agreement and decrease in South–South trade. They find that North–South harmonisation appears to lock Southern partners in hub-and-spoke trade patterns with their Northern partners at the expense of South–South trade.

The contribution by Cadot, Disdier and Fontagné constitutes an ideal transition to the theme of Part IV of the book, which focuses on the linkages between standard-setting and development promotion. In this regard, Stephanie Bijlmakers and Geert van Calster empirically review the creation of the ISO 26000 standard on social responsibility to see how development-related concerns were addressed (or not) during the preparation, drafting and adoption of the standard. The case of ISO 26000 is important when it comes to companies' social responsibility in the absence of any coercive international rules effectively disciplining their activities and operations in the interest of human rights or the environment. Bijlmakers and van Calster find that alternative approaches may lead to change in corporate behaviour more effectively.

Awilo Ochieng Pernet describes the standards development process within CAC and discusses whether development-related considerations are entrenched into the CAC standard-setting process, one of the most influential public-driven international SSBs. Ochieng Pernet identifies the challenges that developing countries face in terms of effective participation even in the presence of administrative tools such as the Codex Trust Fund aiming at improving CAC participation.

Angelos Dimopoulos maps the current (private-driven, for the most part) standards on socially responsible investments and corporate social responsibility and discusses their key characteristics with a view to identifying their contribution to public policy goals. Importantly, Dimopoulos examines the possibility for convergence in the area of responsible investment and corporate social responsibility standards through international investment agreements (IIAs), for instance, by introducing corporate social responsibility (CSR) conditionality and subjecting the conferral of the benefits of such agreements to compliance with relevant CSR standards.

In the last part of the book, the linkages of standard-setting processes with innovation are examined. Pierre Larouche and Geertrui van Overwalle connect the dots among intellectual property protection, standard-setting, innovation and competition policy. Inter alia, they discuss important, topical issues that have arisen (and instigated much mediatised litigation) at the intersection between standard-setting, intellectual property and competition policy such as standard-essential (SEPs) patents and setting fair, reasonable and non-discriminatory (FRAND) terms vis-à-vis licensees; 'patent thickets', 'hold-up' and 'ambush'; or open standards. These are more than buzzwords; they are areas where interoperability, intellectual property and competition policy meet, raising crucial questions of balancing, legitimacy and innovation and revealing a more complicated mix of the 'private' and the 'public' as expressed in standard-setting. Larouche and van Overwalle criticise the current piecemeal approach when examining standardisation because it may lead to an incomplete picture of how standardisation works in practice; instead these authors argue for a more holistic approach when it comes to standard-setting, protection of intellectual property and competition policy.

At this juncture, Neil Gandal and Pierre Régibeau review the existing economic literature on standardisation in general, the standard-setting process and the performance of SSBs. In this respect, they examine both the operating performance of SSBs and their impact on the markets for products, technologies and standards. Crucially, Gandal and Régibeau emphasise the need for more thorough research on empirical evidence regarding the mechanics of standard-setting but also to substantiate the (both positive and/or negative) effects of standard-setting processes on innovation, suggesting that comprehensive research on standard-setting is only at its nascent stage.

Part V of the edited volume concludes with a contribution on technical standards and privacy by Eleni Kosta and Kees Stuurman, who map the existing standards on data protection and privacy and their development processes and discuss the newly emerging standard-setting area of cloud computing. Kosta and Stuurman emphasise the ever-growing importance of technical standards for compliance with data protection regulation in view of the forthcoming general data protection regulation within the EU.

A note offering various avenues for future research concludes this volume on international standardisation. Although this volume improves our understanding of the importance of standard-setting – be it publicly, privately or hybrid driven – and the processes used, it also advocates for more comprehensive research efforts that would allow for testing certain hypotheses and assumptions that are – sometimes light-heartedly – made with respect to the mechanics, dynamics, actors and values of the standard-setting ecology.

PART I

International standard-setting – A multifaceted and evolving area of research

1

Standardisation

A developing field of research

HENK J. DE VRIES

In this chapter, I map the field of standardisation research by confronting a characterisation of standardisation with a systematic listing of sciences and then linking this to the standardisation literature. This enables us to distinguish some missing areas on the map of standardisation research, and it may form the basis for future systematic standardisation research. Academic research on standardisation generally concentrates on one technical topic (for instance, wireless networked control systems),[1] one specific item (for instance, 'relations between construction and standardisation')[2] or one aspect of standardisation, often the economic aspect; for an overview of economic studies, see Swan.[3] Monographs include Gaillard,[4] Glie,[5] Hesser et al.,[6] Hesser and Inklaar,[7] Verman[8] and Choi et al.,[9] and non-academic

This is an updated version of H.J. de Vries, 'Standardization – Mapping a Field of Research' in S Bolin (ed), *The Standards Edge* (2002) 99–121.

1 P Ferrari, A Flammini, M Rizzi and E Sisinni, 'Improving Simulation of Wireless Networked Control Systems Based on WirelessHART'. 35 *Computer Standards & Interfaces* (2013) 605–615.

2 W Hesser, *Untersuchungen zum Beziehungsfeld zwischen Konstruktion und Normung [Study on Relations between Construction and Standardisation]* (1981).

3 GMP Swann, *The Economics of Standardization* (2000); GMP Swann, *The Economics of Standardization: An Update* (2010).

4 J Gaillard, *A Study of the Fundamentals of Industrial Standardization and Its Practical Application, Especially in the Mechanical Field* (1933).

5 R Glie (ed.), *Speaking of Standards* (1972).

6 W Hesser, AJ Feilzer and HJ de Vries (Eds), *Standardisation in Companies and Markets*. 3rd edition (2010).

7 W Hesser and A Inklaar (Eds.) *An Introduction to Standards and Standardization* (1997).

8 LC Verman, *Standardization – A new discipline* (1973).

9 D-G Choi, B-G Kang and T Kim, *Standardization: Fundamentals, Impact, and Business Strategy* (2010).

studies such as Kuppanna,[10] Sanders,[11] Spivak and Brenner[12] and Toth.[13]

A. Standards and standardisation: Fields of research

Standards concern entities or relations between entities. An entity may be (1) a person or group of persons; (2) a 'thing' such as an object, an event, an idea or a process ('things' includes plants and animals[14]); and (3) a combination of the first two kinds of entities (for instance, a car with a driver or a company). Standards provide solutions for matching problems: matching thing–thing (for instance, bolts and nuts), matching man–thing (for instance, safety or ergonomic requirements), or matching man–man (for instance, procedures, management systems).[15] De Vries[16] presents and discusses classifications of standards on this basis. Among other aspects, he distinguishes between performance standards and solution-describing standards. The latter can be distinguished in interference standards, compatibility standards and quality standards. The majority of standards research concerns compatibility standards only, mostly in the field of information and communication technology.

De Vries[17] also classifies standards in relation to the people who develop or use them. In relation to standards development, the main distinction is between formal and de facto standardisation. The academic literature often pays attention to balancing between them (for instance, Egyedi[18]; Schmidt and Werle[19]). Most economic literature is restricted to de facto standardisation. Examples of studies that focus on formal

[10] SMRK Kuppanna, *Monograph on Standardization* (1988).
[11] TBR Sanders. *The Aims and Principles of Standardization* (1972)
[12] SM Spivak and FC Brenner, *Standardization Essentials – Principles and Practice* (2001).
[13] RB Toth (ed.), *Standards Management – A Handbook for Profits* (1990).
[14] Gaillard (1933, p. 33) provides a rather complete list of possible entities.
[15] HJ de Vries, 'Standardization – What's in a name?' (1997) 4 *Terminology – International Journal of Theoretical and Applied Issues in Specialized Communication* 55–83.
[16] HJ de Vries, *Standardization – A Business Approach to the Role of National Standardization Organizations* (1999).
[17] Ibid.
[18] T Egyedi, *Shaping Standardization – A Study of Standards Processes and Standards Policies in the Field of Telematic Services* (1996).
[19] SK Schmidt and R Werle, *Coordinating Technology – Studies on the International Standardization of Telecommunications* (1998).

standardisation are Bonner and Potter,[20] DIN[21] and De Vries.[22] Standards development within companies gets little attention. Exceptions include AFNOR,[23] Adolphi,[24] Nakamura[25] and Van Wessel.[26]

Standards application receives attention in the technical literature, but mostly indirectly. The main exceptions to this are the standards that affect people most – those on quality management. In this case, a substantial literature exists (for reviews on the quality management standard ISO 9001, see Psomas and Fotopoulos[27]; Rusjan and Alič[28]; Sampio et al.[29]; for a literature review of studies on the environmental management standard ISO 14001, see De Vries et al.[30]). Several studies concern formalisation, sometimes without using the terms 'standard' or 'standardisation' (e.g. Bodewes[31]; Foucault[32]; Giddens[33]; Weber[34]).

Between the development of standards and their use are some other activities. At the supplier side, these include publication, announcement

20 P Bonner and D Potter, *Achieving Best Practices in National Standardisation – A Benchmarking Study of the National Standardisation Systems of Finland, Sweden, Denmark and Italy* (2000).

21 DIN, *Economic Benefits of Standardization – Summary of Results. Final Report and Practical Examples.* DIN German Institute for Standardization (2000).

22 See note 19.

23 AFNOR, *La normalisation dans l'entreprise* [Company Standardisation], Association Française de Normalisation (1967).

24 H Adolphi (ed), *Strategische Konzepte zur Organisation der betrieblichen Standardisierung* [Strategic Concepts for the Organisation of Company Standardisation] (1997) 38.

25 S Nakamura, *The New Standardization – Keystone of Continuous Improvement in Manufacturing* (1993).

26 RM Van Wessel, *Toward Corporate IT Standardization Management – Frameworks and Solutions* (2010).

27 EL Psomas and CV Fotopoulos, 'A meta analysis of ISO 9001:2000 research – findings and future research proposals' (2009) 2 *International Journal of Quality and Service Sciences* 128–144.

28 B Rusjan and M Alič, 'Capitalising on ISO 9001 benefits for strategic results' (2010) 27 *International Journal of Quality and Reliability Management* 756–778.

29 P Sampio, P Saraiva and AG Rodrigues, 'ISO 9001 certification research: questions, answers and approaches' (2009) 26 *International Journal of Quality & Reliability Management* 38–58.

30 HJ de Vries, DK Bayramoglu and T van der Wiele, 'Business and environmental impact of ISO 14001' (2012) 29 *International Journal of Quality & Reliability Management* 425–435.

31 WEJ Bodewes, *Neither Chaos nor Rigidity – An Empirical Study of the Role of Partial Formalization in Organizational Innovativeness* (2000).

32 M Foucault, *Surveiller et punir; naissance de la prison* (1975) [English edition: *Discipline and Punish; Birth of the Prison* (1977)].

33 A Giddens, *The Constitution of Society – Outline of the Theory of Structuration* (1984).

34 M Weber, *Wirtschaft und Gesellschaft* (1922).

and distribution of the standard; sometimes additional information is provided. At the user side are the activities of finding, selecting and purchasing standards. I use the term 'standards acceptance' for the combination of these activities. This is a field of interest because different standards may be in competition. Standards battles, such as that between different video systems, receive a lot of attention in economic literature.[35] A final field of research is the impact that standards have on business and society.

In summary, we see five fields of standardisation research: standards themselves, together with their development, acceptance, use, and impact.

B. Fundamental sciences

The five areas of standardisation research may involve many scientific disciplines. For instance, a standard may be studied from the point of view of its technical contents (technical science) or as an expression of language (linguistics). Standards acceptance may be studied by psychologists or economists or by scientists in marketing research that combine these disciplines. Standards impact may be studied by economists or business scholars.

To discuss possible scientific approaches of standardisation, we may consult the philosopher Herman Dooyeweerd,[36] who distinguishes several aspects of phenomena in reality. A table, for instance, may have four legs – an arithmetic aspect. It may have a certain economic value – an economic aspect. It may be beautifully designed – an aesthetic aspect. In all, Dooyeweerd lists fifteen aspects and their respective laws that should be honoured. The first aspects concern laws of nature. For instance, mathematical laws including adding, subtracting and multiplying apply to the arithmetic aspect. For later aspects, these laws are normative principles that humans, in their free will, can obey, ignore or oppose. Each aspect is studied by a fundamental science (see Table 1.1).

[35] G van de Kaa, J van den Ende, HJ de Vries and E van Heck , 'Factors for winning interface format battles: A review and synthesis of the literature' (2011) 78 *Technological Forecasting & Social Change* 1397–1411.

[36] For an introduction to Dooyeweerd's ideas, see L Kalsbeek, *Contours of a Christian Philosophy: An Introduction to Herman Dooyeweerd's Thought* (1975) rather than Dooyeweerd's original works (H Dooyeweerd, *A New Critique of Theoretical Thought* 2: *The General Theory of the Modal Spheres* (1955); H Dooyeweerd, *A New Critique of Theoretical Thought* 3: *The Structures of Individuality of Temporal Reality* (1957).

Table 1.1. *Aspects of phenomena in reality and related scientific disciplines*

Aspect	Science
Arithmetic	Mathematics (algebra)
Spatial	Mathematics (geometry)
Kinematic	Kinematics
Physical	Physics, chemistry
Biotic	Biology (botany)
Feelings	Biology (zoology, human biology), psychology[37]
Analytical	Logic[38]
Historical	History
Linguistic	Linguistics
Social	Sociology[39]
Economic	Economics
Aesthetic	Aesthetics
Juridical	Law
Ethical	Ethics
Faith	Theology

According to Dooyeweerd, an entity or activity functions in all aspects of phenomena in reality and therefore can be studied from the point of view of all fundamental sciences. However, some aspects will dominate others, and these require more attention. This means that each fundamental science can study standards themselves, as well as the activities of developing them, selecting a standard, and using a standard. Therefore, we can create an inventory of contributions that these fundamental

[37] Followers of Dooyeweerd have argued that the feelings aspect should be divided into a perceptive aspect, which can be observed in animals and people, and a sensitive aspect, which is characteristic of people only. The first is studied in zoology and human biology, the second in psychology (WJ Ouweneel, *De leer van de mens – Proeve van een christelijk-wijsgerige antropologie* [Anthropology – Proof of a Christian-Philosophical Anthropology] [1986]).

[38] Logic has not yet developed into a scientific discipline with, for instance, its own faculty. It remains under of the 'umbrella science' of philosophy, although other disciplines also give consideration to it. The same applies to aesthetics and ethics.

[39] It is worth nothing that sociology studies more than simply the social aspect of 'social intercourse'.

sciences have made to the knowledge of these four categories and discuss what else they might contribute. For impact studies, the choice of fundamental sciences is more limited, and economics is the most relevant.

The latter relates to another element of Dooyeweerd's approach: each entity or activity is 'qualified' or, in other words, 'characterised' by one aspect. This aspect takes precedence over the other aspects without making them irrelevant; they should be honoured as they are but get their 'colour' from the dominant aspect. De Vries[40] has argued that the economic aspect qualifies standardisation because it is directed at finding a proper balance between uniformity and variety, and, according to Dooyeweerd, finding such a balance is typical for an economic activity. This applies to standards development and to standards acceptance, the latter of which relates to beneficial application of a standard.

The standard itself, as an entity, is a package of information in which solutions for matching problems have been defined. Because of the informative function of a standard, the linguistic aspect is important. However, as a document resulting from a development or selection process, it serves as an agreement: 'Let's do it this way.' In terms of Dooyeweerd, this is characteristic for the juridical aspect.

C. Fundamental sciences to study standardisation

In this section, I create an inventory of the current and potential contributions of the fundamental sciences to the knowledge of standardisation. I combine standards themselves with their development, selection, application and impact in an effort to keep this chapter to reasonable length.

I. Mathematics, kinematics, physics, chemistry and biology

The arithmetic, spatial, kinematic, physical, biotic and feelings aspects of standardisation concern mainly the following standardisation topics: calculation methods, such as statistical techniques (characterised by the arithmetic aspect); sizes and tolerances thereof (spatial aspect); telecommunication and logistics (kinematic aspect); products (physical

[40] HJ de Vries, 'Standardization – What's in a name?' (1997) 4 *Terminology – International Journal of Theoretical and Applied Issues in Specialized Communication* 55–83.

aspect); crops (biotic aspect); animals; and the physical functioning of people (feelings aspect). Because standards are mostly used for designing or influencing these entities, standardisation is of less interest to these fundamental sciences than to sciences such as engineering (e.g. computer science, civil engineering, mechanical engineering, aerospace science[41]), architecture, agricultural science and medicine. Findings from fundamental sciences are used when designing standards such as those employed in the use of calculation methods. Bongers[42] and Ailleret[43] have provided mathematical studies on standardisation.

Conversely, standardisation contributes to these sciences in such forms as standardisation of numerical systems and statistics (mathematics), quantities, units and symbols[44] (kinematics, physics, chemistry), the periodic system (chemistry), referential materials (chemistry, physics) and test methods (physics, chemistry, biology). Van den Dool[45] is an example of such a study in chemistry.

II. Psychology

Psychologists distinguish between rational and irrational behavior. Standards, as such, represent rationality, but their development, selection and application may include all kinds of irrational considerations. Resistance against standards is a topic that deserves the attention of psychologists. Resistance against standards might be related to the fact that standards, in general, are tools for control that may affect freedom. Brehm[46] and Brehm and Brehm[47] addressed the latter topic. Erkel and Steensma[48] used their

[41] The $125 million NASA *Orbiter* spacecraft crashed because one NASA team had used Newtons, while the other NASA team used English pounds of force.

[42] C Bongers, *Standardization – Mathematical Methods in Assortment Determination* (1980).

[43] P Ailleret (ed), *Dimensional Standardization.* International Electrotechnical Commission (1985).

[44] The International System of Units (SI – Système International d'Unité) was established in International Standard ISO 31, *Quantities and Units.*

[45] H Van den Dool, *Standardisation of Gas Chromatographic Analysis of Essential Oils* (1974).

[46] JW Brehm, *A Theory of Psychological Reactance* (1966); JW Brehm, *Responses to Loss of Freedom: A Theory of Psychological Reactance* (1972).

[47] SS Brehm and JW Brehm, *Psychological Reactance – A Theory of Freedom and Control* (1981).

[48] D Erkel and H Steensma, 'Leidt gevoelde certificeringsdruk tot weerstand tegen TQM? [Does perceived certification pressure cause resistance against TQM?]' (1999) 43 *Sigma* (3) 6–9.

findings to investigate uneasiness related to the ISO 9000 standards for quality management.

III. Logic

There are some examples of studies on standardisation from the point of view of logic. De Vries[49] lists these for the area of standards development and concludes that further research on systematic standards development is necessary, especially directed at linking systematic development to the practices of standardisation committees.

IV. History

Standards are known to have existed as early as 7000 BCE, when cylindrical stones were used as units of weight in Egypt.[50] Muschalla[51] provided a basic historical study on the early history of standardisation. Some of the available historical studies on standardisation are related to the history of national standardisation organisations (e.g. McWilliam[52]; Wölker[53]). Other studies highlight one example of standardisation, for example, intercompany standardisation in the early American automobile industry[54] or, very recently, Internet browsers.[55]

V. Linguistics

Teichmann (2010) provides a basic linguistic study on standardisation. Several other studies have provided underpinnings for linguistic aspects

49 HJ de Vries, *Standardization – A Business Approach to the Role of National Standardization Organizations* (1999).

50 MA Breitenberg, *The ABC's of Standards-related Activities in the United States*, NBSIR 87–3576, National Institute of Standards and Technology (1987).

51 R Muschalla, *Zur Vorgeschichte der technischen Normung* [About the previous history of technical standardisation] (1992).

52 RC McWilliam, *BSI: The First Hundred Years 1901–2001 – A Century of Achievement* (2001).

53 T Wölker, *Entstehung und Entwicklung des Deutschen Normenausschusses 1917 bis 1925* [Genesis and Development of the German Standardization Commission 1917 through 1925] (1992).

54 GV Thompson, 'Intercompany Technical Standardization in the Early American Automobile Industry' (1954) 14 *The Journal of Economic History* 1–20.

55 I Oshri, HJ de Vries and H de Vries, 'The rise of Firefox in the web browser industry: The role of open source in setting standards' (2010) 52 *Business History* 834–856.

of standards development (e.g. Effenberger,[56] Galinski,[57] Interrante and Heymann[58]). Other linguistic topics include standards as an expression of language and language use within technical committees.

Conversely, standardisation contributes to language science by providing standards for transliteration. Moreover, insights from standardisation may be used to describe language canonisation and the battle for acceptation of languages or dialects. In the latter case, languages or dialects themselves could be regarded as standards.

VI. Sociology

The tension between rationality and irrationality is a central topic in sociological studies (e.g. Weber[59]). This has resulted in studies of, for instance, formalisation that add to the standardisation body of knowledge, although the term 'standardisation' is often not used. The development, selection and application of standards can be regarded as social processes that sociologists may study. People act, react, cooperate or compete. Some sociological articles on standardisation were published in a special issue of the French sociological journal *Réseaux*.[60] Cochoy, Garel and De Terssac[61] wrote about the implementation of the ISO 9000 standards for quality management from a sociological perspective. Other academic studies on standardisation that honour social interaction (e.g. Schmidt and Werle[62]; Tamm Halström[63]) are not purely sociological.

[56] D Effenberger, Dietmar, 'Fundamentals of terminology work' (1995) 17 *Computer Standards & Interfaces* 131–137.

[57] C Galinski, 'Terminology Standardization and Standards Information' in Minister of Supply and Services Ottawa, *Proceedings from the International Symposium on Terminology and Documentation in Specialized Communication* (1992) 61–77.

[58] FJ Interrante and FJ Heymann (eds.), *Standardization of Technical Terminology: Principles and Practices* (1982).

[59] M Weber, *Wirtschaft und Gesellschaft* (1922).

[60] *Réseaux* (2000) 'Dossier: La fabrication des normes' [Special issue: The production of standards]. *Réseaux* 18, 102, Hermes Science Publications, Paris.

[61] F Cochoy, J-P Garel and G de Terssac, 'Comment l'écrit travaille l'organisation: le cas des normes ISO 9000' [How the written word operates on organisation: the case of ISO 9000 standards] (1998), 39 *Revue Française de Sociologie* 4, at 673–699.

[62] SK Schmidt and R Werle, *Coordinating Technology – Studies on the International Standardization of Telecommunications* (1998).

[63] K Tamm Halström, 'The production of management standards' (1996) 75 *Revue d'Économie Industrielle* 61–76.

Sociology studies are more than just the social aspect of 'social intercourse' that Dooyeweerd found characteristic of this aspect. This is addressed by Habermas's concept of 'communicative action',[64] which Van Veldhuisen[65] applied to the implementation of the ISO 9000 standard for quality management within a company and Schoechle[66] to standardisation in general.

VII. Economics

Because standardisation is an economic activity, economics is the first fundamental science among all the others that should study it – and it does. The most popular topic in economic studies on standardisation is standards battles. The vast majority of these studies take empirical data, if any, from information and communication technologies and consumer electronics. Van de Kaa et al.[67] provide an overview of this literature and distinguish between evolutionary economics, industrial economics and institutional economics. However, economic studies on standardisation are more apt to address, for instance, the impact of standards and standardisation. A seminal book is *The Economics of Standards – Theory, Evidence, Policy*.[68] Studies by Swann[69,70] provide an overview of economic studies in this field.

VIII. Aesthetics

Artists may talk about individual expression of the most personal emotions; this is the opposite of standardisation. In fact, there are few attempts to apply standardisation to the arts. Exceptions include Müller[71]

[64] J Habermas, *Theorie des kommunikativen Handelns*, 2, *Zur Kritik der funktionalistischen Vernunft* (1981) [English version: *The Theory of Communicative Action*, 2, *Lifeworld and System: A Critique of Functionalist Reason*].

[65] A Van Veldhuisen, *De Macht van het Systeem – Een studie naar kwaliteitssystemen gebaseerd op het werk van Michel Foucault* [The power of the System – A study of quality management systems based on the works of Michel Foucault] (1996).

[66] T Schoechle, *Standardization and Digital Enclosure – The Privatization of Standards, Knowledge, and Policy in the Age of Global Information Technology* (2009).

[67] G van de Kaa, J van den Ende, HJ de Vries and E van Heck , 'Factors for winning interface format battles: A review and synthesis of the literature' (2011) 78 *Technological Forecasting & Social Change* 1397–1411.

[68] K Blind (ed), *The Economics of Standards, Theory, Evidence, Policy* (2004).

[69] GMP Swann, *The Economics of Standardization* (2000).

[70] GMP Swann, *The Economics of Standardization: An Update* (2010).

[71] S Müller, *Kunst und Norm* [Art and standard] (1992).

and the first president of the European Academy for Standardisation (EURAS), who made an initial attempt to create an aesthetic approach to standardisation.[72] De Rijk[73] relates standardisation to the design of products. Of course, the aesthetics of standards, in particular their layout, might be studied as well.

IX. Law

Unfortunately, the most complete juridical study on standardisation[74] is available only in the Dutch language. Stuurman addresses, among other things, differences between formal and de facto standardisation, standardisation as a means of self-regulation, constitutionality, copyright of standards, liability of standards developing organisations, product liability, competition and antitrust law and trade barriers related to standards. Other law studies address intellectual property rights[75] and legislation related to standardisation.[76] Standards can be seen as 'soft law'.[77] There is less attention to topics that might better align with Dooyeweerd's theory: standards' development, choice or use as a result of an agreement between parties, and the consequences thereof.

X. Ethics

Schoechle[78] suggests studying standardisation from the perspective of ethics, using, for instance, Habermas's model of discourse ethics. In discourse ethics, universal ethical principles and the relation between the justification and application of norms can be socially constructed from

72 P Guillet de Monthoux, *Esthétique du management* [Aesthetics of Management] (1999); P Guillet de Monthoux, *Standardization as the Aesthetics of Spectaculum Modernum.* Paper presented at the SCANCORE/SCORE Seminar on Standardization, Arild, Sweden; 18–20 September 1997; SCORE – Stockholm Center for Organizational Research, Stockholm.

73 T De Rijk, *Norm = Form – On standardization and design* (2010).

74 C Stuurman, *Technische normen en het recht* [Technical Standards and the Law] (1995).

75 MA Lemly, 'Intellectual Property Rights and Standard-Setting Organizations' (2002) 90 *California Law Review* 1898–1980.

76 H Schepel, *The Constitution of Private Governance – Product Standards in the Regulation of integrating Markets* (2005); H Schepel and J Falke, *Legal aspects of standardisation in the Member States of the EC and EFTA – Volume 1: Comparative report,* (2000); H Schepel and J Falke, *Legal aspects of standardisation in the Member States of the EC and EFTA – Volume 2: Country reports* (2000).

77 U Mörth, *Soft Law in Governance and Regulation – An Interdisciplinary Analysis* (2004).

78 T Schoechle, 'Towards a Theory of Standards' in *Proceedings Second Interdisciplinary Workshop on Standardization Research, 24–27 May,* 1999, at 84–95.

rational discourse based on language. The increasing interest in business ethics has not yet been reflected in scientific studies on standardisation ethics. Nevertheless, there are several studies on ethical topics. Examples include access to the standardisation arena by less powerful parties (or their delegates); such as developing countries,[79] consumers,[80] small and medium-sized enterprises[81] and their possible influence over the process; decision-making processes[82] or the standards' contents, such as sustainability.[83] More in general, the standardisation discipline can contribute to (business) ethics in the area of codes of behavior,[84] as these can be regarded as standards.

XI. Theology

According to Dooyeweerd, the faith aspect does not concern a religion; rather, it concerns the possibility to believe. In business, this aspect qualifies the organisation's mission statement. By definition, the faith aspect is present in standardisation – a standard is a standard because of its intended *and expected* massive use.[85] The expected, not the factual, use is decisive, and this expectation, although supported by 'facts', is a matter of belief. In general, such topics are not investigated by theology, the scientific discipline most directly related to the faith aspect.

Standardisation theory might contribute to theology at the point of understanding the development and canonisation of religions. This was

[79] The World Bank and ISO, *The Role of Standardization in Economic Development* (1993).

[80] JV Coles, *Standards and Labels for Consumers' Goods* (1949).

[81] HJ de Vries, K Blind, A Mangelsdorf, H Verheul and J van der Zwan, *SME Access to European Standardization – Enabling Small and medium-sized Enterprises to Achieve Greater Benefit from Standards and from Involvement in Standardization* (2009) Brussels: CEN and CENELEC. http://www.cenelec.eu/NR/rdonlyres/88D06BD5-CA51–479D-A416-AB1F3BE67E66/0/SMEAccessReport20090821.pdf.

[82] RW Hawkins 'Standards-making as technological diplomacy: Assessing objectives and methodologies in standards institutions' in R Hawkins, R Mansell and J Skea (eds.), *Standards, Innovation and Competitiveness – The Politics and Economics of Standards in Natural and Technical Environments* (1995), at 147–158; J Riganati, 'Latent benefits of Consensus Processes' (1986) 5 *Computer Standards & Interfaces* 3, at 185–187.

[83] F Wijen, 'Means versus ends in opaque institutional fields: Trading off compliance and achievement in sustainability standard adoption' (2014) 39 *Academy of Management Review* 3, at 302–323.

[84] A Kolk and R van Tulder, 'Setting new global rules? TNCs and codes of conduct' (2005) 14 *Transnational Corporations* 3, at 1–27.

[85] HJ de Vries, 'Standardization – What's in a name?' (1997) 4 *Terminology – International Journal of Theoretical and Applied Issues in Specialized Communication* 55–83.

suggested in the first doctoral thesis on standardisation[86]: religions, 'with all their sects, heresies, and schisms', are standardised to the most minute detail with regard to the time, place, form or conditions of worship, and the doctrines, ritual, beliefs, hopes and expectations of the worshippers.

D. Applied sciences

Next to the fundamental sciences are applied sciences. They study a certain entity, a group of entities or activities and combine several points of view and insights from fundamental sciences. Examples are provided in Tables 1.2 and 1.3.

Table 1.2. *Examples of entity-related sciences*

Entities	Science
People, health of	Medicine
Human teeth	Dentistry
Animals, health of	Veterinary medicine
Companies	Business science
Governments	Public administration
Stars and other celestial bodies	Astronomy

Table 1.3. *Examples of activity-related sciences*

Activities	Science
Education	Pedagogics
Teaching	Didactics
Airplanes (design of)	Aviation science
Buildings (design of)	Architecture
Agriculture	Agricultural science
Marketing	Marketing science
Quality management	Quality management

[86] RA Brady, *Industrial Standardization* (1929).

For marketing and quality management, it may be discussed whether they are independent scientific disciplines or just specialisms within business science. In practice, many sciences do not fit purely in one of the above categories. For instance, business economics studies the economic aspect of (the entities) companies.

Applied sciences may study standardisation in addition to the fundamental sciences and use their results. These applied sciences will primarily be those that are related to the entities and activities involved. The entities include the topics to be standardized, the people concerned and the organisations involved.

E. Applied sciences related to the topics of standardisation

I. Design, development and construction sciences

Most standards concern technical matters and are used by technicians. Technical and biotechnical sciences such as mechanical engineering, civil engineering, architecture, informatics, agriculture and biotechnology may study the role of standards in designing, developing, building, using and maintaining technical artifacts and in growing crops and breeding animals. Examples in the area of mechanical engineering include Henzold[87]; Hesser[88]; and Tenchea, Arghiriade and Lipp.[89]

II. Health care sciences

Medicine, dentistry and veterinary medicine may study the role of standards for areas including products, materials and medical information interchange. Taking the example of dentistry, scientific literature often pays attention to standards, especially to round robin tests. Generally, standardisation is just one of the topics addressed (e.g. Kalenderian[90];

[87] G Henzold, *Handbook of Geometrical Tolerancing – Design, Manufacturing and Inspection* (1995).

[88] W Hesser, 'Untersuchungen zum Beziehungsfeld zwischen Konstruktion und Normung' [Study on Relations between Construction and Standardization] (1981).

[89] P Tenchea, I Arghiriade and G Lipp, *Probleme actuale ale standardizării în construcţia de maşini* [Nowadays problems of standardization in mechanical engineering] (1987).

[90] E Kalenderian, *EZ codes: a diagnostic terminology as the foundational step of quality for the dental profession* (2013).

Lane, Watts and Wilson[91]; Schmalz and Browne[92]; Watts and McCabe[93]).

III. Business science

Standards for management systems and their application are studied by management or business science. Examples include Psomas and Fotopoulos,[94] Rusjant and Alič,[95] Kanji,[96] Sampaio et al.,[97] De Vries et al.[98] and Van Wessel and de Vries.[99]

F. Applied sciences related to people and organisations involved in standardisation

From psychology to theology, the fundamental sciences listed by Dooyeweerd study people, their activities and interrelations. In standardisation some typical groups of people and institutions are involved that can be studied by specific scientific disciplines.

91 DA Lane, DC Watts and NH Wilson, 'Ambient light working times of visible light-cured restorative materials. Does the ISO standard reflect clinical reality?' (1998) 14 *Dental Materials* 353–359.

92 G Schmaltz and RM Browne, 'The biological evaluation of medical devices used in dentistry. The influence of the European Union on the preclinical screening of dental materials' (1995) 45 *The International Dental Journal* 275–282.

93 DC Watts and JF McCabe, 'Aluminium radiopacity standards for dentistry: An international survey' (1999) 27 *Journal of Dentistry* 73–80.

94 EL Psomas and CV Fotopoulos, 'A meta analysis of ISO 9001:2000 research – findings and future research proposals' (2009) 1 *International Journal of Quality and Service Sciences* 128–144.

95 B Rusjan and M Alič, 'Capitalising on ISO 9001 benefits for strategic results' (2010) 27 *International Journal of Quality and Reliability Management* 756–778.

96 GK Kanji, 'An innovative approach to making ISO 9000 standards more effective' (1998) 9 *Total Quality Management* 67–78.

97 P Sampaio, P Saraiva and AG Rodrigues, 'ISO 9001 certification research: questions, answers and approaches' (2009) 26 *International Journal of Quality & Reliability Management* 38–58.

98 HJ de Vries, DK Bayramoglu and T van der Wiele, 'Business and environmental impact of ISO 14001' (2012) 29 *International Journal of Quality & Reliability Management* 425–435.

99 RM Van Wessel & HJ de Vries, 'Business Impacts of International Standards for Information Security Management. Lessons from Case Companies' (2013) 1 *Journal of ICT Standardization* 25–40.

I. *Business science*

Standardisation primarily concerns companies. Their functioning and interrelations are studied by business science. As an example, the research of one of the few chairs on standardisation, at the Rotterdam School of Management, Erasmus University, includes studies on small and medium-sized companies, companies-in-general (that apply standards or are involved in standardisation), companies in specific sectors of industry (chemistry, medical devices, services, retail, professional soccer), specific companies (railways, heating systems factory), standards users, national standardisation organisations, certification and accreditation bodies, industrial consortia, and consumers. Moreover, several disciplines within business science are covered: strategic management, management of technology and innovation, information management, logistics, service management, quality management, marketing management, business and society management and human resource management.

Narayanan and Tianxu[100] claim to provide a review of research on technology standards. They focus on compatibility standards and, more particularly, on standards battles. Due to this limited focus, the paper does not live up to its promise, as stated in the abstract, 'to organize the fragmented management literature on technology standards'.

II. *Science of public administration*

The roles of governments in standardisation may be studied by the science of public administration. These roles include:

- Supporting standardisation as a part of their general role in stimulating business performance and international trade
- Creating a legal foundation for standardisation
- Carrying out standardisation activities themselves
- Supplementing, simplifying or improving the legal system with standardisation by making references to standards in laws
- Using standardisation for specific public sector tasks

[100] VK Narayanan and C Tianxu, 'Research on technology standards: Accomplishments and challenges' (2012) 41 *Research Policy* 1375–1406.

- Using standardisation to improve administration performance in areas that are not specifically governmental

Researchers in the field of law have addressed some of these issues (e.g. Schepel[101]). Scientists in public administration might add their approaches.

G. Applied sciences related to specific standardisation activities

In the foregoing sections, I distinguished at the supplier side among development, publication, announcement and distribution of standards and providing information on standards; and, at the user side, I discussed tracing, selecting, buying, implementing, and using standards, and subsequently assessing their impact on business and society. Scientific disciplines related to some of these activities have emerged and can be helpful in providing a better understanding of standardisation.

I. Policy science

Schoechle[102] argues that policy science may contribute to standardisation knowledge through its focus on decision processes, social processes and problem orientation. Djelic and Sahlin-Andersson et al.[103] and Drezzner[104] address transnational governance of standardisation.

II. Design sciences

Standards can be regarded as products; therefore, insights from product development theories can be applied to standardisation. Examples include

[101] H Schepel, *The Constitution of Private Governance – Product Standards in the Regulation of integrating Markets* (2005).

[102] T Schoechle, 'Towards a Theory of Standards' in *Proceedings Second Interdisciplinary Workshop on Standardization Research, 24–27 May* (1999) 84–95.

[103] ML Djelic, K Sahlin-Andersson et al. *Transnational governance – Instutional Dynamics of Regulation* (2006).

[104] DW Drezner, *All Politics Is Global – Explaining International Regulatory Regimes* (2007).

Schacht[105] and Susanto.[106] The International Organisation for Standardisation's 'high level structure' for management system standards[107] builds on previous research work described by De Vries.[108]

III. *Studies of technology and society*

Studies of technology and society have emerged from the disciplines of philosophy and sociology and offer theoretical concepts that may be of help in analyzing standards making (Schoechle[109]; Egyedi[110]). For example, social constructivism indicates that standards do not suddenly leap into existence as the result of a momentous act by a heroic inventor; rather, they are gradually constructed or deconstructed in the social interactions of social groups. Egyedi[111] used social constructivism to study standardisation.

IV. *Information science*

Publication and distribution of standards, using possibilities of information and communication technology, might be studied by information science. There are several professional publications in this area (e.g. Rössel[112]; Walser[113]). Kuiper[114] wrote a dissertation in this field.

105 M Schacht, *Methodische Neugestaltung von Normen als Grundlage für eine Integration in den rechnerunterstützten Konstruktionsprozeß* [Methodical Standards Redesign as a Basis for Integration in Computer Aided Engineering] (1991).

106 A Susanto, *Methodik zur Entwicklung von Normen* [Methodology for Standards Development] (1988).

107 S Tangen and A-M Warris, 'Management makeover – New format for future ISO management system standards' 3 *ISO Focus+* (2012) (7) 40–41.

108 HJ de Vries, 'Opportunities for better management systems standards' in MJ Holler and E Niskanen (eds.), 15 *EURAS Yearbook of Standardization* 2, Special issue of *Homo oecunomicus* (1999) 381–402.

109 T Schoechle, 'Towards a Theory of Standards' in *Proceedings Second Interdisciplinary Workshop on Standardization Research, 24–27 May* 1999, 84–95, at 89.

110 T Egyedi, 'A Research Autobiography from an STS Perspective' in: J Schueler, A Fickers and A Hommels (eds), *Bargaining Norms, Arguing Standards.* STT 74 (2008).

111 T Egyedi, *Shaping Standardization – A Study of Standards Processes and Standards Policies in the Field of Telematic Services* (1996).

112 T Rössel, 'Intranet – Die neue IuK-Pattform für das ganze Unternehmen' [Intranet – The new ICT platform for the entire company] (1998) 77 *DIN-Mitteilungen* 739–743.

113 A Walser, 'Bedeutung des Internets für DIN' [Internet significance for DIN] (1998) 77 *DIN-Mitteilungen* 750–753.

114 BE Kuiper, *Information systems for the international accessibility of standards* (1975).

For other activities, there are no scientific disciplines in the sense of Kuhn,[115] but there are specific studies devoted to standardisation. Examples include the following:

- In relation to the activity of 'decision making', studies by Ewald,[116] Goerke and Holler[117] and Von Mayntz[118] are available.
- The activity of 'standards selection' is addressed by Weitzel et al.[119]
- From a company's point of view, standards can be seen as a part of knowledge management because standards contain explicit knowledge.[120] Company standardisation can be regarded as a process of transforming an unstructured mixture of tacit and explicit knowledge into available explicit knowledge. Slob[121] showed that scientific insights in knowledge management did not add any issues that had not been previously addressed in professional and scientific literature on company standardisation, such as Adolphi.[122]
- The activities of implementing and using standards – apart from the subject matter content of these standards – are studied by business science.

H. Applied sciences related to the impact of standards

Many standards impact studies stem from the economic discipline. This applies to macro-economic studies (global, regional or country level) and micro-economic studies (company level). Impact studies in business science may take a broader perspective and, for instance, use the Business

[115] TS Kuhn, *The structure of scientific revolutions* (1962).

[116] F Ewald, *Normes, pouvoir, démocratie* [Standards, power, democracy] (1996). La Villette – Centre National de la Recherche Scientifique, Centre de Recherche en Histoire des Sciences et des Techniques, Paris.

[117] L Goerke and M Holler, 'Voting on standardisation' (1995) 83 *Public Choice* 337–351.

[118] R Von Mayntz, 'Entscheidungsporzesse bei der Entwicklung von Umweltstandards' [Decision Processes in the Development of Environmental Standards] (1990) 23 *Die Verwaltung* 137–151.

[119] T Weitzel, D Beimborn and W König, 'A Unified Economic Model of Standard Diffusion: The Impact of Standardization Cost, Network Effects, and Network Topology' (2006) 30 *MIS Quarterly* 489–514.

[120] HJ de Vries and M van Delden, 'Standardisation and Knowledge Management in Services' (2011) 1 *Journal of Standards and Standardization* 70–78.

[121] FJC Slob, *Bedrijfsnormalisatie – De schakel tussen tacit en expicit knowledge* [Company standardization – The link between tacit and explicit knowledge] (1999).

[122] H Adolphi (ed), *Strategische Konzepte zur Organisation der betrieblichen Standardisierung* [Strategic Concepts for the Organization of Company Standardisation] (1997), 38.

Balanced Scorecard[123] to study standardisation impacts.[124] Psomas and Fotopoulos,[125] Rusjan and Alič[126] and Sampio et al.[127] list dozens of academic studies on ISO 9001 impact.

I. Characterisation of the standardisation 'discipline'

The listing of 'fundamental sciences' reviewed here aims at being complete, but the 'applied sciences' listing may require new additions. From the foregoing, it will be clear that the standardisation discipline owes a great debt to the sciences on which it is founded. However, it will never truly establish itself as an academic discipline in its own right until those that profess the subject demonstrate that it is capable of developing, and has developed, its own theoretical foundations. The next question is whether there are, or should be, any typical issues that should be addressed to understand the phenomenon and theoretical foundations related to that.

The most typical for standardisation, given its definition, is the balancing between unity and variety from a business perspective.[128] Within a company, the question is what to do according to standard and what to customize.[129] The same question arises in the relations between the company and its environment: how does the company 'plug in' to its environment in a technical as well as an organisational way? Again, there is the choice between the repeated use of certain solutions versus creating a new solution each time. The environment does not need to be taken for granted – the company can actively be involved in influencing its environment to obtain agreed-on solutions laid down in standards. This

123 RS Kaplan and DP Norton, 'The balanced scorecard – measures that drive performance' (1992). 70 *Harvard Business Review* 71–79.

124 RM Van Wessel, *Toward Corporate IT Standardization Management – Frameworks and Solutions* (2010).

125 EL Psomas and CV Fotopoulos, 'A meta analysis of ISO 9001:2000 research – findings and future research proposals' (2009) 1 *International Journal of Quality and Service Sciences* 128–144.

126 B Rusjan and M Alič, 'Capitalising on ISO 9001 benefits for strategic results' (2010) 27 *International Journal of Quality and Reliability Management* 756–778.

127 P Sampio, P Saraiva and AG Rodrigues, 'ISO 9001 certification research: questions, answers and approaches' (2009) 26 *International Journal of Quality & Reliability Management* 38–58.

128 HJ de Vries, 'Standardization – What's in a name?' (1997) 4 *Terminology – International Journal of Theoretical and Applied Issues in Specialized Communication* 55–83.

129 CAJ Simons and HJ de Vries, *Standaard of maatwerk* (2002).

is the area covered by standardisation, and the typical choice concerns balancing between unity and variety.

Studying such balancing is done in other disciplines as well. Examples vary from fuzzy logic in technical sciences, to variations on a theme in arts, to the development of species in evolutionary biology, and from differences between 'life world' and 'system' in philosophy[130] to controversies between moderates and orthodox in theology. So 'balancing' could be added to Boulding's listing of phenomena that are found in different disciplines and for which general theoretical models might be developed.[131] A characteristic of standardisation, then, is that it concerns this phenomenon in business practice and that the main considerations in the balancing between unity and variety have an economic nature. The standardisation discipline may learn from other disciplines that address this balancing, and they, in turn, may learn from standardisation.

J. Conclusions and discussion

In standardisation, activities occur – namely designing, selecting and applying standards; so if we regard studying standardisation as a scientific discipline, it is an activity-related science. Because the phenomenon is part of the functioning of companies (including their mutual relations and the relations with other parties such as customers and governments), standardisation science can be seen as a specialism within business science, along with marketing, financial management, information management, quality management and so on. This, however, does not exclude the input from fundamental sciences. Four of these appear to be the most important ones:

- technical sciences, because most standards provide solutions for technical (matching) problems;
- economic science, because standardisation is, by definition, an economic activity;
- law, because standards are laid-down agreements; and
- sociology, because of the social interactions in the process of standardisation.

130 J Habermas, *Theorie des kommunikativen Handelns*, 2, *Zur Kritik der funktionalistischen Vernunft* (1981) [English version: *The Theory of Communicative Action*, 2, *Lifeworld and System: A Critique of Functionalist Reason*].

131 KE Boulding, 'General systems theory – The skeleton of science' (1956) 2 *Management Science* 197–208.

Dooyeweerd helped us to classify sciences. However, more important in his philosophy is the observation of restrictiveness in scientific knowledge. An economic analysis, for instance, helps us to understand just one aspect of reality: the economic aspect. This is important, but it is dangerous to draw conclusions for business in the real world on the basis of an approach that is, by its nature, one-sided. Schoechle[132] stresses this: 'The literature of Business Management, Engineering Sciences and Law have treated standards in entirely instrumental and industrial process oriented terms. The social science disciplines of sociology, philosophy, political science and communication have almost entirely overlooked the phenomenon of technical standardisation per se.' Pondy and Mitroff[133] addressed this: 'The problem is that models not only direct attention *to* some phenomena and variables but also *away* from others. And if a model is highly successful in helping a researcher to cope with problems the model says are important, habituation will take place: The researcher will simply not "see" other problems, and he will have no basis for being receptive to competing models.' So we need additional scientific approaches. They may complete each other and business science may try to integrate them.[134] Although standardisation is a professional discipline for some people within companies and standardisation bodies, it is not an academic discipline. At the moment, it can be seen as a research topic that can be studied from several perspectives; it may develop as an independent specialism within business science.

K. Final remarks

The fundamental approach described in this chapter should be completed with a pragmatic approach, listing current problems in standardisation practice that ask for solutions and underpinning thereof. The latter should be offered by academic studies. Then the suitable scientific disciplines should be linked to these issues. For reasons of available space, I cannot carry out that exercise in this chapter.

On the basis of Dooyeweerd's findings, it should be possible to further develop the contours of a systematic theory on standardisation. The

[132] T Schoechle, 'Towards a Theory of Standards' in *Proceedings Second Interdisciplinary Workshop on Standardization Research*, 24–27 May 1999, at 87.

[133] LR Pondy, and II Mitroff, 'Beyond Open System Models of Organization' in BM Staw (ed.), *Research in Organizational Behavior* (1979) 3–39.

[134] A Bryman and E Bell, *Business Research Methods* (2007).

aspects distinguished by Dooyeweerd have a certain sequence, are interrelated and get their 'colour' from the most characteristic aspect, in this case, the economic one. Moreover, his insights into entities and their interrelations should be added. A first attempt, applied to ISO 9000 standards for quality management, can be found in De Vries[135]; a second one, on dissatisfaction with an in-company standardisation project, is described in De Vries and Haverkamp.[136]

A final remark that should be made is that international cooperation is necessary to understand standardisation. In formal standardisation at the international and regional level, Western Europe is leading. In de facto standardisation in the informations and communication technology sector, the United States dominates the scene. Best practices in company standardisation can probably be found in Japan, Germany and Turkey. The classic volume on standardisation[137] originates from India. The standardisation phenomenon as such has different traditions in different parts of the world, and so has academic research. Therefore, researchers from different disciplines as well as different countries should complement each other.

[135] HJ de Vries, *Kwaliteitszorg zonder onbehagen* [Quality Management without Uneasiness] (1999).

[136] HJ de Vries and A Haverkamp, 'Overcoming resistance against quality control – A philosophical-empirical approach' (2015) 32 *International Journal of Quality & Reliability Management* 1, at 18–41.

[137] LC Verman, *Standardization – A new discipline* (1973).

2

ISO standards and authoritative collective action

Conceptual and normative issues

HANS LINDAHL

A. Introduction

The aim of this chapter is to discuss certain fundamental conceptual and normative issues pertaining to the International Organisation for Standardisation (ISO). Conceptually, the question is this: why and how does ISO cast light on the concept of legal order, both in terms of the continuities that link it to earlier forms of legal order – paradigmatically, the state – and the discontinuities that separate it therefrom? This question situates an enquiry into ISO in the more general framework of the debate about the emergence of transnational or, by some accounts, global law. This chapter deals with this question by unveiling a model of legal order that is sufficiently capacious to accommodate a wide variety of putative legal orders while also sufficiently flexible to account for their structural differences: legal order as authoritative collective action, as I call it here. Normatively, the question is this: what understanding of authority would be adequate to the way in which ISO includes and excludes in the very process of constituting and maintaining itself as a legal order over time? If, as I argue, no collective can constitute itself as a legal order without including and excluding, then it is necessary to understand why such is the case, both in general and for ISO in particular, and how this impinges on the ISO's authority to set international standards that might be binding on their addressees.

B. ISO as a conceptual challenge to legal theory

It has become a truism to state that the Westphalian paradigm, premised on the assumption that law true and proper falls on either side of the divide between municipal and international law, has largely forfeited its claim to descriptive completeness. In particular, the emergence of normative orders that are diversely qualified as 'transnational', 'post-national' or 'denationalized', can no longer be described in terms of the canonical features of the state – namely, the exercise of exclusive power over the population within a bounded territory. These orders pose a challenge to the assumption that for there to be a legal order, at least in the full-fledged sense of the term, each of the traditional elements that is associated with statehood must be present: power that is both normative and enforceable; power that is exclusive in the twofold sense of exclusiveness with respect to a population and to a territory. As to the first element, the notion of 'soft' law calls into question the assumption that the enforceability of normative power is a constitutive feature of legal orders. As to the second element, we are witness to the 'unbundling' of normative power, such that emergent orders lay claim to regulating domains of behaviour that had previously been under the aegis of state power. Finally, the claims to normative power of these emergent orders are no longer limited by the borders of national territories; their spatial sphere of validity is transnational and, in some cases, global. These transformations do not merely demand enlarging the concept of legal order to accommodate at least some of these normative orders, in addition to municipal and international law. They also throw into crisis the very concept of a state predicated on the aforementioned distinctive features, as well as the attendant concept of international law, at least to the extent that it takes states, so defined, to be the proper subjects (and objects) of legal power.

ISO is, arguably, one of the emergent normative orders that refuse accommodation on either side of the divide between state and international law. Although composed of national standardisation bodies, ISO is an independent non-governmental organisation in its own right. Moreover, the standards it enacts are in principle voluntary and have the status of 'soft' law. Additionally, the personal and spatial scope of ISO standards is avowedly global: 'It is ISO's aim and expectation that each of its International Standards represents a worldwide consensus and responds to global market needs'.[1] If, initially, the questions about ISO's status as a normative order and the normative authority it exercises could seem merely academic issues best left to the abstract discussions of legal theorists, two developments have lent these questions considerable practical significance. The first is ISO's success in enacting a large number of international standards in response to the extraordinary growth of world

[1] ISO/International Electrotechnical Commission Directives, ix.

trade. Admittedly, this is no more than a quantitative criterion. But the fact that ISO is by far the largest purveyor of international standards becomes qualitatively significant in light of Article 2.4 of the World Trade Organization's (WTO's) Agreement on Technical Barriers to Trade:

> Where technical regulations are required and relevant international standards exist or their completion is imminent, Members shall use them, or the relevant parts of them, as a basis for their technical regulations except when such international standards or relevant parts would be an ineffective or inappropriate means for the fulfilment of the legitimate objectives pursued, for instance because of fundamental climatic or geographical factors or fundamental technological problems.

It would have been possible to interpret this article in such a way that international standards constitute 'a basis' for national technical regulations in that, as Robert Howse argues, 'there must be a reasonable relationship between the international standard and the domestic regulation', thereby 'leav[ing] a great deal to context, in recognition of the wide variety of normative material, and the many different purposes entailed in the broad notion of international standards'.[2] Instead, the Appellate Body, in the *EC – Hormones* and *EC – Sardines* decisions, effectively rendered international standards enforceable, as concerns the WTO's members.[3] Certainly, Article 2.4 allows for putting aside international standards in light of legitimate objectives pursued by a member of the WTO. But whether or not an objective is 'legitimate' is ultimately determined in the framework of the WTO dispute resolution mechanism, and the onus of the proof of such legitimacy rests squarely on the member invoking the objective.

It is against the background of these developments that I explore the question about ISO's status as a normative order, as well as of the nature and extent of its normative authority. My strategy will be to first sketch out the contours of a comprehensive and flexible concept of legal order – that is, one that draws out the common features of states and normative orders such as ISO, while also allowing for their differentiation. The point of this exercise is above all to pinpoint how and why the problem of normative authority is central to the concept of legal order. This will allow me, in

[2] See R Howse, 'A New Device for Creating International Legal Normativity: the WTO Technical Barriers to Trade Agreement and "International Standards"', in C Joerges and E-U Petersmann (eds), *Constitutionalism, Multilevel Trade Governance and International Economic Law* (2011), 385–6.

[3] See *EC Measures Concerning Meat and Meat Products*, Report of the Appellate Body, WT/DS 26 & 48/AB/R, January 16 and February 13, 1998; *EC – Trade Description of Sardines*, Report of the Appellate Body, DS 231/AB/R, September 26, 2002.

a second stage of the argument, to focus on the problem of normative authority confronting ISO in light of Article 2.4 of the TBT Agreement.

C. Authoritative collective action

Legal orders can best be characterised as authoritative collective action, or so I argue here. This thesis will be developed in two steps. First, I provide a nutshell characterisation of collective action, the *genus*. A second step introduces the form of normative authority that constitutes legal orders as a *species* of collective action.

Consider, to get started, a scenario in which several persons happen to be on the platform of a train station, waiting for the train to take them to their respective destinations. And now compare it to a second scenario in which a group of friends is standing on the platform, waiting to catch a train to go on a sightseeing trip to a certain city. Suppose that, in both situations, someone were to ask, 'What are you doing?' In both cases, the answer might be, 'We are waiting for the train'. But the use of the word 'we' is quite different in these scenarios. In the first, 'we' functions as an *aggregative* term: each of a number of individuals happens to be waiting for the train, independently of what the others are doing. Margaret Gilbert deftly characterises this use of the term 'we' as 'we each'. In the second, 'we' functions as an *integrative* term: a manifold of individuals functions as a group of friends engaged in a sightseeing trip; they act together, such that, for example, if one of the group members, who was distracted, hadn't noticed that the train was ready to leave, another group member would call out, 'Step on board!' Here we have a case in which 'we' functions as 'we together'.[4] When acting together, the members of a group take up a first-person plural perspective: we intend, believe, act, etc. as a *unit* even though there can be no intention, belief, or act by a group absent a set of interlocking intentions, beliefs, or acts by its participant agents.

To see the difference between the two scenarios, imagine that, in the first case, one of the persons who happened to be at the platform hadn't noticed that the train was about to leave; no one alerted her to this fact, and she only realised what had happened when it was too late. The would-be passenger would surely be annoyed at having missed the train. But she would have no standing to rebuke the other commuters for not having warned her about the train's imminent departure; they had no *obligation* to do so. In the second scenario, the friend who was left behind would

[4] M Gilbert, *On Social Facts*, 2nd edition (1992), 168.

certainly be entitled to rebuke her fellow sightseers. After all, travelling was something they were doing *together* as part of a joint sightseeing trip; what was the point of going together to visit city X unless they were prepared to help each other out to do so? See here the elementary structure of entitlements between members that attaches to action by social groups – 'plural subjects', to use Gilbert's favoured expression.[5]

Now, my claim is that legal orders are a species of collective action, as described earlier. In other words, legal orders take on the form of collectives in which a manifold of individuals act together over time. By the same token, legal obligations and sanctions are a species of the entitlements and rebukes that emerge between participant agents in the course of collective action. The nature and scope of legally relevant behaviour, as well as the rights and obligations that accrue to it, are internally related to the normative point of joint action: that which our joint act ought to be *about*. Significantly, what it is that we ought to do together – the normative point of joint action under law – may itself be open to discussion and conflict, as is the case with other forms of social groups. However, and in contrast to the latter, legal orders involve second-level authorities that, acting on behalf of the group, *monitor* participant agency with a view to realising the (transformable) normative point of joint action. By 'monitoring', I mean that second-level authorities lay down general and/or individual norms that are a default-setting of the normative point of joint action. Monitoring includes decisions in the face of conflict among the participant agents about the course that their joint action ought to take. If all of this goes into the monitoring of joint action by authorities, so also legal orders typically involve second-level authorities who *uphold* joint action, meaning by such that they take steps to ensure that participant agents comply with joint action and its normative point.

This account of legal order is broad because it includes – but is by no means limited to – municipal law. In particular, the mutually exclusive territoriality of states is but a limiting case of a broader spectrum of possibilities that includes the functional and regional legal orders that emerge with the uncoupling of law and state. Indeed, the emergence of 'overlapping' legal orders is the outcome of the differentiation and multiplication of the normative points and first-person plural perspectives governing the configuration of legal orders as concrete spaces of action. Whereas two or more 'overlapping' legal orders share a given geographical surface, they are different legal spaces because they differentiate and interconnect

[5] M Gilbert, *A Theory of Political Obligation* (2006), 189 ff.

places from different first-person plural perspectives and in view of different normative points. The same holds for membership: the very same set of persons may be members in two entirely different collectives to the extent that joint action under each is governed by different normative points. As concerns authority, the concept of monitoring I have introduced allows for a wide range of normative instruments. It obviously includes legislation, administrative acts, and judicial rulings, no less than international treaties. But it also includes, say, model contracts as deployed in certain sectors of *lex mercatoria*, the regulations drawn up by international accounting organisations, and, importantly, technical regulations and standards. No less importantly, although second-level authorities are required to uphold joint action under law, this need not mean physical coercion. Nor need joint action be upheld by the second-level authorities of the respective legal collective: it suffices that the authorities of another collective take the necessary steps to uphold joint action, as when national authorities enforce the award handed down by an arbitrator or an arbitral tribunal.[6]

D. Means and ends, inclusion and exclusion

What does the account of legal order as authoritative collective action tell us about ISO? Consider, to begin with, Article 2.1 of the Statutes of ISO, which reads as follows: 'The object of the Organization shall be to promote the development of standardization and related activities in the world with a view to facilitating international exchange of goods and services and to developing cooperation in the spheres of intellectual, scientific, technological and economic activity'.

The first point to explore is the development of international standards as a form of collective action. Indeed, standards are the result of a set of interlocking acts by a host of participant agents (among others, a technical committee, a group of experts, and ISO members), all of which are oriented to producing a standard in the course of an intricate, often recursive, procedure. The interlocking set of acts by participant agents allows the members of ISO, if consensus is achieved at the end of the procedure, to say, 'we (as a group) have developed a standard'. Importantly, the individual acts of participant actors, in the course of developing an international standard, are part and parcel of joint action to the extent that they share

[6] For a systematic development of this concept of legal order, see H Lindahl, *Fault Lines of Globalization: Legal Order and the Politics of A-Legality* (2013), esp. 81–90.

a common normative point – namely, to lay down a general rule that 'provides requirements, specifications, guidelines or characteristics that can be used consistently to ensure that materials, products, processes and services are fit for their purpose'.[7] Notice that the definition of a standard advanced by ISO entails that the development of a standard depends on the *purpose* of materials, products, processes, and services. Although the passage refers to purposes, it does so in a way that views it as largely *given in advance*, focusing attention on ensuring that standards are adequate to realising those ends. Clearly, however, the purposes served by standards are not fixed in advance, such that the task of standardisation is merely to establish the conditions that must be met with a view to ensuring that the product is 'fit' for its purpose. Standards also establish, explicitly or implicitly, which purposes are 'fit' for a product, service, material, or process. *To develop a standard is eo ipso to lay down both the end and the proper means to achieve it.* This gives the nay, from the very beginning, to a merely 'technical' – that is, neutral or apolitical – interpretation of standardisation. It is in this full sense of the term, therefore, that it makes sense to understand the development of international standards as a form of collective action: their development calls attention both to the normative point of collective action and to the default-setting thereof. Notice that this calls for an understanding of the concept of rationality that is neither simply reasoning about ends nor reasoning about means but rather about both, together. One way of putting this would be to say that standards call for a *hybrid* form of practical rationality, which is at once instrumental and purposive in scope. Another, perhaps more radical, approach would be to say that standards illustrate the general structure of practical reasoning, which is never merely about means nor merely about ends, but always a recursive form of reasoning in which means are assessed in terms of ends, at the same time that the ends of action are calibrated in terms of available means.[8]

So, prima facie at least, the definition of a standard advanced by ISO has the advantage of inadvertently revealing standardisation as the act of setting means *and* ends, thereby holding open the possibility of a rich normative discussion about the purposes for the sake of which materials,

[7] http://www.iso.org/iso/home/standards.htm (accessed October 13, 2013).

[8] One possible philosophical development of this idea would be to consider whether the kind of practical reasoning called forth by the enactment of (international) standards can be adequately conceptualised in terms of the distinction between what Jürgen Habermas calls 'strategic' and 'communicative' action. This discussion exceeds, however, the scope of this chapter. See J Habermas, *Theory of Communicative Action* (1985).

products, processes and services ought to display certain features. Although this hybrid structure also holds for national standards, the point of the standards developed by ISO is, as posited in Article 2.1 of the ISO Statutes, to 'facilitat[e the] international exchange of goods and services and to [develop] cooperation in the spheres of intellectual, scientific, technological and economic activity'. At one level, this formulation can be seen as involving a shift of the first-person plural perspective whence standardisation takes place: it is no longer only the perspective of a state but rather of the international or even global community with respect to which the means and ends of standards are to be laid down. At another level, the reference to the '*exchange* of goods and services' effectively frames the discussion about the ends and means of standards in terms of *trade*. By the same token, albeit implicitly, it frames the discussion about who, in the first instance at least, is an interested party to the development of (international) standards – namely, *traders*. Indeed, ISO goes on to state, when referring to the 'benefits' of its standards, that these 'ensure that products are safe, reliable and of good quality. For business, they are strategic tools that reduce costs by minimising waste and errors, and increasing productivity. They help companies to access new markets, level the playing field for developing countries and facilitate free and fair global trade'.[9]

This passage inadvertently reveals the ambiguities confronting international standardisation. At one level, one cannot but notice that it grants central stage to businesses as the primarily interested parties in standard-making. This inevitably calls forth the question of whether and how consumer interests might play a role in the determination of the means and ends of standardisation. The ISO/International Electrotechnical Commission (IEC) statement on consumer participation recognises that 'it is essential that consumer representatives participate in the standards development process' for products and services, especially products and services to be used by consumers.[10] However, as the statement acknowledges, 'consumer representatives face difficulties participating in the standards development process, due to a lack of financial resources and technological expertise'. As a result, the statement draws up a series of recommendations, addressed to national and international standards organisations, to promote greater consumer participation. Although the diagnosis of a

[9] http://www.iso.org/iso/home/standards.htm (accessed October 13, 2013).

[10] ISO/IEC Statement on Consumer Participation in Standardization Work, http://www.iso.org/iso/copolcoparticipation_2001.pdf, p. 1 (accessed on December 27, 2014).

lack of technological expertise is surely correct, framing consumer participation in terms of enhanced technological expertise, rather than in terms of the means-ends relation as a whole, tends to entrench a means-centred approach to standards and to downplay the political significance of standardisation.

This concern is confirmed by the seventh of the recommendations of the ISO/IEC statement, which states that national bodies 'should provide consumer representatives with guidance and training on standards procedures and with briefings on technical issues, in order to make their contribution both effective and based on a knowledge of real possibilities'.[11] Although it is certainly the case that well-informed representation is of great importance to effective consumer contributions, the expert-driven approach to standardisation tends to constrain the domain of what counts as a 'real possibility' with respect to the *purposes* to be served by standards. This, in turn, is of considerable importance with respect to the selection of consumer representatives: there is the real risk that only those consumer representatives and groups that are prepared to participate in the standardisation process in ways that are 'effective' and sensitive to 'real possibilities', as determined by experts, will obtain financing and support from national bodies.

But even if the difficulties confronting consumer participation can be addressed, both substantively and procedurally, the question remains which stakeholders are excluded from the process of standardisation to the extent that it focuses on consumers as users of services and products. What approaches to means and ends must remain invisible or are silenced when it is taken for granted that accounting for and balancing the interests of companies and consumers, and of developed and developing countries, is what is most fundamentally at stake in establishing standards conducive to 'free and fair global trade'?

These questions highlight a decisive feature of collective action in general and authoritative collective action in particular. Indeed, if there can be no collective action without a normative *point*, so also can there be no normative point without a normative *closure* – that is, an act of inclusion and exclusion that gets collective action going by establishing what is relevant and important with a view to realising its normative point and what is irrelevant and unimportant thereto. This normative closure is irreducibly ambiguous. On the one hand, exclusion has a positive normative significance: it is a condition for *normative empowerment* in the

[11] Ibid, p. 3.

sense of a domain of practical possibilities that are opened up and made available for participant agency. Without a normative closure, there could be no collective action, in the sense of an interlocking web of participant acts, nor could there be any directed or relational obligations, nor, finally, would there be any standing by a participant agent to demand of the other agents that they contribute to realising the point of their joint act. On the other hand, exclusion has a negative normative significance because it brings about *normative disempowerment* by placing other practical possibilities beyond the realm of what is deemed relevant and important for participant agency. In other words, the closure wrought by the normative point of collective action establishes what is deemed important and relevant to collective action, pushing all the rest beyond the pale and thereby marginalising it.

By these lights, the development of an (international) standard is inevitably an act of inclusion and exclusion: unless there is at least a minimal determination of what purposes are to be served by products, materials, services, and processes, standardisation is literally pointless; but including the purposes that products and such ought to serve involves excluding other possible purposes they could realise. No less important, the normative closure also involves the inclusion of members in, and the exclusion of, non-members from collective action, where 'membership' means those who are deemed to have an interest therein and who may, accordingly, claim standing in the collective, both to invoke rights and obligations and to rebuke participant agents who do not contribute to realising the normative point of collective action. So by approaching the purposes to be served by international standards in terms of facilitating international trade, ISO also determines who has an interest in (re-)negotiating the default-setting of standards and their purposes: primarily producers and, secondarily, consumers.

E. Authority and representation

The time is ripe to develop more fully the problem of authority. I noted in Section C that a legal order can best be understood as authoritative collective action. Legal orders involve, or so I argued, authority in the twofold sense of the authoritative monitoring of collective action and the authoritative upholding thereof. But the nature of authority presupposed herein remains in the dark. The question we must now address is how authority plays a role with respect to ISO and the specific problems of authority this plural subject confronts when developing international standards.

Let us begin with the authoritative monitoring of joint action. Notice that this form of authority has already been anticipated, albeit in passing, when I noted that the authority of (international) standards turns on the relation between means and ends that they posit. In effect, the authority of standards – their claim to being binding for their addressees – is never only about identifying the means necessary to realising a given end; their authority also and always depends on the ends posited as binding for the addressees of the standards. The development of standards always confronts ISO with a fundamental practical – in fact, *political* – question: what ought our collective action be about? Certainly, it would be reductive simply to qualify standardisation as a political process oriented to establishing the ends to be served by products, services, and some such; standardisation has an irreducibly *technical* dimension – that is, a dimension concerning the means adequate to achieving an end – that cannot be filtered out of its monitoring authority. My argument is not an argument against technical considerations in standardisation. Instead, my argument is that it would be no less reductive to filter out the political dimension of standardisation from the claim to authority raised by standards. Standardisation is a prime example of the intertwinement of epistemological and normative authority. To repeat the crucial point, to posit a standard is to posit the ends for the sake of which products and the like ought to display certain features. More precisely, at issue is the *collective* ends, the ends pursued by the collectivity and which define it *as* a collective. This entails that the authority of international standards depends on mechanisms that ensure a rich normative discussion about the ends they ought to serve, not least because ISO and other such entities effectively claim authority to posit standards on behalf of the world community *as a whole*.

When assessed in this light, ISO's extant standardisation procedure is clearly *not* conducive to the development of authoritative international standards in the full sense of the term indicated heretofore. Indeed, there is a strong bias in favour of a technical interpretation of standardisation. To begin with, ISO's statutes make no bones about the fact that the key organs involved in the development of standards are technical: a 'technical managing board', '(joint) technical committees', 'working groups (of experts)', and 'liaison organizations'.[12] To be sure, it is possible to appoint advisory

[12] So, for example, Category D liaisons shall include 'organizations that make a technical contribution to and participate actively in the work of a working group. This can include manufacturer associations, commercial associations, industrial consortia, user groups

groups within a technical committee, 'bearing in mind the requirement for sufficient representation of affected interests'.[13] However, not only does a *technical* board decide what counts as affected interests, but the members of these advisory boards are also nominated by national standardisation bodies, which are themselves technical organs. Not surprisingly, although ISO procedures allow for the maintenance of standards, that is, a renegotiation of standards as default settings of authoritative collective action, the recursivity of standard-setting is viewed, as is the case with their development, as a primarily technical matter. This is not to say, of course, that the procedure for the adoption of ISO standards curtails the possibility of a discussion about the purposes to be served by products and such, but the procedure is not set up in a way that explicitly recognises the strongly political dimension of standard-setting and, accordingly, invites a critical and broadly based discussion thereof. In effect, Article 2.1 of the ISO Statutes entails that the discussion about purposes is framed in terms of facilitating international trade. It is not obvious that the focus on international trade, even when it is construed in a way that includes consumer interests and the interests of 'developing' countries, facilitates a rich normative debate that integrates, say, social justice and protection of the environment as part and parcel of the purposes to be served by *all* international standards. By the same token, it is not obvious that the procedure could make room for other potentially interested parties, who, procedurally and substantively speaking, remain beyond the pale of collective action oriented toward setting international standards.

This lopsided approach to standardisation is particularly worrisome in light of the claim to global authority raised on behalf of ISO standards. Indeed, although in the first instance, these standards are issued on behalf of the first-person plural perspective of the ISO, the 'we' uttered on behalf of ISO by its second-level authorities, it is also uttered on behalf of a much broader 'we', the global community of interested parties in the development of international standards: 'It is ISO's aim and expectation that each of its International Standards represents a worldwide consensus and responds to global market needs'.[14] An important difference comes into

and professional and scientific societies'. Art. 1.17.2.2.2, ISO/IEC Directives, Part 1 – Consolidated ISO Supplement – Procedures specific to ISO.

[13] Ibid, Art. 1.13.2.

[14] ISO/IEC Directives, Part 1 – Consolidated ISO Supplement – Procedures specific to ISO, at: http://isotc.iso.org/livelink/livelink?func=ll&objId=14883169&objAction=browse&viewType=1 (accessed on June 18, 2015).

view between national and international standardisation. If the political dimension of standardisation can at least be accounted for in the framework of *national* standardisation, to the extent that state organs, including the legislature and the judiciary, can in principle monitor not only the procedure but also the choices of ends underpinning standards enacted by national standardisation organisations, no such political assessment on behalf of the global community seems possible for the standards issued by ISO. Whereas the exercise of state authority is organised in such a way that, at least in principle, potentially competing interests in facilitating trade, social justice, the environment, and other potential interests can be balanced, the structures of authority and the scope of the purposes served by the ISO systematically conceal the fact that there are such competing interests, let alone allowing for their balancing.

The problems encountered by ISO as concerns its monitoring authority spill over into the authoritative upholding of international standards. At least to a certain extent, the deficient claim to monitoring authority of ISO standards could be compensated by the fact that the adoption of these standards by their addressees is voluntary. Certainly, states may decide to declare ISO standards obligatory for one or more domains, such as health or safety. Moreover, ISO's addressees, such as businesses that claim that their products and services meet those standards, can in principle be held accountable for not fulfilling the conditions stipulated by the apposite standards, even though action would presumably need to be brought against them in the state in which they violate the conditions stipulated in the standards. As is the case with *lex mercatoria*, so also here the authorities that uphold international standards need not be the authorities of the collective that enacts them. But it remains the case that ISO standards are in principle voluntary in that states are not obligated to give them legal force, nor are businesses or other users of standards obligated to comply with them.

The problem of the authoritative status of ISO standards becomes acute with the rulings proffered in the framework of the dispute resolution mechanism of the WTO in the *EC – Hormones* and *EC – Sardines* cases, as these rulings effectively render international standards enforceable with respect to the WTO's members. Crucially, both rulings avoid any kind of scrutiny with respect to the authority of such standards, simply taking for granted that they are authoritative by dint of having been issued by an international standardisation body. The *Sardines* decision, in particular, is striking, given that it declared nonbinding the TBT Committee's decision, issued in 2000, that six principles should be taken into account in the process of developing international standards: transparency, openness, impartiality, consensus, effectiveness and relevance, coherence, and due attention given to the concerns of developing countries.[15] A shift is marked by the *US – Tuna II* decision, which held, on the one hand, that the Technical Barriers to Trade (TBT) Committee's decision is binding with respect to the interpretation of TBT rules as concerns standards, and, on the other hand, that international standardisation must be open and transparent.[16] In short, it would seem that the WTO is poised to engage in what one might call third-level authoritative monitoring and enforcement of international standardisation.

To what extent does this development contribute to enhancing the authoritativeness of ISO standards? A first and crucial point to bear in mind is that it will not do to isolate procedure and substance from each other, such that the authoritativeness of standards is made to turn on their satisfying certain procedural qualities, regardless of the content of the enacted standards. This caveat follows from the structure of authoritative collective action, as outlined heretofore. In effect, if the content of standards can be viewed as a default-setting of the normative point of collective action, so also the procedure through which standards are developed. More forcefully, procedure and substance are connected to each other by way of the normative point of collective action. This means concretely, in the case of ISO, that the reduction of standardisation to a largely technical procedure goes hand in hand with a reduction of the normative point of ISO to facilitating international trade. As long as it is taken for granted that what is procedurally at stake in ISO standards is giving adequate voice and weight to interests concerning international trade, then it will be possible to devise a procedure that meets the six criteria laid down by the TBT Committee yet blocks a wide-ranging normative debate that could include purposes such as social justice and environmental protection as part of the substantive criteria that should be met by all international standards. This spills over, of course, into the problem of inclusion and exclusion in the process of drawing up international standards: who is to partake in the procedure, hence what is to count as an open, transparent, impartial, and consensus-related procedure, cannot be isolated from a determination of the ends to be served by ISO standardisation. *What counts as a good procedure is always ultimately a substantive judgment – that is, a judgment about what ought to be the normative point*

[15] WTO, 'Decisions and Recommendations adopted by the WTO Committee on Technical Barriers to Trade since 1 January 1995', G/TBT/1/Rev. 10, June 9, 2011.

[16] *United States – Measures Concerning the Importation, Marketing and Sale of Tuna and Tuna Products*, WT/DS381/AB/R, of May 16, 2012.

of collective action. This means that there is no purely 'procedural' understanding of rationality in which an independent and neutral procedure can accommodate any and all substantive claims and counter-claims. Not only are procedures thoroughly imbued by substantive values, but they also presuppose a prior closure that determines who is to count as the authority that monitors whether the procedure is being followed – that is to say, what normative claims and counter-claims are to be allowed to enter the procedure and be settled within its framework.[17] Although I recognise that the aforementioned criteria of transparency and some such have a certain value, what is most urgently needed is to expose the strongly reductive nature of the current default-setting of the ISO as concerns its normative point and the procedure oriented to realising it. No listing of procedural qualities to be met by ISO standardisation can substitute for this essentially *political* task. More precisely, what is required is to give substantive interpretations of procedural principles that reveal, for the one, what is politically at stake therein and, for the other, what gets concealed in an allegedly 'content-neutral' interpretation thereof.

Crucially, a substantive reading of procedures is by no means limited to ISO: it goes to the heart of the WTO itself. Indeed, one may wonder to what extent an organisation such as the WTO has the normative wherewithal to critically monitor ISO standardisation. Inevitably, the question arises whether the WTO, which is, after all, oriented to fostering international trade, could be the proper institutional platform for a critical engagement with the ISO's avowed aim of 'facilitating the international exchange of goods and services'. More pointedly, should we expect the WTO's dispute resolution mechanism to do anything other than to assess the aforementioned six criteria in light of the normative point of promoting global trade, thereby effectively entrenching the normative point that requires critical engagement? This question seems especially apposite in light of studies that expose the producer-bias of WTO law and the impact of its functional logic on human rights.[18]

[17] Here again, the question is whether the 'procedural' interpretation of legal rationality that has been advanced by Habermas could deal with this problem – *quod non*. But that is another matter. See J Habermas, *Between Facts and Norms* (1998).

[18] See, among others, E-U Petersmann, 'Human Rights, Cosmopolitan Democracy, and the Law of the WTO', in IF Fletcher, M Cremona and L Mistelis (eds), *Foundations and Perspectives of International Trade Law* (2001), 79–96, and S Leader, 'Trade and Human Rights II', in P Macrory, AE Appleton and MC Plummer (eds), *The World Trade Organization: Legal, Economic and Political Analysis*, Vol. II (2005), 663–696.

Everything turns on how one interprets the task of promoting global trade – that is to say, in what way trade might count as the realisation of a *global good*. The Preamble to the Agreement Establishing the WTO is emphatic. The Parties thereto recognise, among other concerns,

> that their relations in the field of trade and economic endeavour should be conducted with a view to raising standards of living, ensuring full employment and a large and steadily growing volume of real income and effective demand, and expanding the production of and trade in goods and services, while allowing for the optimal use of the world's resources in accordance with the objective of sustainable development, seeking both to protect and preserve the environment and to enhance the means for doing so in a manner consistent with their respective needs and concerns at different levels of economic development.

This passage amounts to the normative point of the WTO, or more precisely, a default-setting thereof. Global trade is not an end in itself; it is worthwhile promoting to the extent that global trade is a means to realising these (and other) higher ends. Crucially, these ends provide a *critical perspective* on the current default-setting of trade as a means to realising global purposes. In this vein, the normative point of the WTO, as laid down in the Preamble to the Agreement, allows for a critical perspective on ISO's reductive interpretation of the purposes served by international standardisation. The WTO should assess procedural criteria such as the openness, coherence, transparency, and consensus-orientation of ISO *substantively* – that is, in terms of their capacity to realise its normative point, a normative point that has obtained a default-setting in the Preamble of the Agreement.

This means that the capacity of WTO to enhance the authoritativeness of ISO standards is internally linked to its own capacity to engage in a rich substantive debate about its own normative point, and the extent to which the rules it enacts to govern global trading are capable of realising its point. The current institutional setting of the WTO, which mirrors its highly reductive reading of the ends to be realised by promoting global trade, gives little cause for optimism about its capacity to enhance the authority of international standards.

3

From standards to quality infrastructure

A review of impact studies and an outlook

KNUT BLIND

A. Introduction

Standards fulfil various economic and social functions, which influence and improve the efficient allocation of resources. Therefore, the increased economic efficiency generated by the implementation of standards generates economic benefits not only for the supplying industry by increasing its profits, but also for the users, who pay lower prices or can implement goods and services of higher quality. In addition to these economic benefits, the implementation of standards generates social benefits for the consumers, the environment and the labour force by reducing possible damages and risks. Consequently, standards are able to improve both the welfare of the actors in the economy and the situation for society as a whole by better protection of the consumers, the labor force and the environment. In general, these economic impacts benefit both the supply and the demand side and are not in contradiction but are beneficial to the objective of improving the overall welfare of society. However, in some constellations, standards may change the distribution of welfare between actors in the economy, for example, by resulting in somewhat lower profits for the supplying industry, but higher social benefits for consumers.

The classification of standards can be performed in several ways. Economists proposed a categorisation based on the economic effects of standards, which is most suitable for the illustration of the benefits of standards for the global economy. This is a useful approach to analyse

This article is derived in part from K Blind, 'The Benefits of Standards for National Economies', in WSC Newsletter No. 03, October 2011. Available at: http://www.worldstandardscooperation.org/newsletters/003/newsletter03.html.

Table 3.1. *General effects of standards*

Type of standard	Positive effects	Negative effects
Compatibility/Interface	• Network externalities, i.e. frictionless communication • Avoiding lock-ins in old technologies • Increased variety of systems products	• Monopoly
Minimum Quality/Safety	• Correction for adverse selection by low-quality suppliers • Reduced transaction costs • Correction for negative externalities	• Restriction of competition
Variety Reduction	• Economies of scale • Building focus and critical mass	• Reduced choice • Market concentration
Information	• Reduced information asymmetry • Reduced transaction costs	• Restriction of competition

Source: K Blind, 'The economics of standards: theory, evidence, policy.' Edward Elgar Publishing (2004), modified after Swann, 'The economics of standardization', in Report for Department of Trade and Industry, Standards and Technical Regulations Directorate, 2000, p. 90. Available at http://www.dti.gov.uk/strd/fundingo.htm#swannrep

both economic driving forces for standardisation and the economic impact dimensions of standards.

In the classification, we use the dimensions of compatibility, quality, variety-reducing and information standards. However, even if standards are developed just to serve one purpose they often fulfil multiple functions. Therefore, it is unlikely that each standard will fall exactly and exclusively into a single category, but several economic aspects are often touched by a single standard. However, a distinction is important for the theoretical discussion, because standards have different economic effects.

Table 3.1 summarises these four purposes of standardisation and highlights their positive and negative effects. The usefulness of this distinction lies in what it contributes to our understanding of the variety of standard types and their sometimes ambivalent effects. Compatibility and interface standards are crucial for the development of network industries such as information and communication technology because they allow the communication between different users or the interplay between hardware and software. Minimum quality and safety standards prohibit inferior and even risky goods and services from being offered by suppliers, which increases the confidence of consumers in new, innovative products and services. Because standards often define a certain specification, they automatically exclude other options and reduce variety, if the standard succeeds, but represent critical masses attracting new investors and companies contributing to the development of new markets. Standards related to information and product description are usually treated as a distinct category, but for many purposes, it is sufficient to treat these as a hybrid of the foregoing three categories. A standardised product description often has the function of a quality standard, eases compatibility with other products and is an expression of product variety.

The likelihood of negative effects of standards can be reduced if the standardisation process follows the World Trade Organization (WTO) principles of transparency, openness, impartiality and consensus, effectiveness and relevance, and coherence because then the potential to misuse standards by specific stakeholders in the standardisation process can be reduced or avoided.[1] Here I note the phenomenon of 'regulatory capture',[2] which describes the situation in which some producers may lobby so effectively in the standardisation development processes that standards are eventually defined in the interest of specific producers rather than of all interested stakeholders, including the customer (as originally intended). This can also be applied to the strategy of referencing patents to protect companies in the standardisation process, which can lead to an even stronger monopolisation of technologies and consequently of markets, especially in the field of information and communication technologies.

In the following subsections, I roughly follow this typology to illustrate and discuss the benefits of standards for the global economy as a whole, starting with their impact on innovation, global trade and value

[1] World Trade Organization Committee on Technical Barriers to Trade. *Decisions and Recommendations Adopted by the WTO Committee on Technical Barriers to Trade since 1 January 1995*. Geneva (2011).

[2] JJ Laffont and J Tirole, 'The politics of government decision-making: A theory of regulatory capture.' *The Quarterly Journal of Economics* 106, no. 3 (1991), 1089–1127.

chains, economic growth and social objectives. Finally, I widen the focus from the impacts of standards to those of the wider concept of quality infrastructure.

B. Innovation

Innovation is the major component of countries' competitiveness and growth but is also the key to solving the greatest challenges to today's societies, such as climate change, scarcity of natural resources, health and security. Consequently, we must expand the role of standards for innovation.[3]

On the basis of the various functions of standards displayed in the Introduction, I describe here an additional dimension, which makes explicit that standards play a crucial role in the research and innovation process with its various feedback loops (see Figure 3.1). More specifically, I distinguish the various roles of different types of standards within and between the different phases, particularly in the research process, bearing in mind the innovation cycle as a whole.[4]

Research in emerging fields of science and technology often leads to the creation of new terms and concepts. Consequently, terminology standards are already required in basic research investigating new technologies, such as nanotechnology, to allow or facilitate efficient communication, but they also play a crucial role in the transfer of knowledge from basic to oriented-basic and applied researchers. The transfer of knowledge from basic to applied research also requires measurement and testing of standards, which allows progress toward initial product-related developments. The gap between applied research and experimental development of new products and processes is facilitated by interface standards, which allow interoperability of components integrated into product or process technology. Finally, the transition of pilot products into mass markets requires, in addition to compatibility standards, ensuring interoperability between products or whole systems, and, above all, quality standards that guarantee the products comply with some minimum safety regulations. The role of compatibility standards as a mediator between research contexts is additionally moderated by the scientific and technological field of the

[3] K Blind, 'The impact of standardisation and standards on innovation.' In *Handbook of Innovation Policy Impact*. Edward Elgar (2016).

[4] K Blind and S Gauch, 'Research and standardisation in nanotechnology: evidence from Germany.' *The Journal of Technology Transfer* 34 (2008), 320–342.

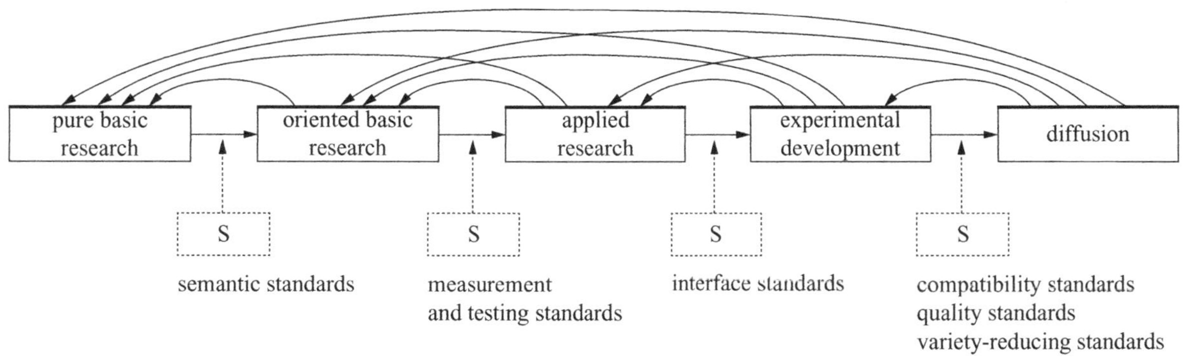

Function of Standards	Reduction of information cost Reduction of transaction cost	Interoperability between components Savings in adaption cost	Increased quality Reduced health, safety, privacy risks Building critical mass Economies of scale Creation of network externalities Interoperability between products

Figure 3.1. Various Roles of Different Types of Standards in the Innovation Process

Source: K Blind and S Gauch, 'Research and standardisation in nanotechnology: evidence from Germany.' *The Journal of Technology Transfer* 34 (2008), 320–342

researchers. Fields that incorporate an underlying logic of networking of products, information and communication technology being the prime example, or other fields that facilitate large technical systems moderate the relative importance of compatibility and interface standards at early stages in the research process.

Requirements such as quality and safety standards are either set by a regulator,[5] who refers in his or her regulation to a set of specific standards, as in the New Approach[6] within the European Union, or are demanded by lead users and early adopters who are interested in the new products but not willing to accept rather large risks by using these unknown products. Finally, the diffusion of new products is fostered by variety-reducing standards that allow the exploitation of economies of scale and by compatibility standards generating positive network externalities among users.

In general, the self-regulation of the actors in the innovation system via standardisation promotes existing trajectories of technologies but is not an instrument to push radical innovations, that is, completely new technologies. The involvement of all relevant stakeholder groups leads to a consensus of interests taking all preferences of the involved parties into account. This procedure has implications for the released specifications and consequently for the impacts on innovation. If we assume, that in addition to the technology providers, i.e., the supply side, users and consumers, for example, are adequately represented, as well as representatives from environmental groups and other stakeholders in society, then we can expect standards that promote incremental innovations balancing the interests of the involved stakeholders. This means that they promote user-driven or customer-oriented solutions and also take environmental concerns into account, which is different from simple technology-driven specifications. However, despite the openness of standardisation processes, weak stakeholders such as consumers, but also small and medium-sized enterprises, have difficulty participating adequately in these processes. Therefore, they require special support to avoid the development of specifications biased to the preferences of dominant players with possible negative implications on innovation.

In summary, standards are, on the one hand, crucial infrastructure for research, development and innovation. On the other hand, they are also

[5] K Blind, B Bührlen, K Menrad, S Hafner, R Walz and C Kotz, 'New products and services: Analysis of regulations shaping new markets.' Study financed by the European Commission (2004), Luxembourg.

[6] Available at: www.newapproach.org.

a channel and outlet for the broad dissemination and therefore implementation of new insights from science and technology. If both functions are effectively coordinated, for example, by integrating standardisation in research programs and projects,[7] the performance of the global innovation system will be significantly improved in the long term. So far, there is little empirical evidence of the innovation fostering impact of standardisation (see Blind 2015; Shin et al. 2015[8]). In general, positive interactions are identified. For example, Lim and Prakash (2014)[9] find a positive influence of implementing the environmental management system ISO 14000 on countries' innovation performance measured by patents in environmental technologies.

C. Global trade and value chains

Classical trade across borders is increasingly substituted by global value chains; within international production, trade and investments are organized across the globe. Returning to the general economic functions of standards, the same arguments elaborated in the introduction to this chapter apply to international trade and global value chains. Hence, any standard is trade-enhancing and reduces the frictions in global value chains because standards reduce information asymmetries, signal quality to customers and create a common language for potential trading partners, thus reducing the overall transactions costs.

This line of argument applies both for exports and imports. However, countries focus especially on exports because they are an indicator for international competitiveness and an important source for growth and employment. Nevertheless, imports are under the new paradigm of global value chains that are not only beneficial for a country if the imported products and services are cheaper than domestic production and the potential importers can put positive competitive pressure on the domestic industry; they are also an opportunity for intermediate products to act as an indicator of a country's competitiveness as locus for production and distribution.

[7] Available at: http://www.cencenelec.eu/research/news/publications/Publications/STAIR.pdf

[8] K Blind, 'The impact of standardisation and standards on innovation.' In *Compendium of Evidence on the Effectiveness of Innovation Policy Intervention.* Edward Elgar (2015) and D-H Shin, H Kimb and J Hwang, 'Standardisation revisited: A critical literature review on standards and innovation.' *Computer Standards & Interfaces* 38 (2015) 152–157.

[9] S Lim and A Prakash, 'Voluntary regulations and innovation: The case of ISO 14001.' *Public Administration Review* 74, no. 2 (2014), 233–244.

In addition to the different impact of standards on exports and imports, we must differentiate between national and international standards. International standards facilitate imports, whereas national standards might be used to protect domestic suppliers but can also reduce information asymmetries for foreign competitors. In contrast, national but especially international standards promote the export performance of companies because both are able to reduce production costs and signal quality to customers. In addition, the development of global value chains, which are benefitting not only single economies but the global economy, crucially depends on international standards. Finally, countries trying to integrate their separate economies into common markets (e.g. the Single Market within the European Union) must completely remove technical barriers to trade. Here, merging and consolidation of the individual national stocks of standards into a common standardisation union have proven to be an effective and efficient strategy within the European Union and elsewhere.

A precondition for the general trade-enhancing impacts of standards is a level of compliance cost and openness that does not restrict market entry of and competition between companies.

The development of international standards has long been driven by stakeholders of the industrialized nations, especially by the United States and Germany. Recently, Asian countries, particularly Japan and South Korea and now also China, are taking the initiative in international standardisation, measured by the increasingly chairing of technical committees in international standardisation organisations such as the International Standardisation Organisation (ISO). Obviously, the development processes of international standards also have to be taken into account when analysing their trade impacts.

Swann[10] reviewed the various studies assessing the impact of standards on trade. Overall, the findings show that there is often, but not always, a positive relationship between international standards and exports or imports, which is in line with the arguments elaborated in this chapter. For national (i.e. country-specific) standards, the results are ambivalent, partially supporting the commonly held view that national standards can be barriers to trade. Obviously, there is no single answer to the question of the net trade effects of standards caused by their multiple economic effects and differences in their elaboration and implementation.

[10] GP Swann, 'International Standards and Trade: A Review of the Empirical Literature.' OECD Trade Policy Working Papers, No. 97. OECD Publishing (2010).

In addition, the role of standards for the development of global value chains has hardly been empirically investigated yet,[11] although international standards are obviously promoting global sourcing and distributed production.[12] As argued earlier, the development process of international standardisation has not yet been adequately addressed in empirical analyses of the impact of international standards on trade flows and global value chains.

D. Economic growth

The tradition of studies on the impact of standards on economic growth started in principle with the study of the German Institute for Standardisation (DIN) on the Economic Benefits of Standardisation.[13] In these studies, standards are perceived as the output of research and development or even broader innovation activities; consequently they can be used as other science and technology indicators[14] for assessing their impact on economic growth. The basic assumption is that the total output of an economy is based on the input of capital (i.e. machinery) and labour (i.e. employees). A third factor, which is increasingly driving economic growth, is technical progress or, even more broadly, the pool of knowledge, which explains more than the half of economic growth. From the section on innovation, we learned that standards help promote research and innovation. Consequently, standards are also a driving factor for economic growth via innovation. In addition, the broad implementation of standards comprising the most recent state of the art in science and technology is an efficient channel of technology and knowledge transfer that facilitates the development of new markets by building critical mass on

[11] See a first analysis by F Ramel, A Mangelsdorf and K Blind 'The Effects of Standards on Value Chains and Trade in Europe.' Proceedings of the 20th EURAS Annual Standardisation Conference, 'The Role of Standards in Transatlantic Trade and Regulation' (2015), eds. K Bergh Skriver, K Jakobs and J Jerlang, 341–358. An overview on global value chains can be found in Organisation for Economic Co-operation and Development (OECD), World Trade Organization (WTO) and World Bank Group, 'Global Value Chains: Challenges, Opportunities, and Implications for Policy.' Report prepared for submission to the G20 Trade Ministers Meeting, Sydney, Australia (2014).

[12] For information on communication and technology, see K Blind, S Gauch and R Hawkins. 'How stakeholders view the impacts of international ICT standards.' *Telecommunications Policy* 34, no. 3 (2010), 162–174.

[13] DIN, 'Economic Benefits of Standardisation – Summary of Results: Final Report and Practical Examples.' DIN German Institute for Standardisation (2000).

[14] K Blind, *The Economics of Standards: Theory, Evidence, Policy*. Edward Elgar (2004).

one hand while also increasing companies', and consequently economies', productivity on the other. Higher productivity is also achieved by lowering production costs, particularly by the implementation of standards in more mature industries, but also by promoting the development of global value chains. Finally, standards level the playing field between companies, promoting competition by requiring higher productivity via process innovation and even more innovative products. Companies can also improve their productivity by outsourcing specific tasks to more efficient external producers or service providers within global value chains. Here, interface standards in particular facilitate outsourcing (i.e. a further division of labour between companies) by lowering the transaction costs and further differentiating global value chains.

So far, we have elaborated the positive impacts of standards on the supply side, particularly enhancement of productivity. However, economic growth is crucially dependent on sufficient demand for products and services. Here, we can refer back to the trade-enhancing role of standards: reducing barriers to trade and facilitating the integration of the global economy. In addition, economies can improve their international competitiveness by setting and implementing international standards. They increase the acceptance of a company's products and services by global customers but also help to reduce costs and therefore prices, leading to higher demand for domestic products and services – a necessary condition for economic growth.

The economic impact of standards on economic growth has been assessed only since the last decade. On the basis of the assumption that codified technological know-how, like patents but also standards, contributes to economic growth, Jungmittag et al. estimated a production function based on macroeconomic data for Germany ranging from 1960 to 1996 within the pioneering DIN 2000 study.[15] The main result of this pioneering exercise is the significant contribution to German growth of the active number of standards, which is similar to the impact of the number of granted patents. Recently, an update of the study confirmed the results of the original work.[16]

[15] A Jungmittag, K Blind and H Grupp, 'Innovation, standardisation and the long-term production function: A co-integration approach for Germany.' *Zeitschrift für Wirtschafts- und Sozialwissenschaften* 119 (1999), 205–222.

[16] K Blind, A Jungmittag and A Mangelsdorf, *The Economic Benefits of Standardisation. An Update of the Study Carried Out by DIN in 2000*, DIN (2011).

A first follow-up study for the United Kingdom[17] replicate the approach and find significant, but lower, impacts of standards on total factor productivity and economic growth.[18] The same is true for a Danish,[19] a Canadian[20] and an Australian study.[21] One explanation for the lower contribution of standards on economic growth can that these countries mainly use international standards and significantly reduced their production of idiosyncratic national standards beginning in the 1990s. France, the third large EU country producing standards,[22] replicated the German approach and compares their results with those of the German research, finding almost identical impacts of approximately 1% of the annual gross domestic product. Blind and Jungmittag[23] performed a comparative study for Germany, France, Italy and the United Kingdom based on pooling data from twelve industry sectors. The empirical results confirm that both the stock of patents and the stock of technical standards contributed significantly to economic growth in the 1990s. Although the results among the four countries are similar, significant differences between the twelve sectors are observed, indicating that standards are more important for growth in less R&D-intensive industries, compared with patents taking precedence in R&D-intensive industries.

E. Social objectives

So far this chapter has focused on the economic benefits of standards for the global economy, that is, the economic dimension of sustainability. However, many standards have a direct and beneficial impact on the

17 UK Department of Trade and Industry DTI. 'The Empirical Economics of Standards.' DTI Economics Paper No. 12 (2005).

18 Recently, the updated study by Cebr 'The Economic Contribution of Standards to the UK Economy.' published by BSI in 2015 confirmed the results of the studies commissioned by DIN.

19 Centre for Economic and Business Research (CEBR), 'Standards and Economic Growth in Denmark' (2007).

20 J Haimowitz and J Warren, 'Economic Value of Standards.' Standards Council of Canada (2007).

21 Centre for International Economics (CIE), 'Standards and the Economy.' Standards Australia (2006).

22 Association Française de Normalisation, 'The Economic Impact of Standardisation – Technological Change, Standards Growth in France.' AFNOR Group (2009).

23 K Blind and A Jungmittag, 'The impact of patents and standards on macroeconomic growth: a panel approach covering four countries and 12 sectors.' *Journal of Productivity Analysis* 29, no. 1 (2008), 51–60.

global society, that is, on the environmental and social dimension of sustainability. These include 'traditional' consumer or work force–related standards such as those for domestic appliances, machine safety or signs and symbols or quality (e.g. ISO 9000), as well as those newer types of standards for sustainability (e.g. ISO 14000), social responsibility (e.g. ISO 26000) or services. If the participation of those stakeholders interested in these topics is ensured, the content of standards will reflect the needs and proper expectations of the general public with regard to factors such as safety and security, labelling, accessibility, fairness and sustainability not only in its economic sense, but also in both its social and environmental dimensions. Overall, the economic impacts related to the implementation of ISO 9000 and ISO 14000 are mainly positive, but no evidence has been collected yet related to ISO 26000.

The protection of health, environmental and safety issues can and has been achieved by governmental organisations, such as regulatory bodies, to avoid market failures that result in risky products and production processes or generating significant damages to the environment. However, standards play an important role in achieving better regulation in two dimensions. First, the involvement of all relevant stakeholder groups in the standardisation process leads to a consensus of interests taking all preferences of the involved parties into account. This procedure has positive implications for the acceptance and implementation of the released specifications compared with regulations set by regulatory bodies in a more top-down approach. Second, those companies that voluntarily adopt standards could be entitled to a lighter regulatory regime, such as that implemented in the Technical Barriers to Trade Agreement. One extreme version of this could be that the implementation of standards, also required for participating in global value chains,[24] would become a substitute for compliance with a given regulation, making inspections by regulatory bodies superfluous.

Finally, standards can be referenced by public procurers not only to make public procurement more efficient[25] but also to promote demand-driven innovation policies,[26] which again and eventually shows the close interaction of the various functions of standards for whole economies.

[24] OECD, WTO and World Bank Group, supra note 11.

[25] K Blind, 'Driving innovation – Standards and public procurement.' *ISO Focus* (September 2008), 44–45.

[26] OECD, 'Demand-side innovation policies.' OECD Publishing (2011).

F. An interim summary

Thus far, we have elaborated the various benefits of standards for the global economy. However, the economic focus must be widened. Tackling tremendous future challenges such as global warming; tightening supplies of energy, water and food; ageing societies; public health; pandemics; and security, cooperation is required. Here, international standards can help enable and improve international collaboration, especially because we lack a global regulatory framework to achieve the various social objectives. However, the development and implementation of common international standards benefit not only our ability to find solutions to these various global challenges, such standards would also allow better collaboration among companies in global value chains. First, today innovations are only feasible and successful through the close collaboration of partners with complementary expertise and functions within the innovation systems, which are no longer located around the globe. Furthermore, the economic impact of innovative products and services is only fully exploited if implemented on a global scale. Finally, the social impact of innovations is greatest if all citizens worldwide can make a use of or benefit from them. Consequently, international standards are an important infrastructure for global research, development and innovation processes (as proved for information and communication standards by the stakeholders' assessment in Blind's work[27]). Likewise, the outcome of all these processes should be reflected in further development of international standards to sustainably fulfil the various functions for research, development and innovation.

As already elaborated, international standards can help promote innovative products and services. However, from trade theory, we know that eliminating trade barriers completely, which can be achieved by replacing different national standards with common international ones, is beneficial for all involved economies – at least in the long run, because the further division of labour between countries according to their production capacities and resource endowments allows higher output of products at lower costs.

In summary, international standards are pushing the growth of the global economy by providing a common infrastructure for global research, development and innovation; by integrating national markets,

[27] K Blind, S Gauch and R Hawkins, supra note 12.

which allows a better division of labour between economies and therefore increases productivity at the global level; and finally, by helping tackle the growing future global challenges that all nations face.

G. From standards to quality infrastructure

The focus on analyzing the impacts of standards has to be widened to all dimensions of sustainability, with an eye toward recognizing all opportunities within the entire system of quality infrastructure. A quality infrastructure is a system of institutions that jointly ensure that products, processes and services meet certain predefined specifications. Most notably, a quality infrastructure enables companies to improve not only their production processes, but also their products, facilitating compliance with regulations or international requirements, particularly international standards. The elements of quality infrastructure are, in addition to standardisation and the produced standards, conformity assessment (i.e. certification) and accreditation, but also metrology.[28] These components are highly complementary to standards. To illustrate, it is only possible to formulate a standard regarding the properties of a product if the respective units of measurement are defined and adequately calibrated measurement instruments exist. The standard is only made effective when producers comply with it, which is more likely if there are incentives to do so – for instance, if compliance can be indicated to potential customers through a certification scheme. The impacts of certification are more significant the greater the trust customers have in the certification institute, which is higher if an accreditation body has evaluated the competences of the certifier. The high interdependencies between the various elements of quality infrastructure require a comprehensive or holistic approach by perceiving these elements and their activities as integral parts of a complex system. They cannot be tackled as independent activities isolated from the functioning of the whole. This is an important consideration when setting up a new or improving an existing quality infrastructure. It requires comprehension of how the components support each other,

[28] JL Guasch, JL Racine, I Sánchez and M Diop. 'Quality Systems and Standards for a Competitive Edge.' The International Bank for Reconstruction and Development/The World Bank (2007); C Sanetra and RM Marbán, 'The Answer to the Global Quality Challenge: A National Quality Infrastructure.' Physikalisch-Technische Bundesanstalt, Braunschweig (2007).

the embedding of each element within the whole system, and the framing of this infrastructure as a pillar of the national economy or the national innovation system.[29]

Conformity assessment is the demonstration that what is being supplied actually meets the requirements that are specified or claimed. Conformity assessment can be applied to products or services, a process, a system, a body or persons and includes activities such as testing, inspection and certification.[30] Within the system of quality infrastructure, conformity assessment has the function of differentiating those goods and services that conform to a standard and those that do not. Therefore, the importance of the conformity assessment is to increase the information that customers have regarding the products, including services, and processes that are certified, thereby creating incentives for producers to upgrade their production processes to eventually meet the requirements of standards. It establishes a visible link between standards and products.[31] Conformity assessment comprises the following activities:[32]

- Testing and inspection: testing serves as a means of studying the characteristics, contents or quality-determining parameters of products, components, substances or processes. Depending on the respective testing field (chemical, microbiological, physical testing, etc.), the methods of analysis, test or inspection will differ along with their respective analytical equipment and test apparatus. All tests of whatever type are decisively dependent on universally applicable, recognized guidelines defining how and under what circumstances the test in question is to be implemented. In this connection, standards have a major role because many test methods are standardized to guarantee that the results obtained will be mutually comparable and reproducible.
- Certification: certification is a third-party attestation declaring that specified requirements pertaining to a product, person, process or management system have been met. The certification of management systems is sometimes referred to as registration. In addition to being

[29] J Gonçalves and J Peuckert, The contributions of quality infrastructure to national innovation systems: Implications for development policy. In: W Siyanbola (ed), *Innovation Systems and Capabilities in Developing Regions*. Ashgate Publishing (2012), pp. 35–48.

[30] Guasch et al., supra note 28.

[31] UK Department of Trade and Industry DTI, Conformity assessment policy in the United Kingdom. DTI, London (2007), Vol. 07/594.

[32] C Sanetra and RM Marbán, supra note 28.

conducted by an independent party, a characteristic of certification is that it often involves an element of surveillance. Once something is certified as compliant, it may be subject to the continuing verification by the certifier of its compliance both before and after it reaches the market (premarket and postmarket surveillance).[33]

Whereas conformity assessment focuses on companies and other organisations complying with specific requirements, especially international standards, accreditation is related to those organisations that perform conformity assessment. The accreditation itself is a statement from an accreditation body – an independent third-party entity – declaring that specified requirements related to conformity assessment bodies have been met and that the accredited body is competent to perform certain functions. Accreditation bodies conduct evaluations based on the series of international standards, ISO 17000. The accreditation bodies themselves must comply with international standards to demonstrate that their competence may also be subject to peer review, performed by regional and international accreditation organisations.

Whereas conformity assessment and accreditation refers to the existing body of standards, metrology has a different function. Metrology is the technology or science of measurement. The effective implementation of many standards depends on measurement instruments, on the accuracy of the existing tools and procedures, on the dissemination of the measurement knowledge and access to measurement technology throughout the productive agents. Consequently, new standards require the development of new measurement approaches, or the latter will allow more advanced standards to be developed, that is, we have a reciprocal relationship between standardisation and metrology.

The ability to bring quality into new domains of social, economic and environmental realities and to benefit from a better correspondence between the actual characteristics of a product and formalized expectations depends on the functioning of metrology, which can be subdivided as follows:[34]

- Scientific metrology: the organisation and development of the highest level of measurement standards. The role of scientific metrology is

[33] American National Standards Institute (2014). Available at: http://www.standardsportal.org/usa_en/conformity_assessment/3party_conformity_assessment.aspx

[34] Sanetra and Marbán, supra note 28.

to establish a measurement system to maintain, develop and diffuse measurement standards for basic units and to disseminate metrological expertise to the economy. Moreover, the scientific metrology institutes develop secondary standards, which are the means for the comparison of a country's national standards with other countries' standards.
- Legal metrology: the accuracy of measurements where these have an influence on the transparency of economic transactions, health and safety. Legal metrology involves the technical procedures that cover regulated areas where there is a public interest in the correctness of units, measurements and measuring devices.
- Industrial metrology: proper maintenance and control of industrial measurement equipment, including calibration of instruments and working measurement standards.
- Calibration: the activities of checking and adjusting (by comparison with a measurement standard of higher accuracy) the exactness of a measuring tool through the regular repetition of measurements.

Turning to the empirical evidence of the impacts of the quality infrastructure as a whole and its single components with the exception of standards, we identify a large gap. On one hand, the impacts of conformity assessment have been studied in depth with the examples of ISO 9000 and ISO 14000 finding positive effects on innovation, trade (e.g. Swann[35]) and economic performance. On the other hand, in relation to accreditation, Frenz and Lambert[36] performed an initial assessment of the general economic impacts of accreditation relying on survey data. Blind and Mangelsdorf[37] provide empirical evidence of how being a member of the international accreditation organisation, especially for the developing countries, promotes exports, while the influence of certification loses importance.

Finally, the insights related to the impact of metrology are limited and mainly based on indirect effects. Temple and Williams[38] found a positive

[35] GP Swann, supra note 10.

[36] M Frenz and R Lambert, 'The Economics of Accreditation.' United Kingdom Accreditation Service (2013).

[37] K Blind and A Mangelsdorf, 'The trade impact of ISO9000: certifications and international cooperation in accreditation', *Proceedings of the 17th Annual EURAS Standardization Conference 'Standards and Innovations'* (2012), eds. M. Orviska and K Jakobs, 21–33. Available at: http://www.gbv.de/dms/tib-ub-hannover/716889994.pdf.

[38] P Temple and G Williams, 'Infra-technology and economic performance: evidence from the United Kingdom measurement infrastructure.' *Information Economics and Policy* 14, no. 4 (2002), 435–445.

impact of research and development in measurement technologies for the economic growth, whereas Choudhary et al.[39] found empirical evidence for the trade-enhancing impact of metrology within the European Union.

The overall impact of quality infrastructure illustrated by case studies can be found in Gonçalves and Peuckert,[40] whereas Harmes-Liedtke and Oteiza[41] find positive correlations between the quality infrastructure as such and the economic level and growth of economies. Finally, Gonçalves and Peuckert[42] and Harmes-Liedtke[43] underline the contributions of quality infrastructure to national innovation systems, particularly in developing countries.

H. A future research agenda

Reviewing the numerous impact studies related to standards and the few analyses about the effects of quality infrastructure reveals a rather large and complex research agenda. First, the focus of the studies of the economic impact of standards has to be expanded to new complex phenomena, such as global value chains, but also to the service sectors, which have only recently started to produce standards. This is also relevant for some new technologies, such as biotechnology. Second, the effects of standards must be extended to the other two dimensions of sustainability – environment and society. Initial studies have begun to address the impact of standards on the environment, but the social dimension of standards has been largely neglected.

Third, the few studies focusing on the impact of quality infrastructure reveal the complexity of impact dimensions, which also put the influence of standards into an adequate context. Although we have gained initial insight into the function of conformity assessment, the role of the accreditation system needs broader, in-depth approaches. Even more important,

39 MA Choudhary, P Temple and L Zhao, 'Taking the measure of things: the role of measurement in EU trade.' *Empirica* 40, no. 1 (2013), 75–109.

40 J Gonçalves and J Peuckert, 'Measuring the Impacts of Quality Infrastructure: Impact Theory, Empirics and Study Design', Guide No. 7/2011, Physikalisch-Technische Bundesanstalt, Braunschweig (2011).

41 U Harmes-Liedtke and JJ Oteiza Di Matteo. 'Measurement of Quality Infrastructure.' Discussion Paper 5/2011, Physikalisch-Technische Bundesanstalt, Braunschweig (2011).

42 Gonçalves and Peuckert, supra note 40.

43 U Harmes-Liedtke. 'The Relevance of Quality Infrastructure to Promote Innovation Systems in Developing Countries.' Discussion Paper 3/2010, Physikalisch-Technische Bundesanstalt, Braunschweig (2010).

the role of metrology has been underestimated and deserves much more attention. Finally, the interactions between the various components of the quality infrastructure have to be disentangled to be able to address its complex impact on the economy, the environment and society as a whole – that is, on the three dimensions of sustainability.

PART II

International standardisation and the World Trade Organization

4

Global food safety standards

The evolving regulatory epistemology at the intersection of the SPS Agreement and the Codex Alimentarius Commission

ALESSANDRA ARCURI

A. Introduction

Global food safety standards are one of the paramount examples of the unfolding process of global standardization. These standards emerged through the interaction of the Codex Alimentarius Commission (hereafter referred to as 'Codex')[1] and the World Trade Organization (WTO). It is uncontested that the WTO has contributed to the transformation of Codex from a rather obscure standard-setting institution into a powerful global regulatory agency in the field of food safety. The adoption of the Sanitary and Phytosanitary Measures (SPS) and the Technical Barriers to Trade (TBT) agreements transformed Codex standards from merely voluntarily standards to quasi-binding obligations.[2] Sociolegal scholars have amply studied the transition by which Codex standards have become quasi-binding. What has gone less studied, however, is how the WTO has transformed the philosophy underpinning the standardization work of

The author thanks Geert Van Calster, Panagiotis Delimatsis, Petros Mavroidis, Alexia Herwig and the participants of the International Conference, 'International Standardization, Marrying Public and Private, Global and Local, Law and Economics', Tilburg Law and Economics Center (TILEC), Tilburg University, 7–8 November 2013, Tilburg. Thanks also to Carlos Esposito and the participants of the Agora on International Law and Human Sciences hosted by the 10th Conference of the European Society for International Law, 4–6 September 2014, Vienna. A slightly different version of this paper has appeared in the EUI Working paper series (RSCAS 2014/97).

1 In this chapter, the term 'Codex' is used to refer to the Codex Alimentarius Commission. Codex can also refer to the set of standards adopted by the Commission (the Code itself); in fact, the Latin term *Codex* means 'code'. In this chapter, it is specified when the term is used specifically to refer to the set of standards rather than the commission.

2 See also Delimatsis's contribution in this volume (Chapter 5).

Codex. At a closer scrutiny of the history of Codex empowerment, it becomes clear that Codex was not simply vested with regulatory powers by the WTO, but its modus operandi was profoundly influenced by the WTO. In particular, it has been argued that the WTO has infused Codex with a technocratic ethos.

Building on this scholarship, this chapter sets out to investigate the dynamic relationship between the WTO and Codex and shed light on the evolving role of expert knowledge in the global regime for food safety standards. The chapter's first thesis is that technocracy (as the rule of the knower) is an unsustainable regulatory paradigm in the field of global food safety standards. The technocratic paradigm rests on an idealized conceptualization of science, according to which science (and scientific experts) can solve international disputes over food safety policy. Such vision of science – or better Science, capital S – has been criticized by a wide scholarship. The controversial ractopamine case, discussed in the chapter, is chosen because it clearly bears witness of the limits of a purely technocratic approach to global food safety standards. The chapter's second thesis is that the global food safety regime is shifting toward a paradigm that marries science with democratic values.

The chapter is organized as follows. The next section introduces the reader to the area of international food safety standards by tracing the evolution of the international food standards regime from the early twentieth century to the present. Section C analyzes the regulatory philosophy underpinning this evolving regime. It will be shown that the initial Codex regime was not built around the Science paradigm but was aimed at creating food standards following what is called a multiple approach. It is only after the establishment of the WTO that the practice of risk assessment became central within the organization of Codex. Section D zooms in on the ractopamine case; this case, which echoes the previous hormones case, illustrates well the problematic features of a regime based on a naive conception of science in which values apparently play no role. To further substantiate this thesis, this section briefly reviews the social science scholarship, shedding light on how values unavoidably shape science. It is further argued, in Section E, that the recent evolution of WTO jurisprudence departs from the technocratic ethos that allegedly underpins the global food safety regime. In the context of the global food safety regime coproduced by WTO and Codex institutions, this section also looks at the vision of science endorsed by key documents of Codex, such as the Working Principles for Risk Analysis and finds some coherence within the regime. Section F draws conclusions.

B. The history of the international food standards regime: A bird's-eye view

I. The pre-Codex regime (in a nutshell)

In the early twentieth century, a variety of transnational initiatives on food standardization emerged to facilitate trade in food products. Standardization initiatives intensified in the 1950s. At the international level, the Food and Agriculture Organization (FAO) and the World Health Organization (WHO)[3] began joint work on a number of issues related to food quality and safety. In 1955, they established the Joint WHO/FAO Expert Committee on Food Additives (JECFA), which produced the General Guidelines Governing the Use of Food Additives at its first meeting; today, it is still operating as one of the scientific committees advising Codex.[4] Building on the work that the International Dairy Federation started in the 1930s, FAO also convened the committee of government experts on the Code of Principles concerning milk and milk products.

In addition to these initiatives, two regional standardization projects were set up.[5] In 1958, the Council of the Codex Alimentarius Europaeus was first proposed by Austrian Minister for Agriculture Hans Frenzel, and it was created under the aegis of the International Commission on Agricultural Industries and the Permanent Bureau of Analytical Chemistry.[6] In 1959, several Latin American countries also created their own Code, the Código Latino-Americano de Alimentos.[7] To overcome the fragmentation of this system and to better address the challenges of international trade, in 1962, the FAO and WHO launched the joint FAO/WHO food standard programme in Geneva.[8] One year later, the Codex Alimentarius

[3] FAO and WHO were created on 16 October 1945 and 7 April 1948, respectively.

[4] N Rees and D Watson (eds), *International Standards for Food Safety* (2000), p. 4.

[5] For an overview of the early history of Codex, see http://www.codexalimentarius.org/about-codex/codex-timeline/en/ (last visited 21 March 2014).

[6] This code was in turn the evolution of the Code and standards developed under the Austro-Hungarian Empire, which were consolidated in the Codex Alimentarius Austriacus; for a more detailed account, see M Masson-Matthee, *The Codex Alimentarius Commission and Its Standards* (2007), pp. 13–15.

[7] CA Guajardo, Código Alimentario Argentino: su valoración juridical (1998). Available at http://www.worldcat.org/title/codigo-alimentario-argentino-su-valoracion-juridica/oclc/39456182/viewport (last visited 8 May 2014).

[8] The proposal to 'internationalize' the existing standards was made by the Council of the Codex Alimentarius Europeaus (see Masson-Matthee, supra note 6, for an overview of this process). From this angle, Codex can be seen as the continuation of the Codex Alimentarius Europeaus. This may explain the preponderant influence exercised by European actors on Codex in the first years of its existence.

Commission was born as a subsidiary body of FAO and WHO, with the mandate of implementing the programme.[9] Codex would have also subsumed and continued the work of the Codex Europaeus.

II. The birth of Codex (the pre-WTO regime)

Codex is a multilayered international body, composed of a multitude of committees dealing with both horizontal and vertical issues: next to an Executive Committee, there are 11 commodity committees (e.g. Codex Committee on Milk and Milk Products, Codex Committee on Cocoa products and Chocolate) and 10 General Subject committees (e.g. Codex Committee on Food Additives and Contaminants).[10] Each committee is administered, organized and financed by a Codex Member (so-called host country); most commonly, countries that have a specific interest on a certain Codex issue are the host of that specific committee (e.g. Switzerland is the host country for the Codex Committee on Cocoa products and Chocolate). The scientific meta-analysis that is used in Codex decision-making process is produced by one of the several expert committees, which were jointly established by the WHO and FAO and provide the studies that are used to adopt standards (e.g. the JECFA and the Joint FAO/WHO Expert Meetings on Microbiological Risk Assessment [JEMRA]).

The decision-making process of Codex consists of an eight-step procedure, starting with the proposal of a standard by a Committee or a Member, followed by several evaluations and discussions (steps 2 to 7) and ending with the submission of the standards for adoption by the Codex Commission.[11] At step 2, a draft text is prepared, and an assessment by an expert body (e.g. JECFA) with recommendations of Acceptable Daily Intakes (ADIs) and/or Maximum Residue Levels (MRLs) is prepared. This is arguably one of the most important phases of the decision-making process because experts' recommendations are typically endorsed in the final standards.

[9] Codex was first approved by the FAO and later by the WHO Assembly. Cfr. Resolution No. 12/61 of the FAO Conference, "Codex Alimentarius," adopted at the 11th Session of the FAO Conference, Report of the 11th Session of the FAO Conference, Rome 4–24 November 1961, para. 263. Resolution WHA 16.24 "Joint FAO/WHO Programme on Food Standards (Codex Alimentarius)," adopted at the 16th Session of the World Health Assembly, Geneva 7–23 May 1963, Off. Rec. World Health Organization, 124, 74.

[10] For a more detailed explanation of how these committees operate, see Masson-Matthee, supra note 6, 2007, at 31–50.

[11] Codex Alimentarius Commission, Procedural Manual (2013), 21st ed., at 31.

The 1962 Report of the joint FOA/WHO Conference that led to the establishment of Codex sheds light on the purpose and scope of Codex and on the nature and types of standards. The Guidelines for Codex standards, part of the 1962 Report, clarify that food standards have dual purposes: to protect consumers' health and to ensure 'fair practices in food trade'.[12] The Guidelines draw a distinction between 'minimum platform standard' and 'trading standards'. The former are to be 'acceptable on as wide basis as possible (on the understanding that acceptance of the minimum standard in no way limits the existence or establishment of higher national standards)'. Trading standards, by contrast, are characterized as 'higher' standards. These standards are not per se superior but are a 'matter of consumer preference'.[13] These standards imply that members accepting them should not restrict the importation of those products on the basis of food safety/quality considerations. The latter standards were mainly conceived for more integrated communities, such as regional areas. With few exceptions (e.g. the European standard on fresh fungus, chanterelle), the development of regional Codex standards never materialized, and global standards became the norm in Codex.[14]

III. The post-WTO regime

The pre-1995 Codex regime has been described as a 'gentlemen's club'.[15] Standards were mainly voluntary, and Codex members followed the norm of not obstructing the adoption of standards, even if these were not in their interests. The number of delegates attending committee meetings was overall small, which facilitated a process of consensus building. Accordingly, standards were always adopted by consensus, even if the procedural manual provided for voting, in cases when such consensus could not be reached.[16]

The voluntary nature of the standards and the small size of meetings to reach agreements were radically changed with the establishment of the WTO. The principal legal framework that has enabled this transformation is the WTO Sanitary and Phytosanitary Agreement (SPS Agreement),

[12] Report of the Joint FAO/WHO Conference on Food Standards, Geneva, 1–5, October 1962, at para 1, Part I.

[13] 1962 Report, supra note 12, p. 7.

[14] Masson-Matthee, supra note 6.

[15] F Veggel and SO Borgen, 'Negotiating International Food Standards', The World Trade Organisation's Impact on the Codex Alimentarius Commission' (2005), 18(4) *Governance* 675.

[16] Codex Procedural Manual (2013), supra note 11.

which provides that Members shall base their measures either on international standards or on science and risk assessment. Articles 3.1 and 3.2 SPS Agreement provide that 'to harmonize sanitary and phytosanitary measures. . . . Members shall base their sanitary or phytosanitary measures on international standards, guidelines or recommendations, where they exist;' and that (2) 'Sanitary or phytosanitary measures which conform to international standards, guidelines or recommendations shall be... presumed to be consistent with the relevant provisions of this Agreement and of GATT 1994'.[17] Because Codex is listed among the international standardization bodies, it is clear that conformity with Codex standards almost invariably implies conformity with WTO law. *A contrario*, the decision to set standards higher than the Codex ones imposes on Members a higher burden to defend its own measures.

Moreover, the WTO is endowed with a uniquely strong dispute settlement system. Under the General Agreement of Tariffs and Trade (GATT), disputes were initiated only on the basis of consensus. Equally, rulings by panels were adopted by consensus. Under the WTO, disputes and the adoption of rulings are quasi-automatic. In case of noncompliance, complaining Members are authorized to retaliate (suspension of concessions).[18] The WTO enforcement system is accordingly one of the most powerful in the international legal arena. For this reason, the nature of post-1995 Codex standards has been characterized as 'semi-binding'.[19]

The new normativity gained by Codex standards may explain why Codex has resorted to voting in the post-WTO era.[20] By virtue of the TBT Agreement, Codex is also relevant for technical barriers to trade, such as labeling. Although this agreement does not explicitly refer to Codex, it does refer to international standards as benchmark for legality of regulatory measures that may adversely affect trade, and it is highly likely that Codex standards qualify as international standards.

The fact that, as of today, virtually no measures disputed at the WTO and diverging from Codex standards has been found to be

[17] *Agreement on Sanitary and Phytosanitary Measures*, Final Act Embodying the Results of the Uruguay Round of Multilateral Trade Negotiations, General Agreement on Tariffs and Trade, Annex 1A, 15 April 1994 (hereafter 'SPS Agreement').

[18] For a more detailed explanation, see P van den Bossche, *The Law and Policy of the World Trade Organization* (2013), 3rd ed.

[19] Veggeland and SO Borge, supra note 15.

[20] Yet the prediction by some that voting would become the main decision-making mechanism did not materialize. In fact, consensus remains the customary practice, and voting has only been used in few controversial cases, such as the approval of standards for growth-promoting hormones and for ractopamine, discussed in section D.

WTO-compatible may be indicative of the new role played by the post-SPS Codex standards. The words of a European Commission representative, commenting the new regime, capture this transition well: 'In the past, if we disagreed with Codex Standards . . . we could ignore and take our own legislation. Now we can't.'[21]

C. Global food safety standards: coproduction and the emergence of a technocratic ethos

Several scholars in the past have tried to clarify the relationship between Codex and the WTO.[22] This article adheres to the theory that the international regime for food safety standards is being coproduced by both Codex and WTO institutions.[23] One of the central arguments of the coproductionist approach is that the WTO has influenced the regulatory philosophy underpinning the global regime for food safety standards, by setting 'Science' at the center of the regime.[24] To better understand this argument, it is worth first identifying the approach underpinning Codex in the pre-WTO era.

I. Regulatory epistemology in the pre-WTO era: the multiple approach

> 'It is due to its scientific basis that Codex texts are considered by WTO as the international reference for food safety standards.'[25] – Codex website

The narrative of the official Codex website suggests that the international trade regime has chosen Codex for its adherence to a scientific paradigm.

[21] From an interview reported by F Veggeland and SO Borge, supra note 15, at 683, conducted by the authors in 2000 with an employee of DG Consumer and Health Protection at the European Commission.

[22] T Buthe, 'The Globalization of Health and Safety Standards: Delegation of Regulatory Authority in the SPS Agreement of the 1994 Agreement Establishing the World Trade Organisation' (2008) *Law and Contemporary Problems*, 219; F Veggeland and SO Borge, supra note 15, DE Winickoff and D Bushey, 'Science and power in global food regulation: The rise of the Codex Alimentarius' (2010) 35 *Science, Technology, & Human Values* 356.

[23] Codex institutions comprise the many committees and subsidiary bodies operating within Codex realm. WTO institutions include the rules as drafted in the final Agreements (most importantly the provisions of the SPS and the TBT Agreement) and the evolution of these rules through the process of interpretation by the WTO quasi-judicial bodies.

[24] Winickoff and Bushey, supra note 22.

[25] Text available at the official Codex website under the section 'Scientific basis for Codex work'. Available at http://www.codexalimentarius.org/scientific-basis-for-codex/en/ (last visited 21 March 2014).

This narrative, however, may be somewhat reductionist, and a glimpse into Codex's history sheds lights on the various factors that have played a role in the food standardization processes developed by Codex. If it is normal today to associate Codex standards with science, the early years of Codex were not characterized by a science-centric approach.

The 1962 Guidelines, briefly discussed in the previous section, did not endorse science as the main paradigm underlying the new standardization project. Although science is undoubtedly one of the factors that should be resorted to when adopting Codex standards, the Guidelines refer to a multiplicity of criteria, as evidenced by the following paragraph: 'The Conference drew attention to the many problems involved in setting up such standards and emphasized the need to study them from the health, scientific, technological, economic and administrative points of view. Only by following this *multiple approach* would it be possible to make the widely desired progress in this fields'.[26] Such an approach appears reconcilable with legal pluralism and more at odds with a universalistic vision of science that enables the formation of uniform global standards. This is further exemplified in the possibility granted by the 1962 Guidelines of adopting different standards for the same product.

The Guidelines in fact clarify that when there is a division of views, 'two or more standards could be proposed, each with its own area of application'.[27] This is better understood because Codex was initially conceived as a forum where also regional standards could have been adopted. Hence, in no case could standards desired by one region have been rejected by outside countries. In this context, it is interesting to read in the Guidelines that 'a food standard . . . does not intend to affect consumer preference, but aims at ensuring that the consumer can know what he is buying'.[28] By granting a status to consumer preferences, the Guidelines implicitly concede that factors other than science may play a role in the determination of standards. Overall, the initial regime was not framed within the boundaries of the science/risk grammar that figures prominently in the SPS Agreement.

II. How the WTO has transformed Codex: toward 'sound science' imperialism?

According to science and technology studies scholars David Winickoff and Douglas Bushey, the WTO has transformed the regulatory epistemology

[26] 1962 Report, letter f, p. 2, emphasis added, supra note 12.

[27] Ibid, para 32, p. 12. [28] Ibid, para 7.

underpinning much of the Codex standard-setting processes.[29] These scholars focus especially on the SPS Agreement, which they characterize as the 'most extreme example' of the technocratic paradigm promoted by the GATT/WTO regime. According to the authors, together with the new normative status, Codex was also endowed with the technocratic aspiration of the SPS Agreement. As they put it:

> The central coproductionist point is this: although the trading regime claimed to be adopting pre-existing science based standards at the international level, the WTO's legal and executive power was necessary to *transform* the Codex into a global agency that could generate such standards. The Codex had been an international body with fairly low visibility. As its increase in power became imminent, the Codex began acting with an invigorated mandate and a sense of itself as 'science-based' organization. It was the rising trading system that drove the development of new norms and practices for the management of knowledge, expertise and evidence in regulatory decision making at the Codex – in short its regulatory epistemology.[30]

Winickoff and Bushey discuss a number of initiatives launched with the aim of creating the scientific framework allegedly necessary for the new scientific identity of Codex. For instance, the 1995 *Joint FAO/WHO Expert Consultation on the Application of Risk Analysis to Food Standard Issues* was convened to enhance consistency and transparency in these practices. The fact that the consultations resulted in recommending changes in Codex practices to promote 'harmonized approaches, consistent with science-based risk assessment' is a clear indication that the pre-existing Codex practices were lacking such consistency.[31] In short, they contend that a well-established regime of 'risk analysis' did not exist under the pre-WTO era; it was envisaged and standardized only in the post-1995 era. Other scholars have noted how the discussion about the role of science within Codex became topical because it became clear that the normativity of Codex standards would have been strengthened by the adoption of the SPS Agreement.[32] Several reports of joint FAO and WHO conferences, starting from 1991 when the negotiations of the SPS Agreement were already concluded,[33] discuss the role of science in Codex.

[29] Winickoff and Bushey, supra note 22. [30] Ibid at 360, emphasis added.

[31] 1995 Joint FAO/WHO Expert Consultation on the Application of Risk Analysis to Food Standard Issues, at 1.

[32] D Jukes, 'The role of science in international food standards' (2000) 11(3) *Food Control, Volume*. Available at http://www.reading.ac.uk/foodlaw/science.htm

[33] For an analysis of the negotiation process that led to the adoption of the SPS Agreement see T Buthe, supra note 22.

The Codex Committee on General Principles (CCGP) was requested to provide guidance on the role science should play in the process of standard setting. After lengthy negotiations, the CCGP produced a text, eventually adopted by Codex in 1995, with four 'Statements of Principle concerning the Role of Science in the Codex Decision Making Process and the Extent to Which Other Factors are Taken into Account' (Statements of Principle).[34] The first statement provides that 'the food standards, guidelines and other recommendations of Codex Alimentarius shall be based on the principle of *sound scientific analysis* and evidence'.[35] The next statement allows 'other legitimate factors' to be taken into account in Codex decision-making process. The Statements of Principle are valid until today as they have been integrated in the Procedural Manual and are still referred to by the Working Principles for Risk Analysis.[36]

Although it is clear that Codex started to profile itself as a science-based institution, the question of the relation between 'sound science' and 'other legitimate factors' remained open. The debate became heated in the controversial process that led to the adoption of standards for growth-promoting hormones. In 1987, the JECFA assigned Acceptable Daily Intakes (ADIs) for a number of growth promoters used in farm animals (particularly cattle) and maintained that it was not necessary to establish Maximum Residue Levels (MRLs).[37] Following the JECFA opinion *sic et simpliciter* would have implied the adoption of Codex standards for the controversial growth-promoting hormones. However, a number of countries, led by the European Communities, opposed these standards because of concerns relating to consumer protection as well as concerns over the practical difficulties of controlling and enforcing the administration of the hormones in meat.[38] The Codex Committee on Residues of Veterinary Drugs in Food (CCRVDF) considered the standards in October 1989[39] and put them forward for a Codex decision.

[34] FAO/WHO 1995, supra note 31, also discussed by Jukes, supra note 32.

[35] Emphasis added.

[36] Codex Alimentarius Commission, 21st edition, Joint FAO/WHO Food Standards Programme, FAO, Rome 2013.

[37] The hormones for which standards were adopted are the following: estradiol 17β, testosterone, progesterone, trenbolone and zeranol.

[38] For the record of the debate in which different arguments were reported, see FAO/WHO (1991). Codex Alimentarius Commission – Report of the 19th Session. (ALINORM 91/40). FAO/WHO Food Standards Programme, Rome (arguments of the EEC at para. 155).

[39] CCRVDF (1989). Report of the Fourth Session of the Codex Committee on Residues of Veterinary Drugs in Foods, 24–27 October 1989. (ALINORM 91/31) Codex Alimentarius Commission.

As the standard reached step 8 in 1991, when the draft for the SPS Agreement was ready, Codex members were split as to the adoptions of these standards. In this context, 'sound science' and 'other legitimate factors' have been used as arguments to defend the opposite positions of the US and the European Communities. Besides the divisions on the questions of whether standards should have been adopted, Codex members seemed to agree that adopting standards would have meant endorsing a sound science approach, whereas the decision of not adopting the standards could have been based on 'other legitimate factors'. In a 1991 discussion of the matter, the Codex Secretariat informed the Commission that the 'FAO Legal Counsel had noted that the Statutes, Rules and Procedures of the Commission *did not bind the Commission to science* as the basis of the decision-making process'. In later documents, the United States objected to this view arguing that 'Codex . . . must show that its standards . . . rested on a sound scientific basis', meaning that a departure from JECFA opinion would have corresponded to a scientifically unsound decision.[40] This use of the term 'sound science' has proliferated ever since. But is it analytically accurate and intellectually honest to juxtapose sound science to other legitimate factors? The next section argues that it is not.

D. Why does my heart feel so bad? Ractopamine and the limits of scientism

On 5 July 2012, Codex has adopted standards on MRLs for ractopamine hydrochloride (ractopamine), a beta-agonist used in meat (mainly in pigs and cattle) for achieving fast growth and leanness of the meat.[41] The process that has led to the adoption of the standards is unusually controversial, and it reminds the hormones episode. The Codex standard was not adopted by consensus, as it is normally the case, but by a vote in which a thin majority voted in favor of the standard (69 in favor vs. 67 opposing and 7 abstentions).[42] About 160 countries, including

[40] Tenth Session of the Codex Committee on General Principles, 1992. As quoted in Jukes, supra note 32.

[41] There are three commercial products derived from ractopamine: Paylean, Optaflexx and Tomax, administered respectively to pigs, cattle and turkeys. Ractopamine is widely used in the United States. For a more in-depth description of the use of the drug in factory farming, see A Alemanno and G Capodieci, 'Testing the limits of global food governance: The case of ractopamine' (2012) 3 *European Journal of Risk Regulation*. Available at http://papers.ssrn.com/sol3/papers.cfm?abstract_id=2133908.

[42] For early comments on the ractopamine case see, A Alemanno and G Capodieci, supra note 41; C-F Lin, 'Scientification of politics or politicization of science: Reassessing

major meat importers and exporters, such as China and the European Union, do not allow the use of ractopamine. The risks related to the administration of ractopamine in animals were assessed a number of times by the JECFA: in 1993, 2004, 2006 and 2010. In its first 1993 evaluation, the JECFA did not establish Acceptable Daily Intakes (ADIs) for ractopamine because of a lack of data to determine safe levels.[43] In its subsequent evaluations in 2004, the JECFA assigned both ADIs and MRLs on the basis of new data provided by Elanco, the major producer of ractopamine.[44] The recommendations were later confirmed, also when revised on the basis of new data.[45]

The adoption of MRLs for ractopamine has been hailed as a victory of science-based regulation. The adoption of the standard has been characterized by Kathy Simmons, the National Cattlemen's Beef Association (NCBA) chief veterinarian, as 'the Codex decision to move forward with *science-based standards*'.[46] Likewise the American delegate, in his oral statements at the 35th Codex session concluded that 'by not adopting MRLs . . . we risk other factors to overshadow science . . . when the science is the core competency of Codex'. Countries opposed to the adoption of the standards have lamented the fact that '[a]s an international organisation seeking to harmonise standards across the globe, Codex should respect consensus-based decision-making, one of the fundamental principles of the organisation. It is clear that for standards to be universally applicable, they also need to be universally accepted.'[47] Interestingly, the two arguments used to defend opposite positions seem to echo the schism between sound science and other legitimate factors. To understand the troubling feature of this schism, it may be useful to briefly introduce

the limits of the Codex Alimentarius Commission' (2013) *Columbia Science and Technology Law Review*. Available at the Social Science Research Network: http://ssrn.com/abstract=2227539.

43 This was decided during JECFA 40th meeting.

44 Elanco is a subsidiary of Ely Lilli.

45 The JECFA monographs on ractopamine and the respective standards are available at: http://www.fao.org/food/food-safety-quality/scientific-advice/jecfa/jecfa-vetdrugs/en/. The MRLs established by JECFA related to different parts of the meat; they are 10 μg/kg for muscle and fat for cattle and pigs, 40 μg/kg for liver and 90 μg/kg for kidney; the ADI are 0–1μg/kg body weight.

46 Emphasis added; as reported in Rutherford, 'Codex Commission Adopts Global Standards For Ractopamine Hydrochloride' (July 5, 2012), *Beef*. Available at http://beefmagazine.com/health/codex-commission-adopts-global-standards-ractopamine-hydrochloride (last visited 8 May 2014).

47 Codex Alimentarius Commission (35th Session) Rome, 2–7 July 2012, EU Statement on ractopamine. Available at http://ec.europa.eu/food/fs/ifsi/eupositions/cac/cac_index_en.html (last visited 8 May 2014).

some of the science underpinnings the arguments that have been used to oppose the use of ractopamine for factory farming.

I. The Science of Ractopamine Politics

Ractopamine, a phenethanolamine β-adrenoceptor agonist (beta-agonist), is administered to animals to increase the rate of weight gain and the leanness of the carcasses.[48] Such effects, in turn, increase the economic value of the animal.[49] Ractopamine works by accelerating the heartbeat and causing the relaxation of blood vessels. The drug has not been approved for human use, and producers acknowledge its negative effects. For example, the label of Paylean (the commercial version of ractopamine administered to pigs) explicitly states that the drug is not for human use and that 'individuals with cardiovascular disease should exercise special caution to avoid exposure'.

There are various concerns over the use of ractopamine. Some relate to human health and others to animal health and welfare. In relation to human health, the concern is rather straightforward: given its properties of accelerating the heartbeat, exposure to residues of the substance may increase risks related to cardiovascular diseases. JECFA, by recommending MRLs, has clearly considered these risks are negligible.[50] However, the risk assessment conducted by the JECFA has been criticized by, among others, the scientific food safety body of the European Union, the European Food and Safety Authority (EFSA). According to a lengthy Opinion of the EFSA Panel on Additives and Products or Substances used in Animal Feed (FEEDAP), administering ractopamine to animals could possibly be dangerous to humans exposed to traces of it. The main argument of the FEEDAP is that data are not sufficient to derive safe residue levels, particularly in relation to the most vulnerable groups, such as people with

[48] See the label of Paylean (the commercial version of ractopamine used in finishing swine).

[49] According to data from Elanco, the increased gain is $2 per head, as reported by H Bottemiller, 'Dispute over Drug in Feed Limiting US Meat Exports' (January 25, 2012) *Food and Environmental Reporting Network*. Available at http://thefern.org/2012/01/dispute-over-drug-in-feed-limiting-u-s-meat-exports/ (last visited 8 May 2014).

[50] Another question, not discussed in this section, is the residue levels of ractopamine in different body organs of the animal (particularly offal). China was especially concerned with higher concentration of the substance in offal, given the consumption of offal by Chinese, and submitted new studies to JECFA. JECFA reviewed the studies and concluded that, with the exception of concentration in lungs, her previous assessment, recommending ADIs and MRLs were not to be reviewed; see JECFA Monograph, No. 9. (2010).

cardiovascular diseases or children. Methodologically, the FEEDAP contested the appraisal by JECFA because 'the safety factor applied by JECFA to derive the ADI from the NOEL does not sufficiently take into account population subsets at higher risk of adverse events after ß-adrenergic stimulation'.[51]

Without entering into the merits of the dispute, one thing is clear. Two scientific panels reached different conclusions on the safety of a certain substance. The panels are both institutional bodies, composed of arguably well-established 'regulatory scientists'. Arguing that following JECFA recommendations is sound science and all the rest is not means disqualifying EFSA science as unsound. The boundary work implicit in this logic should be puzzling for legal scholars and policy makers. Labeling as sound science only the recommendation coming from Codex expert bodies means entrusting the monopoly of science to a few dozen embedded scientists. The risk is that global trade wars will be transformed into turf wars among few experts on narrowly defined scientific issues. But who would subscribe to such a model? The first lesson from the ractopamine case is that a purely technocratic paradigm is reductionist at best and illiberal at worst.

The other concerns on the use of ractopamine are related to animal welfare and health. Several studies have shown that animals treated with ractopamine become more aggressive and difficult to handle[52] and as a consequence are 'more likely to be subjected to rough handling and increased stress during transportation'.[53] In light of these studies, the FEEDAP expressed concerns in relation to the safety of the pigs. Although there may remain some doubts as to the extent of the phenomenon, it is undeniable that ractopamine has some negative effects on animal

[51] Scientific Opinion of the Panel on Additives and Products or Substances used in Animal Feed (FEEDAP) on a request from the European Commission on the safety evaluation of ractopamine. *The EFSA Journal* (2009) 1041, 1–52.

[52] JN Marchant-Forde, DC Lay Jr., EA Pajor et al., 'The effects of Ractopamine on the behavior and physiology of finishing pigs' (2003) 81(2) *Journal of Animal Science* 416; R Poletto, HW Cheng, RL Meisel et al., 'Aggressiveness and brain amine concentration in dominant and subordinate finishing pigs fed the β-adrenoreceptor agonist ractopamine' (2010) 88(9) *Journal of Animal Science*, 3107. As reported by Marchant-Forde et al., 'Pigs fed ractopamine were more difficult to handle and had elevated heart rates and catecholamine levels after 4 weeks of administration. Pigs that are more difficult to move are more likely to be subjected to rough handling and increased stress during transportation, implying reduced welfare, increased workload for the handlers and, potentially, poorer meat quality.'

[53] *The EFSA Journal*, supra note 51, at 1041.

welfare.[54] Even producers have a disclaimer in this regard on the label; Paylean label reads: 'Caution: Ractopamine may increase the number of injured or fatigued pigs during marketing. Not for use in breeding swine.'[55]

The question of animal welfare has been often listed among the 'other legitimate factors' that ought to be taken into account in the context of Codex decision making. For the purpose of our analysis, it is worth emphasizing that no matter which category we may want to place animal welfare, the studies showing the effects on the animals can be plausibly classified as scientific. Thus, even when animal welfare is considered under the rubric of 'other legitimate factors', to stigmatize as unscientific a decision taken on such basis is incorrect and possibly demagogic.

Some commentators have pointed to the ractopamine case, which is echoing the previous hormones case, as evidence of the limits of Codex.[56] However, rather than Codex's limits, the ractopamine case points at the limits of a rigid technocratic model of legitimacy that proves untenable before a political and scientific divided community.

II. The sound of science: Science vs scientism

As illustrated here, decisions diverging from the recommendations of Codex expert bodies may be scientifically sound as much as those portrayed as science-based. Disagreement among scientists may legitimately exist because science does not provide univocal answers to policy problems, in contrast to what an overly simplistic view of science may suggest. Science is fraught with uncertainties and value choices: models of extrapolation (from animals to humans) can be more or less conservative; decisions to study certain variables (and certain hazards) instead of others may differ; some studies may look at potential effects on average people, and others may consider vulnerable groups (e.g. children, pregnant women, people with diseases, etc.); studies can consider cumulative

[54] According to one report, 'when Elanco studied the drug in pigs for its effectiveness, it reported that "no adverse effects were observed for any treatments." But within a few years of Paylean's approval, the company received hundreds of reports of sickened pigs from farmers and veterinarians, according to records from the FDA's Center for Veterinary Medicine.'

[55] Available at: http://www.elanco.us/products-services/swine/feed-efficiency-finishing-swine.aspx (last visited on 27 March 2014).

[56] Lin, supra note 42.

effects or not and so on. This means that within the realm of scientific studies, it is possible to reach different conclusions, all plausible according to generally agreed standards of how to conduct scientific research.

Moreover, the assessment of factors outside the scope of the reports of Codex expert committees, such as JECFA, is not in itself unscientific. The assessments of effects on animal welfare and/or on the environment are as scientific as the assessment of the carcinogenic properties of a certain substance and the relative risks to humans. Symmetrically, the judgments of whether it is acceptable to let animals suffer or to pollute the environment or to allow the marketing of a substance with a carcinogenic potential are all equally political. It is thus unfortunate that the term 'sound science' is often used to stigmatize all the standards not based on the recommendations of Codex expert bodies. Using the jargon 'sound science' to characterize only specific assessments of certain types of risk and discount all other considerations as unscientific is either banal or parochial.

Political scientists have coined the term 'scientism' to refer to a 'discourse or framework for discussion that excludes considerations of distributional and other social impact criteria in the determination by a regulatory agency that a product is or is not suitable for markets.... In its neoliberal form, scientism tends to restrict democratic participation and weakens the option for governments to regulate'.[57] This concern is echoed by Winickoff and Bushey, who worry that in the context of Codex, the adoption of 'risk as the single dominant grammar of global food regulation' may create certain governance biases: '[r]isk discourse implicitly empowers some people as experts while marginalizing others as inarticulate or irrelevant'.[58] To better understand how a reductionist view of the risk analysis paradigm can degenerate into scientism and create a regime of exclusion, it is worth a brief excursion into the academic literature on risk.[59]

[57] K Moore, DL Kleinman, D Hess and S Frickel, 'Science and neoliberal globalization: A political sociological approach' (2011) 40 *Theory and Society*, at 517.

[58] Winickoff and Bushey, supra note 22, 364.

[59] Although the scholarship on risk relevant for this debate is vast, this chapter deals only with a limited subset of studies. Important contributions not discussed in this article are the following: SO Funtowicz and JR Ravetz, 'Science for the post-normal age' (1993) *Futures* 739; M O' Connor, S Faucheux, G Froger, S Funtowicz and G Munda, 'Emergent complexity and procedural rationality: Post-normal science for sustainability' in R Costanza and O Segura (eds), *Getting Down to Earth: Practical Applications of Ecological Economics* (1996); K Shrader-Frechette, 'Reductionist approaches to risk' in DG Mayo and RD Hollander (eds), *Acceptable Evidence – Science and Value in Risk Management* (1991);

Social scientists have amply shown that, beyond the conclusions of so-called experts, there are many factors that could legitimately be taken into account when deciding about risks. Several studies in the field of cognitive psychology have shown that people value different qualitative characteristics of risks and do not merely focus on the probabilities related to morbidity and mortality.[60] For instance, people prefer voluntary over involuntary risks, they are sensitive toward perceived inequities of distribution of risks, and so on.[61] Preferences toward risk are also shaped by culture.[62] Several studies have shown that people sharing similar values fear similar risks; for instance, egalitarians tend to be more afraid of environmental risks and the risks of gun possession, whereas individualists may be more concerned about security issues and disfavor gun controls. The different attitudes of Americans and Europeans toward biotech food products may then be seen as pertaining to 'cultural cognition' rather than stemming from different attitudes towards science. As Kahan et al. put it, 'if risk disputes are really disputes over the good life, then the challenge that risk regulation poses for democracy is less how to reconcile public sensibilities with science than how to accommodate different visions of the good within a popular system of regulation'.[63]

Against this background, it may be clear why the risk analysis paradigm enlisted by the WTO/Codex coproduced regime could be accused of being a form scientism, if it is not entrenched in democratic values. Acknowledging the role that values play within scientific processes does not mean to corrupt science with values; to the contrary, it means aspiring to a more transparent scientific process that can enable dialogue on divisive issues.

E. An evolving regime: marrying science and democracy?

If technocracy may have been the driving paradigm in the post-WTO phase of Codex, the incapability of this paradigm to solve international

B Wynne, 'May the sheep safely graze? A reflexive view of the expert-lay knowledge divide' in S Lash, B Szerszynski and B Wynne (eds), *Risk, Environment and Modernity: Toward a New Ecology* (1996).

[60] P Slovic, 'Perception of Risk' (1987) 236 *Science*, 280; P Slovic, 'Beyond numbers: A broader perspective on risk perception and risk communication', in DG Mayo and RD Hollander (eds), *Acceptable Evidence – Science and Value in Risk Management* (1991), 48.

[61] P Slovic, supra note 60.

[62] O Renn and B Rohrmann (eds), *Cross-Cultural Risk Perception – A Survey of Empirical Studies* (2000); DM Kahan, P Slovic, D Braman and J Gastil, 'Fear of democracy: A cultural evaluation of Sunstein on risk', (2006) 119 (4) *Harvard Law Review*, 1071.

[63] Kahan et al., supra note 62, at 1073.

trade disputes is becoming apparent. The main claim of this chapter is that the 'regulatory epistemology'[64] that has driven the erection of the global regime, coproduced by Codex and WTO institutions, is changing. The emerging paradigm is not a radical departure from technocracy, but it could be seen as an attempt to reconcile technocratic and democratic approaches to food safety regulation. Such an approach is more in line with a model advocated by a rich and diverse body of social science scholarship, briefly presented in the foregoing sections.

I. WTO jurisprudence

As already explained, with the establishment of the WTO, the international trade system has been endowed with a unique dispute-settlement mechanism. This system not only confers normative bite to WTO obligations, it also greatly contributes to the clarification and evolution of trade rules. Paying attention to how the rules of the SPS and TBT Agreements are being interpreted is thus crucial for a better understanding of the global regime for food safety standards.

The SPS Agreement allows WTO Members to adopt more stringent sanitary and phytosanitary measures than international standards (in our case, Codex standards), only when these measures are justified by risk assessment or when there is no sufficient scientific evidence on which to base the risk assessment measures.[65] The concepts of 'risk assessment' and '(in-)sufficiency of scientific evidence' are then crucial in understanding the role of Codex standards in the international regulatory space. A broad interpretation of these conceptual categories will imply that WTO Members have more freedom to diverge from Codex standards and vice versa. It follows that the degree of normativity of Codex provisions depends on how the provisions of the SPS Agreement are interpreted. At a more fundamental level, the interpretation of concepts such as a risk assessment reflects the regulatory epistemology endorsed by WTO institutions.

[64] The term is used by Winickoff and Bushey to refer to the 'norms and practices for the management of knowledge, expertise, and evidence in regulatory decision making' and it draws on the concept of civic epistemology previously elaborated by Sheila Jasanoff; see S Jasanoff, 'Ordering knowledge, ordering society', in S Jasanoff (ed), *States of Knowledge: The Coproduction of Science and Social Order* (2004).

[65] Cfr. Article 2 and 3 of the SPS Agreement. For a more articulate discussion of the basic structure of the SPS Agreement, see A Arcuri, 'Food safety at the WTO after "Continued Suspension": A paradigm shift?' in A Antoniadis, R Schütze and E Spaventa (eds), *The European Union and Global Emergencies: A Law and Policy Analysis* (2011).

A key case to understand the turn taken by WTO jurisprudence in the interpretation of the SPS Agreement is *Canada/ US – Continued Suspension*. This case is the follow-up of the *Hormones* case. Let me briefly summarize the main steps that led to the *Continued Suspension* dispute for readers not familiar with this body of case law. After the Appellate Body ruling in *EC – Hormones* in which Europe was faulted for not having based its measures on risk assessment, the United States and Canada were allowed to retaliate against the noncompliant European Union. In the meanwhile, Europe asked its food-science body – at the time, the Scientific Committee on Veterinary Measures relating to Public Health (SCVPH) – to conduct a new risk assessment of the disputed hormones.[66] On the basis of the new SCVPH opinion, Europe amended its laws, adopting Directive 2003/74/EC, which prohibits the use of estradiol-17β for growth promotion purposes because of its carcinogenic and genotoxic properties, and it prohibits, on a provisional basis, the other disputed hormones because the available data, although indicative of risk, were not sufficient to conduct a proper risk assessment. At this point, Europe argued that it was complying with the Appellate Body ruling in *EC – Hormones*. The United States and Canada did not agree and continued to retaliate.

As the WTO dispute settlement understanding (DSU) is silent on what do in these cases, Europe initiated a new dispute against the United States to challenge the legality of their retaliatory measures. In *Canada/US – Continued Suspension*, the Appellate Body did not resolve the saga about the use of hormones for growth promotion purposes[67] but, in reversing the Panel's Report,[68] has shed light on the interpretation of crucial norms for the global governance of food safety. Most important, the concepts of 'risk assessment' and 'sufficiency of scientific evidence' have been interpreted by the Appellate Body as 'relational' concepts; this implies

[66] The first SCVPH 'Assessment of Potential Risks to Human Health from Hormones Residues in Bovine meat and Meat Products' was produced on April 30, 1999. The SCVPH conducted two new assessments on May 3, 2000, and on April 10, 2002, respectively. The latter assessments confirmed the conclusions of 1999. This body has been replaced by EFSA.

[67] In May 2009, the European Union and the United State signed a Memorandum of Understanding (MoU) 'regarding the importation of beef from animals not treated with certain growth promoting hormones and increased duties applied by the United States to certain products of the European Communities', of which they notified the WTO in September of that same year.

[68] Panel Reports, *Canada/United States – Continued Suspension of Obligations in the EC – Hormones Disputes*.

that issues, such as 'the appropriate level of protection' chosen by a government, can shape the methodology and questions studied in the risk assessment.[69] Such a conceptualization subscribes to a vision of science that is shaped by social values. The Appellate Body in *Continued Suspension* has unequivocally embraced this approach, as evidenced by the following quote:

> The risk assessment cannot be entirely isolated from the appropriate level of protection. There may be circumstances in which the appropriate level of protection chosen by a Member affects the scope or method of the risk assessment. This may be the case where a WTO Member decides not to adopt an SPS measure based on an international standard because it seeks to achieve a higher level of protection. In such a situation, the fact that the WTO Member has chosen to set a higher level of protection may require it to perform certain research as part of its risk assessment that is different from the parameters considered and the research carried out in the risk assessment underlying the international standard.[70]

The quote shows how the Appellate Body has relativized the centrality of international standards. Hence, international standards may be less central than previously thought. In the ractopamine case, in which the EFSA report focused on risks left unaddressed by JEFCA (e.g. risks to people with cardiovascular diseases), it may well be the case that Codex standards will not be decisive in the determination of compliance with WTO law.

The turn taken by the Appellate Body in *Continued Suspension* can be read as a more general sign that the regulatory epistemology underpinning the global food safety regime rests on a vision of science that acknowledges its value-laden nature. This turn may have been favored by the experience gained during the development of SPS jurisprudence. The evolution of the *Hormones* case is exemplary of the limits of scientism and thus, it may be no coincidence that in *Continued Suspension* the Appellate Body has clarified that the WTO regime needs to deal with the value-laden nature of science. Moreover, in another salient case, the *EC – Biotech*, in which the European regulatory regime of Genetically Modified Organisms was at stake, the WTO was directly exposed to the reasoning of social scientists. In fact, a group of well-known social scientists submitted an *Amicus Brief*

[69] For a more detailed analysis of this issue, see A Arcuri, supra note 65.

[70] Appellate Body Report, *Continued Suspension*, para 534 (emphasis added).

to the Panel in *EC – Biotech* in which the arguments showing the value-laden nature of science are lucidly presented.[71] Although the *EC – Biotech* case never reached the appeal stage, it is likely that the Appellate Body has been exposed to the *Amicus Brief* submitted in the Biotech case. Is this approach also reflected by Codex?

II. Codex general principles for risk analysis

Despite the emphasis placed on science-based decision making, the new approach endorsed by the Appellate Body is also compatible with the general principles for risk analysis negotiated and agreed at Codex. Two key documents define these principles: 1) the Working Principles for Risk Analysis for Food Safety for Application by Governments[72] (*Working Principles for Governments*) and 2) the 'Working Principles for Risk Analysis for Application in the Framework of the Codex Alimentarius' (*Working Principles for Codex*).[73] Both sets of principles have been negotiated for several years and both have been adopted in the new millennium (in 2007 and 2003, respectively) and may thus be seen as representing the existing consensus on the constitutive elements of risk analysis. They are therefore an important piece of the puzzle in understanding the regulatory epistemology underpinning the current global food safety regime.

Both sets of principles share similar features. The principles endorse an Anglo-Saxon model of risk analysis originally articulated in the United States in the 1983 National Research Council (NRC)'s report on *Risk Assessment in the Federal Government – Managing the Process*, also known as the Red Book.[74] According to this report, risk assessment is separated from risk management. This functional separation is clearly endorsed by

[71] L Busch, R Grove-White, S Jasanoff, D Winickoff and B Wynne, 'Amicus Curiae Brief Submitted to the Dispute Settlement Panel of the WTO in the case of EC-Biotech' (WT/DS291, 292 and 293) (2004). Also published as a scholarly article, D Winickoff, S Jasanoff, L Busch and RB Grove-White, 'Adjudicating the GM food wars: Science, risk, and democracy in world trade law' (2005) 30 *Yale Journal of International Law* 81.

[72] These principles are the result of seventeen years of negotiations and were finally adopted in 2007, Cfr Codex Alimentarius Commission, Working Principles for Risk Analysis for Food Safety for Application By Governments (2007), *CAC/GL 62–2007.*

[73] These principles were agreed on in 2003 and are now included in the Procedural Manual, supra note 11, at 107.

[74] National Research Council, Risk Assessment in the Federal Government: Managing the Process (1983).

Codex principles, where it is established that 'there should be a functional separation of risk assessment and risk management'.[75] The debate on the functional separation between risk assessment and risk management has evolved over time and, within the United States, several NRC reports succeeding the Red Book asserted the importance of having an iterative process where risk assessment and management are in a dialogical relationship.[76] The acknowledgment of the importance of a linkage between the two phases is premised on the idea of value-laden science.

In line with the evolution of this debate, Codex principles also acknowledge the need of linking risk assessment and risk management and set limits to the functional separation of the two. In the first place, risk assessment and risk management are to be separated 'to the degree practicable'.[77] In addition, the risk analysis process is framed as an iterative one: 'it is recognized that risk analysis is an iterative process, and *interaction* between risk managers and risk assessors is *essential* for practical application'.[78] Even more important, both sets of principles recognize that 'Precaution is an inherent element of risk analysis'.[79]

[75] Codex (2007), supra note 72, para. 11.

[76] For an overview of the evolution of the debate, see E Fisher, 'Framing risk regulation: A critical reflection' (2013) 2 *European Journal of Risk Regulation* 125. Important Reports for the evolution of these debate are: National Research Council *Science and Judgment in Risk Assessment* (1994); National Research Council, *Understanding Risk: Informing Decisions in a Democratic Society* (1996), and Royal Commission on Environmental Pollution, *Setting Environmental Standards,* 21st Report (1998).

[77] This sentence, which is only included in Codex 2007, was the result of intense negotiations; for a recollection of the negotiations on these principles, see D Demortain, 'Enabling global principle-based regulation: the case of risk analysis in the Codex Alimentarius' (2012) *Regulation and Governance* 207, at 215–217.

[78] Codex 2007, supra note 72, para. 11; 'Working Principles' in Procedural Manual, supra note 73, para. 9, second sentence. The importance of this interaction is further recognized in other Codex documents. For instance, Principle 3 of the 'Risk Analysis Principles Applied by the Codex Committee on Food Additives' establishes that 'CCFA and JECFA recognize that *continuous interaction* between risk assessors and risk managers is critical to the success of their risk analysis activities', Codex Procedural Manual, supra note 11, at 116.

[79] Codex 2007, supra note 72, para. 12, and 'Working Principles' in Procedural Manual, supra note 73, para. 11. Despite the general recognition of the role of precaution in risk analysis, no consensus was found on including a reference to the precautionary principle in the section of risk management of the Working Principles (e.g. cfr. para. 8, ALINORM01–33). For a discussion of the negotiating history of the Working Principles for Codex and questions related to the precautionary principle, see C Gerstetter and M-L Maier, 'Risk regulation, trade and international law: Debating the precautionary principle in and

Asserting the importance of the interaction of risk assessment and risk management, as well as recognizing the role of precaution in risk analysis, means acknowledging the value-laden nature of science. In other words, the 'technē' or 'episteme' that serves the decision-making process is itself shaped by values. Although other Codex documents and general principles may still be inspired by a narrow vision of science and display a technocratic aspiration, the conceptualization of risk analysis in the *Working Principles* contrasts with an eminently technocratic approach. It can thus be concluded that the current guidelines for risk analysis are not informed by a purely technocratic approach, which is commonly associated with Codex practices.

III. Representation and participation

Enabling indirect representation through a progressive interpretation of the WTO-Codex legal framework may contribute to reduce the democratic deficits associated with regimes of global governance.[80] The jurisprudential turn of the WTO Appellate Body in *Continued Suspension* can be said to realize a form of vertical representation in the global arena. This is because the preferences of domestic constituencies get a voice under the interpretative framework in which conceptions of 'risk assessment' and 'uncertainty' are more organically integrated with 'the appropriate level of protection'. Whereas under the previous interpretative framework, consumers' preferences may have been more easily disregarded, governments are now in a better position to defend risk analysis endorsing the values cherished by its people. As discussed earlier, Codex guidelines for risk analysis, by recognizing the importance of values in risk regulation, also enhance the potential for a more democratic decision-making process for the adoption of global food safety standards. From this perspective, the new regulatory epistemology enhances the legitimacy of the WTO-Codex regime for food safety standards.

It remains to be seen whether, in practice, room is left to direct representation when standards are proposed, negotiated and adopted at the

around the WTO' (2005), TranState Working Paper, 018/2005. Available at http://www.sfb597.uni-bremen.de/pages/pubApBeschreibung.php?SPRACHE?de&ID?20

80 For an overview of the debate see R Keohane and JS Nye, Jr., 'Between Centralization and Fragmentation: The Club Model of Multilateral Cooperation and Problems of Democratic Legitimacy' (February 2001). *KSG Working Paper* No. 01–004. Available at SSRN: http://ssrn.com/abstract=262175.

global level. In this context, it may be useful to distinguish issues of equal representation and participation of different countries from issues of participation of different constituencies of the civil society. In relation to representations of different countries, Codex has been criticized for not sufficiently empowering developing countries.[81] As discussed by Ochieng Pernet in this book (Chapter 13), Codex has launched some laudable initiatives (e.g. the Codex Trust Fund) to fix some of these deficiencies.[82] The jury is still out as to whether, through these institutions, developing countries will be able to more effectively participate in Codex decision making. Codex has also been criticized because of the overrepresentation of industrial groups vis-à-vis public interest groups. Various studies have provided ample evidence of this asymmetry.[83] The *de facto* limited participation of developing countries as well as the overrepresentation of industry groups suggests that the WTO-Codex regime has not solved all of its legitimacy problems. Although the evolving regulatory epistemology described in this chapter does lend more legitimacy to the system, institutions capable of effectively rebalancing representation and participation of different stakeholders in the standard setting process are equally needed.

F. Concluding remarks

The locution 'sound science' has dominated the global battlefield of food politics. It has been often used to defend measures following from the recommendations of Codex expert bodies and to discount all the rest as unscientific. The ractopamine case, in which various scientific bodies have produced diverging opinions, shows that science cannot be the ultimate arbiter on divisive issues fraught with uncertainties. As beautifully put by Bruno Latour: 'When scientists add their findings to the mix, they do not put an end to politics; they add new ingredients to the collective process.'[84] The challenge for the twenty-first century is thus to rethink

[81] For an overview of arguments, see Masson-Matthee supra note 6, 241–245.

[82] See A. Ochieng Pernet's contribution in this volume (Chapter 13).

[83] E Smythe (2009), 'In whose interests? Transparency and accountability in the global governance of food: Agribusiness, the Codex Alimentarius, and the World Trade Organization', in J Clapp and D Fuchs (eds), *Corporate Power in Global Agrifood Governance* (1993), at 93; N Avery, M Drake and T Lang, *Cracking the Codex: A Report on the Codex Alimentarius Commission*. London: National Food Alliance; Masson-Matthee supra note 6, 248–254.

[84] B Latour, 'From the World of Science to the World of Research?' (10 April 1998) 280 no. 5361 *Science* 208.

our legal and political institutions so that they can better cope with the value-laden nature of science. This article suggests that the coproduced regime by the WTO and Codex is starting to take up this challenge by forging a new regulatory epistemology. Although it may be premature to declare the end of the technocratic era of global food safety standards, seeds of a more democratic regime have been sown.

5

'Relevant international standards' and 'recognised standardisation bodies' under the TBT Agreement

PANAGIOTIS DELIMATSIS

A. Introductory remarks

International standards have a prominent position in the Agreement on Technical Barriers to Trade (TBT) of the World Trade Organization (WTO). The TBT preamble commences with an orthodox assumption: that international standards improve efficiency of production and facilitate the conduct of international trade.[1] Similar assumptions are made under the Agreement on Sanitary and Phytosanitary Measures (SPS), although the latter appears to be more flexible by acknowledging the right of Members to choose their level of protection.

Standards constitute a form of codified technical knowledge that enables the development of products and processes. Standards regularise and constrain behaviour (regulative function); lend a taken-for-granted quality to certain technologies and modi operandi (cognitive function); and favour cooperative strategies over adversarial ones (normative function).[2] In the absence of standards, technological progress would miss an important instrument for benchmarking and capitalising on advances in the field of technology. In addition, first-mover advantages in standardisation[3] are substantial incentives for firms to

I am grateful to the participants of the 10th TILEC Anniversary conference, 'International Standardisation', held 6–7 November 2013 in Tilburg and Christian Häberli for their comments and suggestions. All hyperlinks were updated in June 2015.

1 G. Swann, P. Temple and M. Shrumer, 'Standards and Trade Performance: The UK Experience', 106(438) *Economic Journal* (1996), pp. 1297–1313.

2 C. Lane, 'The Social Regulation of Inter-Firm Relations in Britain and Germany: Market Rules, Legal Norms and Technical Standards', 21 *Cambridge Journal of Economics* (1997) 197.

3 Art. 1.1 of he ISO/IEC Guide 2:2004 defines 'standardisation' as the 'activity of establishing, with regard to actual or potential problems, provisions for common and repeated use,

innovate.[4] In that sense, standards are constitutive of markets[5] and a decisive instrument for economic growth.[6]

The WTO does not have a standard-setting capacity. The Committee on Technical Barriers to Trade (hereinafter, the TBT Committee) does not develop nor adopt standards itself. Rather, the TBT exerts a high level of deference toward technical rationality as expressed through international standard-setting activities *outside* the WTO. Standards developed within international standard-setting bodies (ISSBs) acquire a prominent role within the WTO through the very text of the TBT, which requires that WTO Members use 'relevant international standards' and presumes compliance with the TBT when such standards are used as a basis for domestic technical regulations.

Is this regulatory outsourcing justified? Indeed, it is one thing to say that the TBT Committee would be unable to develop any standards whatsoever. It is yet quite another to say that certain non-WTO rules are WTO compatible as long as they are relevant to the product at issue in a WTO dispute *regardless* of the process that led to their adoption (which process, by the way, is totally out of the control of the WTO). Quite astonishingly, the TBT entails such delegation of regulatory power[7] without any inquiry as to the actual processes used throughout the development of international technical standards. This is even more surprising if one considers that such delegation of international standard-setting activities is directed toward private actors, thereby creating an alternative to formal international law.[8]

aimed at the achievement of the optimum degree of order in a given context. Note 1: in particular, the activity consists of the processes of formulating, issuing and implementing standards.'

4 Again, and more generally, if we consider standardisation as infrastructure, it can promote but also hamper innovation. See also D. Acemoglu, G. Gancia and F. Zilibotti, 'Competing Engines of Growth: Innovation and Standardization', 147 *Journal of Economic Theory* (2012) 570.

5 H. Schepel, *The Constitution of Private Governance – Product Standards in the Regulation of Integrating Markets* (Hart Publishing, 2005).

6 K. Blind and A. Jungmittag, 'The Impact of Patents and Standards on Macroeconomic Growth: A Panel Approach Covering Four Countries and 12 Sectors', 29 *Journal of Productivity Analysis* (2008) 51.

7 For a similar observation under the SPS, see T. Büthe, 'The Globalization of Health and Safety Standards: Delegation of Regulatory Authority in the SPS Agreement of 1994 Establishing the World Trade Organization', 71 *Law and Contemporary Problems* (2008) 219.

8 E. Benvenisti, '"Coalitions of the Willing" and the Evolution of Informal International Law' in C. Callies, G. Nolte and P.-T. Stoll (eds), *'Coalitions of the Willing' – Avantgarde or Threat?* (Carl Heymanns Verlag, 2007).

The vantage point of this chapter is that such a deferential approach is untenable. Attributing to international standards developed elsewhere automatic legal force in the WTO is contrary to contemporary demands for more transparency and due process within global governance institutions more generally, and openness in international standard-setting in particular.[9] One key hypothesis to be tested is whether the WTO can play an instrumental role in improving standard-setting processes within ISSBs, because its very foundational treaty (and the TBT in particular through the presumption of WTO compatibility mentioned earlier) makes it a prominent promoter of such standards and indeed a high-level diffusion mechanism. Conceptually, this chapter aims at the tentative identification of an analytical framework that would lead to a gradual democratisation of international standardisation via a more critical approach by the WTO judiciary.

This chapter goes into the heart of the problem of legitimacy of international standardisation processes used within ISSBs and identifies the WTO as a potential drive for change in transnational standard-setting. Section B describes the nature of international standards, tilting between public and private, in a grey area of law, whereas Section C analyses the position of international standards and the bodies that create them from a positive and normative viewpoint. Section D concludes the chapter.

B. The peculiar nature of international standardisation

I. *At the interstices between public and private*

Increased international standardisation by the private sector results from an ever-increasing demand of consumers for better and safer products, technological advances, the expansion of global trade and the increased attention paid to leveling the playing field when it comes to social and sustainability issues.[10] International standards, although voluntary, affect our everyday life in many ways. Standards are about consumer welfare because they bring about and solidify technological evolution, innovation and diffusion of knowledge. They play a decisive role as to whether the

[9] Cf. A. von Bogdandy, 'The European Lesson for International Democracy: The Significance of Articles 9–12 EU Treaty for International Organizations', 23: 2 *European Journal of International Law* (2012) 315.

[10] M. Egan, *Constructing a European Market: Standards, Regulation, and Governance* (Oxford University Press, 2001).

business and market environment will be conducive to increased innovation and trade. They form an important condition for doing business and affect access to markets, determining the profitability, growth and ultimately the survival of entrepreneurs. Hence, standards have a crucial *trade facilitation function.*

Yet standards can also be adopted with a view to restricting access to a given market, thereby neutralising any trade concessions made in other fora, including the WTO or preferential trade agreements (PTAs). When standards are diverse and applied in a thoughtless manner, trade is negatively affected, and economies of scale become more difficult to attain. For small entrepreneurs in particular, the costs of compliance with this heterogeneous set of standards may be prohibitive, de facto precluding any attempt to gain access to foreign markets from the outset.[11]

Standardisation can also be captured by the industries involved, which exploit the presence of asymmetric information and organisational differences. Industries organise themselves more efficiently than consumers and manage to capture norm-setting institutions.[12] This can also be the result of structural bias:[13] for instance, the International Organisation for Standardisation (ISO) is a non-governmental, mainly industry-driven, international standard-setter. More recently established SSBs such as the ISEAL Alliance claim to be more inclusive, yet even in such bodies full membership is reserved to those presumed to have the necessary technical expertise – that is, standard-setting organisations and accreditation bodies.[14] Consumer associations or NGOs are barely involved in the development of the standards; however, they may be more engaged in the consultation stage later and shortly before the standard becomes final.[15]

Standard-setting resembles law-making because standards, like laws, are the outcome of discussion, bargaining, deliberation and compromise

[11] P. Messerlin, *Measuring the Costs of Protection in Europe: European Commercial Policy in the 2000s* (Institute for International Economics Press, 2001).

[12] R. Baldwin, 'Regulatory Protectionism, Developing Nations, and a Two-Tier World Trade System' in S. Collins and D. Rodrik (eds), *Brookings Trade Forum: 2000* (Brookings Institution Press, 2001).

[13] Cf. M. Koskenniemi, 'The Politics of International Law – 20 Years Later', 20:1 *European Journal of International Law* (2009) 7.

[14] See the contribution by Axel Marx and Jan Wouters in this volume (Chapter 10).

[15] See Stephanie Bijlmakers and Geert van Calster's contribution in this volume (Chapter 12).

among non-state actors.[16] However, standards established by ISSBs like the ISO are not law per se, but rather serve a clear regulatory function prescribing rules for others to follow.[17] An important distinction, again, is that, whereas domestic standardisation can encompass both binding and voluntary technical specifications and standards, international standardisation, as noted previously, typically involves standards with which compliance is voluntary. Once international standards are adopted, it is a country's prerogative to adopt these standards in the form of domestic technical regulations (compliance with which is mandatory), or as standards (compliance with which is optional). Depending on the domestic constitutional structures, such standards may be adopted through the relevant national SSBs or public regulatory agencies. A third option for a country is to maintain or adopt its own standards or allow for some leeway to domestic companies to decide whether they would like to comply voluntarily with the international standards at stake.[18]

II. Standard-setting within ISO

ISO is the most important international standard-setting organisation internationally. ISO is not an intergovernmental organisation, and thus its output is deprived of any formal coercive force; it is a network (or federation) of national SSBs, composed of both governmental and industry representatives. ISO currently comprises 160 members involved in the development of standards,[19] which predominantly are national standardisation bodies, which in turn are of a hybrid nature but primarily composed of representatives from the private industry, whereas government

[16] Cf. B. Kingsbury, N. Krisch and R.B. Stewart, 'The Emergence of Global Administrative Law', 68 *Law and Contemporary Problems* (2005) 15.

[17] See also J. Black, 'Legitimacy, Accountability and Polycentric Regulation: Dilemmas, Trilemmas and Organisational Response' in A. Peters, L. Koechlin, T. Förster and G. Fenner Zinkernagel (eds), *Non-State Actors as Standard Setters* (Cambridge University Press, 2009), 241, at 246.

[18] See J. Nakagawa, *International Harmonization of Economic Regulation* (Oxford University Press, 2011), at 109.

[19] According to the ISO Membership Manual, ISO Members are the national organisations most representative of standardisation in their respective country. ISO has one Member per country. See *ISO Membership Manual*, August 2013, p. I-7, available at: http://www.iso.org/iso/iso_membership_manual_2013.pdf. See also S. Bernstein and E. Hannah, 'Non-State Global Standard Setting and the WTO: Legitimacy and the Need for Regulatory Space', 11:3 *Journal of International Economic Law* (2008) 575.

staff experts act as members in the national bodies.[20] The ISO Secretariat is relatively small (154 full-time employees), serving a coordination function.

Standardisation within ISO starts with the creation of a technical committee (TC) or the approval of a standard's development proposal by ISO Technical Management Board (TMB), which then assigns the development of a particular standard to an existing TC. TCs may be composed of working groups and sub-committees, which are essentially run by the national standardisation bodies. There are more than 230 TCs within ISO.[21] Because the work of a given TC is very technical, having a bird's-eye view of developments in every TC is highly unlikely. Typically, national SSBs are called on to run the Secretariat of a given TC. Secretariats are distributed unevenly, with the American National Standards Institute (ANSI) and DIN (*Deutsches Institut für Normung*, ANSI's German counterpart) maintaining a disproportionally large number of such secretariats.[22]

ISO works in close collaboration with the International Electrotechnical Commission (IEC) on all matters relating to electrotechnical standardisation. In 1987, the first Joint Technical Committee (JTC 1) was created between ISO and IEC to prepare standards in the areas of ICT, including multimedia, ICT security or cloud computing.[23] Along with ISO and IEC, a third important standardisation body is the International Telecommunication Union (ITU). In fact, these three organisations are regarded as the only international standardisation bodies in the field of technical standards for certain Members such as the European Union.[24] In 2001, ISO, IEC and ITU established the World Standards Cooperation

[20] This varies depending on the country's (centralised or decentralised) approach to technical standards. For the United States, for instance, where the American National Standards Institute (ANSI) is a private entity, ISO standards are regarded as standards adopted by and addressed to private parties. Cf. D. Wirth, 'The International Organization for Standardization: Private Voluntary Standards as Swords and Shields', in G. van Calster and D. Prévost (eds), *Research Handbook on Environment, Health and the WTO* (Edward Elgar, 2013), 139, at 141. In Japan, on the other hand, the Japanese Industrial Standards Committee (JISC), an advisory council of the Ministry of Economy, Trade and Industry, is the ISO Member Body for Japan. While under the auspices of the government, JISC is in reality a multi-stakeholder body.

[21] A list of the ISO TCs can be found at http://www.iso.org/iso/home/standards_development/list_of_iso_technical_committees.htm.

[22] For the current figures, see ISO in figures 2014, available at http://www.iso.org/iso/home/about/iso-in-figures.htm.

[23] See http://www.iso.org/iso/jtc1_home.html.

[24] See also EU Regulation 1025/2012 on European Standardization [2012] OJ L 316/12, Art. 2 (10).

(WSC) 'to strengthen and advance the voluntary and consensus-based international standard-setting systems'[25] that they created through the avoidance of duplication and overlap of work.

Generally, ISSBs choose consensus as the decision-making mode par excellence, which the ISO defines as 'general agreement, characterised by the absence of *sustained opposition* to *substantial issues* by *any important part of the concerned interests* and by a process that involves seeking to take into account the views of all parties concerned and to reconcile any conflicting arguments' (emphasis added). However, it is made clear that consensus need not imply unanimity.[26] In addition, the ISO Directives consider sustained oppositions as a peculiar category of objection. They are defined as 'views . . . maintained by an important part of the concerned interest and which are incompatible with the committee consensus'.[27] The Directives call on the leadership of the relevant body to solve the issue based on certain guidelines; for example, the leadership of the committee must ensure that the opposition is sustained by an important part of the concerned interest, which will vary depending on the dynamics of the committee; if so, it should be dealt with in good faith. However, and crucially, a sustained opposition is not akin to a right to veto.

Appeals can be filed on the condition that they are against the Statutes and Rules of Procedure, the ISO/IEC Directives or detrimental to trade, safety, health or the environment. They can relate not only to technical but also to administrative issues. However, appeals with respect to decisions concerning draft standards are possible only on condition that questions of principle are involved or the contents of the draft may be detrimental to the reputation of ISO. Crucially, only P-members (those actively participating in the TC) are permitted to file an appeal. O-members (members who have requested observer status within the TC) or other categories of ISO members do not have this possibility. Finally, appeals do not have suspensive effect until the approval stage.[28]

Although unanimity is not required, most international organisations aim at consensus building and have those mechanisms in place in their

[25] See http://www.worldstandardscooperation.org/about.html.

[26] See ISO/IEC Directives (ISO/IEC Dir 1), Part 1, Edition 10.0, 2013–10, clause. 2.5.6. This definition is generally accepted as reflecting the understanding of what consensus entails in SSBs. See also the EU Regulation 1025/2012, Annex II, para 3(b).

[27] See ISO/IEC Directives, Part 1, Consolidated ISO Supplement – Procedures specific to ISO, Fourth edition, 2013, clause 2.5.6.

[28] Ibid, clause 5.

constitutions and secondary law.[29] Consensus however can cause delays, in which competitors argue for their preferred solution or simply hold out until one side concedes or withdraws to the benefit of the other.[30] Endorsement of a given standard at the end of the process can generate substantial rents, which make the effort quite worthwhile,[31] but also confirms the value of (and, in the end, legitimises) the standard-setter as a stabilising factor in its capacity as a coordinating authority. In addition, it was shown that in areas of rapid technological change and innovation and thus important rents being at stake (distributional conflicts), the standard-setting process may be slower in a consensus-based SSB, but delays will be efficient when the underlying technology improves with the time. Thus, and quite importantly, at the end of the lengthy process, higher-quality outcomes will be produced.[32] This means that, contrary to conventional belief and somehow counter-intuitively, striving for consensus may have a limited impact on the technical and scientific excellence of a given standard. However, when vested interests are strong, relaxing the way consensus is required or identifying a neutral participant to break deadlock (i.e. binding arbitration) may be preferable to increase the effectiveness of a given standard.[33]

In practice, ISO does not decide by unanimity but rather has adopted qualified majority voting rules in the various stages (from the preliminary stage to the enquiry draft and up to the publication stage) that lead to the adoption of an international standard.[34] According to the ISO/IEC directives, within ISO, if there is doubt as to whether consensus was reached for registration as an enquiry draft (that is, the TC draft), a two-thirds majority of the actively involved members in the TC (the P-members) approving it would suffice.[35] The 'two-thirds rule' seems to

[29] See J. Steffek, 'Sources of Legitimacy beyond the State: A View from International Relations', in C. Joerges, I.-J. Sand and G. Teubner (eds), *Transnational Governance and Constitutionalism* (Hart Publishing, 2004), 81, at 94.

[30] Farrell and Saloner first described this tactic as a 'war of attrition', suggesting that it may lead to the technically best solution, but with a significant delay. See J. Farrell and G. Saloner, 'Coordination through Committees and Markets', 19 *RAND Journal of Economics* (1988) 235.

[31] M. Rysman and T. Simcoe, 'Patents and the Performance of Voluntary Standard-Setting Organizations', 54:11 *Management Science* (2008) 1920.

[32] T. Simcoe, 'Standard Setting Committees: Consensus Governance for Shared Technology Platforms', 102:1 *American Economic Review* (2012) 305.

[33] J. Farrell and T. Simcoe, 'Choosing the rules for consensus standardisation', 43:2 *RAND Journal of Economics* (2012) 235.

[34] See ISO/IEC Dir 1, Art. 2.3ff.

[35] Ibid, Art. 2.5.6.

be generally applicable. Thus, rather than unanimity, the two-thirds rule should be regarded as the general benchmark expressing the multilateral scientific consensus in the international standardisation community.

In line with this observation, a final draft international standard circulated by a TC is approved if two-thirds of the votes cast by the ISO members actively participating in the TC are in favour and not more than one-quarter of the total number of votes cast are negative. The two conditions are cumulative. Abstentions do not count, and the same goes for negative votes that are not based on technical reasons.[36] This applies to both the enquiry stage and the approval stage. In other words, objections with respect to procedural defects, for instance, would need to be raised at an early stage, while the relevant TC or subcommittee is still discussing drafting the standard. This is also made clear in the ISO/IEC Directives of 2013, which provide that[37]

> to avoid re-discussion, national bodies have the responsibility of ensuring that their technical standpoint is established taking account of all interests concerned at national level, and that *this standpoint is made clear at an early stage of the work* rather than, for example, at the final (approval) stage. Moreover, national bodies need to recognise that substantial comments tabled at meetings are counter-productive, since no opportunity is available for other delegations to carry out the necessary consultations at home, without which rapid achievement of consensus will be difficult. [emphasis added]

However, for this even to be possible in the first place, effective participation is a prerequisite. Effective participation would include the ability to be at all meetings; to follow several meetings simultaneously, which may be organised in different places around the world; and to find the way through a large number of technical documents in a short period of time. Under such conditions, much of the work within TCs is dominated by industry-driven developed-country interests.[38] Such transnational corporate domination[39] of the process of drafting standards may be worrisome in various respects, not only in terms of fairness and participation but, more fundamentally, of legitimacy.

[36] Ibid, Art. 2.7.3. [37] Ibid, p. 7. See also Art. 2.5.1.

[38] See also N. Roht-Arriaza, '"Soft Law" in a "Hybrid" Organization: The International Organization for Standardisation' in D. Shelton (ed), *Commitment and Compliance – The Role of Non-Binding Norms in the International Legal System* (Oxford University Press, 2000), 263, at 267.

[39] Ibid, 267.

C. The TBT and international standards

The impressive reduction of tariffs since the inception of the General Agreement on Tariffs and Trade (GATT) in the late 1940s and the outright prohibition of the use of quantitative restrictions that the GATT requires has a significant consequence: WTO Members used (and continue to use) 'behind-the-border', domestic instruments to protect their industries.[40] These non-tariff barriers (NTBs) have ultimately emerged as important obstacles to trade.[41] This situation has recently been exacerbated due to the emergence of non-state actors that increasingly impinge on economic action, particularly in the area of transnational standardisation, which domestic antitrust laws may be unable to capture. Thus, delegation of regulatory power to private actors and the rise of private government and transnational networks are not unproblematic from an international law viewpoint because these activities may fall outside the scope of traditional inter-state rules.[42]

The GATT is a negative integration contract. It is essentially based on one obligation that is imposed on the WTO Membership: every domestic policy that has an impact on trade has to be applied in a non-discriminatory manner, thereby ensuring equality of opportunities between domestic and foreign products, but also among foreign products only. On the other hand, policies affecting trade are to be unilaterally prescribed at a national level, and thus there is no predefined set of policies by which all WTO Members must abide. This means, inter alia, that the GATT does not imply any compulsory adherence to international standards.

However, with the adoption of the TBT in the mid-1990s and the establishment of the WTO, the WTO drafters decided to go beyond non-discrimination to promote regulatory efficiency domestically in areas

[40] D. Kono, 'Optimal Obfuscation: Democracy and Trade Policy Transparency', 100 *American Political Science Review* (2006) 369 (showing that democratic regimes have asymmetrical influence over the various types of protection, i.e. while leading to lower tariffs, they lead to higher NTBs).

[41] J. Wilson, 'Standards, Regulation and Trade: WTO Rules and Developing Country Concerns' in B. Hoekman, A. Mattoo and R. English (eds), *Development, Trade and the WTO : A Handbook* (The World Bank, 2002); also P. Messerlin, *Measuring the Costs of Protection in Europe: European Commercial Policy in the 2000s* (Institute for International Economics Press, 2001).

[42] See, among many others, C. Donnelly, *Delegation of Governmental Power to Private Parties – A Comparative Perspective* (Oxford University Press, 2007); A. Boyle and C. Chinkin, *The Making of International Law* (Oxford University Press, 2007).

such as technical or food safety regulations.[43] As we will see subsequently, this shift of focus inevitably included voluntary international standards developed in ISSBs such as the ISO or the Codex Alimentarius Commission (CAC) as expressions of international technical consensus. With this change in the multilateral trading system, the processes used within ISSBs, neglected by many up to that point in time, came to the forefront, leading to ensuing transformations of their modus operandi.[44]

I. The TBT preamble

The TBT, just like the SPS, substantiates Members' attempt to effectively deal with NTBs and harness badly designed and badly applied technical regulations, specifications, standards and procedures domestically. However, it also acknowledges Members' prerogative to autonomously pursue public policy objectives such as those relating to the protection of the environment or consumers, provided that the non-discrimination principle is observed.[45] Early on in its preamble, the TBT points to the importance of harmonisation of compatibility standards by acknowledging the role of international standards as trade facilitators and technology transfer vectors. In the latter case in particular, the TBT explicitly refers to international standardisation as a process *in abstracto*, thereby endorsing by reference the importance of the work of several decades done within ISSBs to advance technological progress. The fact that the TBT, contrary to the SPS, does not explicitly refer to particular ISSBs does not alter this observation. As a result, the TBT encourages the development of such harmonised standards at the international level.

This is no coincidence but rather a considerate choice in both the TBT and the SPS. If one is to strike a reasonable balance between concealed

43 See G. Marceau and J. Trachtman, 'The Technical Barriers to Trade Agreement, the Sanitary and Phytosanitary Measures Agreement, and the General Agreement on Tariffs and Trade: A Map of the World Trade Organization Law of Domestic Regulation of Goods', 36(5) *Journal of World Trade* (2002) 811; also R. Howse and J. Langille, 'Permitting Pluralism: The Seal Products Dispute and Why the WTO Should Accept Trade Restrictions Justified by Non-Instrumental Moral Values', 37 *Yale Journal of International Law* (2012) 367.

44 For instance, Motaal unequivocally describes the transformation that ISSBs, which prepare, draft and adopt SPS standards, have undergone since 1994 due to the entry into force of the *SPS Agreement* at the conclusion of the Uruguay Round. See A. Motaal, 'The "Multilateral Scientific Consensus" and the World Trade Organization', 38(5) *Journal of World Trade* (2004) 855.

45 See also Appellate Body Reports, *US – Tuna II*, para. 213; and *US – Clove Cigarettes*, para. 95.

protectionism and the well-meant pursuit of legitimate objectives by a benign government, benchmarks (or proxies) are needed. Under the GATT, the necessity test of the general exceptions provision enshrined in Article XX GATT exerts this role,[46] allowing to identify measures that exceed what is necessary to achieve the degree of contribution that a given measures makes to the pursuit of a legitimate objective.[47] A necessity test as a proxy in the case of the TBT and the SPS would be necessary (and indeed such a test is to be found in both agreements), but it is not sufficient for a treaty that aims at *positive* integration through harmonisation. Additional common denominators or heuristic devices are needed. Because of their unambiguous technical expertise, ISSBs and the instruments they create (i.e. international standards) are deemed appropriate for this role.

At the same time, and perhaps contrary to the mainstream literature about TBT,[48] there is no carte blanche for international standard-setting processes enshrined in the TBT preamble. Rather, the approach is much more critical when one looks carefully at the preamble. Indeed, the TBT preamble, after hailing the importance of harmonising international standards and conformity assessment systems, it refers to Members' desire to ensure that technical regulations and standards do not create unnecessary trade barriers. Arguably, this relates not only to domestic measures but also to international standards. This interpretation is reasonable due to the use of the word 'however' in the fifth recital of the TBT preamble. It is only plausible to suggest here that 'however' refers to what preceded under the third and fourth recitals, which exclusively refer to *international* instruments.

II. The substantive scope of TBT: textual ambiguities and jurisprudential mishaps

Ratione materiae, the TBT distinguishes between technical regulations and standards. More specifically, it covers mandatory technical regulations and voluntary product standards (such as those relating to size, quality, composition or labelling) as well as conformity assessment procedures.

46 See P. Delimatsis, 'Protecting Public Morals in a Digital Age: Revisiting the WTO Rulings on *US – Gambling* and *China – Publications and Audiovisual Services*', 14:2 *Journal of International Economic Law* (2011) 257.

47 Cf. Appellate Body Report, *US – Tuna II (Mexico)*, para. 319.

48 See, among many others, A. Appleton, 'The Agreement on Technical Barriers to Trade' in P. Macrory, A. Appleton and M. Plummer (eds), *The World Trade Organization: Legal, Economic and Political Analysis* (Martinus Nijhoff, 2005), 371.

Despite the reference to ISO/IEC Guide 2:1991, which is by now revised twice within ISO, the TBT offers definitions for the three main categories of measures coming under the TBT ambit in Annex 1 of the agreement. For our purposes, I focus on the first two categories. Thus, a technical regulation is defined as a

> document which lays down product characteristics or their related processes and production methods, *including the applicable administrative provisions*, with which *compliance is mandatory*. It may also include or deal exclusively with terminology, symbols, packaging, marking or labelling requirements as they apply to a product, process or production method. [emphasis added]

Furthermore, it defines a standard as a[49]

> document *approved by a recognised body*, that provides, for common and repeated use, rules, guidelines or characteristics for products or related processes and production methods, with which *compliance is not mandatory*. It may also include or deal exclusively with terminology, symbols, packaging, marking or labelling requirements as they apply to a product, process or production method. [emphasis added]

The two definitions are not free from ambiguity; overlap is clear already in their very wording. More strikingly, the second sentence in both definitions is identical. Importantly, both definitions make clear that they cover labelling requirements and production and process methods (PPMs). After the Appellate Body decision in *EC – Sardines*, the traditional view that allowed drawing the line between the two categories of products was whether non-compliance with a given measure in fact prohibits access to a given domestic market (technical regulation) or whether access to this market was possible irrespective of compliance with a given technical specification (standards).[50]

In *EC – Asbestos* and *EC – Sardines*, and, more recently, in *EC – Seal Products*, the Appellate Body examined the constitutive elements of a technical regulation under the TBT. First, it suggested that a technical regulation may lay down one or more *binding* product characteristics

[49] For the negotiating history of this definition, see WTO (TBT Committee), 'Negotiating History of the Coverage of the Agreement on Technical Barriers to Trade with Regard to Labelling Requirements, Voluntary Standards, and Processes and Production Methods Unrelated to Product Characteristics', WT/TBT/W/11, 29 August 1995.

[50] Cf. P. van den Bossche, *The Law and Policy of the World Trade Organization – Text, Cases and Materials*, 2nd edition (Cambridge University Press, 2008), at 807. As discussed later, the Appellate Body decoupled the two in *US – Tuna II*, para. 196.

in a *positive* or *negative* form – that is, it may require that a product possesses or does not possess particular characteristics. Thus, the document at stake must have a certain normative content.[51] According to the Appellate Body, this includes any objectively definable features, qualities, attributes or other distinguishable mark of a product. Those product characteristics may be intrinsic, or they may be related to the product such as the means of identification, the presentation and the appearance of a product.[52] Furthermore, it should be applicable to an identifiable product or group of products. Finally, compliance is mandatory, suggesting that non-compliance would allow an enforcement mechanism to sanction a particular producer.[53]

In *EC – Sardines*, the Appellate Body found that compliance with the relevant EC Regulation was mandatory because the legislative instrument used was binding and directly applicable in all EU Member States. In view of the many similarities among the definition of technical regulation and that of standard, two appear to be the main differences between the two types of measures: the first is that, whereas the technical regulation is adopted by a governmental body and thus is a State measure, a standard is typically issued by private or semi-private standardising bodies, that is, bodies with standardisation activities. A standard can later become a technical regulation if adopted or used as a basis for a legislative act by the State. The second difference is that a compliance with a standard, contrary to a technical regulation, is voluntary.

The actual contours of the definition of a technical regulation and its relationship with the definition of a standard under the TBT was discussed in the controversial *US – Tuna II* dispute relating to a dolphin-safe labelling scheme for tuna products.[54] Under the measure at issue, tuna products sold in the United States could be labelled 'dolphin-safe' only if certain requirements were met – in particular, with respect to the way tuna was harvested. However, for tuna to be imported, no 'dolphin-safe' label was required. First the Panel and subsequently the Appellate Body found that the measure at issue was a technical regulation within the

[51] Appellate Body Report, *EC – Seal Products*, para. 5.10. For instance, a measure only consisting of a prohibition on a product in its natural state may not constitute a technical regulation. Ibid, para. 5.34–5.

[52] Appellate Body Report, *EC – Asbestos*, para. 67.

[53] See Appellate Body Reports, *EC – Asbestos*, para. 76; and *EC – Sardines*, para. 176.

[54] See also P. Mavroidis, "Driftin" Too Far from Shore – Why the Test for Compliance with the TBT Agreement Developed by the WTO Appellate Body Is Wrong, and What Should the AB Have Done Instead', 12:3 *World Trade Review* (2013) 509.

meaning of Annex 1.1 TBT. Both considered that whether the measure at issue was imposing conditions on the access to the 'dolphin-safe' label rather than the US market was important in this respect. Decisive in the judiciary's view appears to have been that the US measure covered the 'entire field' of what 'dolphin-safe' meant in relation to tuna products in the US.[55] In the Appellate Body's words:[56]

> In effect, the measure at issue establishes a single definition of 'dolphin-safe' and treats any statement on a tuna product regarding 'dolphin-safety' that does not meet the conditions of the measure as a deceptive practice or act.

After observing that the measure at issue consisted of legislative and regulatory acts of the US federal authorities and that included administrative provisions, the Appellate Body found that the 'dolphin-safe' labelling requirement was a technical regulation for the purposes of the TBT.

The analysis of the Panel and the Appellate Body is not convincing[57] and makes much of Annex 1.2 TBT superfluous by blurring the distinction between mandatory and voluntary compliance. More importantly, it deprives Annex 1.2 TBT of its *effet utile*, rendering redundant any distinction between standards and technical regulations. Systemically, it may have important repercussions for various sustainability-related labelling schemes, organised and applied by both governmental or hybrid and private standardising bodies. This is particularly so in cases where: (i) no governmental scheme is present in the market and (ii) some connection with the government or administrative guidance can be proven, thereby linking the standardisation body with the State. If none of the two happens, then only a very marginal set of rules could come within the definition of a standard under Annex 1.2 TBT, taking into account the similarity of the second sentence of Annex 1.1 and 1.2 TBT.[58]

Both the Panel and the Appellate Body failed to give meaning to the fact that the US measure was not concerned with setting characteristics for a product to be regarded as tuna. Rather, the measure at issue

55 See Appellate Body Report, *US – Tuna II*, para. 193. 56 Ibid, 195.

57 Note that one of the panelists in the panel stage filed a separate opinion on this particular matter. See Panel Report, *US – Tuna II*, paras. 7.146ff.

58 Taking *US – Tuna II* as example, if the same dolphin-safe scheme was promulgated by a private standardising body, then the scheme would have been classified most likely as a standard and not a technical regulation. However, is it only the governmental involvement that determines whether we apply Annex 1.1 or 1.2 TBT? The focus on compliance and exclusion of the market for a given product seems to be safer a criterion.

set the traits of *dolphin-safe* tuna products. Tuna products, no matter how they were harvested, could still enter the US market. In addition, there is a fundamental difference with the situation in *EC – Sardines* that both the Panel and the Appellate Body invoked to corroborate their findings: the EC regulation allowed only one particular species of sardines to be marketed as preserved sardines within the EU. Thus, it was clear that no other sardines could enter the market and be called 'preserved sardines'.

However, in *US – Tuna II*, traders were free to access the tuna market without any labels relating to their 'dolphin safety'. Note, in stark contrast, that Peruvian sardines could not enter the EU market. In the facts of *US – Tuna II*, however, tuna products could still be marketed and sold as tuna even if they do not comply with the 'dolphin-safe' labelling requirements.[59] Thus, the WTO judiciary conflated the requirements for access to the *label* with the requirements for access to the *market*. Here it is at the discretion of the producer to comply or not with the requirements of the label. Conversely, in *EC – Sardines*, that was definitely not the case.

To show the problematic character of this interpretation, one can juxtapose it to the US measure that was at stake in the Certain Country of Origin Labelling (COOL) dispute, imposing an obligation on retailers selling specific meat products in the US downstream market to label those products with their country of origin.[60] In this case, such products could not get to the final point of sale without such label. If the *US – Tuna II* interpretation holds, then the COOL measure and the 'dolphin-safe' label would be labelling systems producing the same effects. In fact, however, they clearly do not, as in the case of country of origin labelling for meat, the meat product would not be allowed to reach consumers unless such label is affixed to the product. This result is absurd. The same line of reasoning holds when reading the recent *EC – Seal Products* Appellate Body Report. In this case, the Appellate Body underscored that the EU-relevant legislation prescribes rules relating to the *placing* on the EU market of seal product 'in a binding and compulsory fashion'.[61]

For the sake of comparison, long-established case law in the EU would most likely accept the legality of the measure at issue. Such a labelling

[59] See also the EU arguments in Appellate Body Report, *US – Tuna II*, para. 155; and the panelist's dissenting opinion in Panel Report, *US – Tuna II*, para. 7.164.

[60] See Appellate Body Report, *US – COOL*, para. 239.

[61] Appellate Body Report, *EC – Seal Products*, para. 5.22.

scheme would most likely pass the *Keck* test applied by the Court of Justice of the European Union (CJEU): the US measure applies to all relevant traders operating in the market and affects in the same manner, both in law and in fact, the selling of domestic products and of those from other Members.[62]

From a systemic point of view, the WTO judiciary, quite importantly, brought under the definition of technical regulation of the TBT non-incorporated PPMs, that is, PPMs that are not observable on the product itself.[63] This may be a welcome development for the WTO judiciary because it would like to keep for itself intrusive judicial review of such measures. From a legal point of view, it messes up the different categories of TBT measures and arguably the balance among the WTO agreements. Indeed, in that case, a correct analysis of the facts should have led the WTO judiciary to find that the TBT was not applicable and thus roll back to the application of Article III:4 GATT. However, one problematic feature of this option was that the Appellate Body would be unable to complete the analysis because the Panel exercised judicial economy with respect to the Mexican claims under the GATT, after finding that the TBT is *lex specialis* and after making findings under the latter agreement. However, you cannot correct a wrong with a wrong, and that is what the Appellate Body did.

III. The Code of Good Practice relating to standards

The structure of the TBT reveals a hierarchical relationship between the two types of instruments that I discussed earlier. Whereas the obligations relating to technical regulations are in the main body of the TBT, the obligations regarding the development, adoption and application of standards are to be found in Annex 3 incorporating a Code of Good Practice (hereinafter 'the Code'). This is in stark contrast with the previous, plurilateral Standards Code adopted in the Tokyo Round where rules for technical regulations and standards belonged to the main body of the agreement without differentiation between the two categories of measures. This does

[62] See CJEU Case C-267/91 and C-268/91, *Keck and Mithouard* [1993] ECR I-6097, para. 16.

[63] In *EC – Seals*, the Appellate Body noted (para. 5.12): 'In order to determine whether a measure lays down related PPMs [within the meaning of Annex 1.1 TBT], a panel . . . will have to examine whether the processes and productions methods prescribed by the measure have a sufficient nexus to the characteristics of a product in order to be considered related to those characteristics.'

not alter the fact that the Annex is equally binding because it is an integral part of the TBT by virtue of Article 15.5 TBT.

However, it appears that the Code is not incorporated in the TBT but in a separate annex because the Code includes obligations that are mainly addressed to self-regulated and self-governed national SSBs, whereby many of those bodies, as noted earlier, are private bodies composed of industry representatives. For this reason, the TBT foresees the possibility for the Code to be accepted by standardising bodies – be it governmental or non-governmental; central; regional or local – established in the territory of any WTO Member. Indeed, the Code is open for acceptance and the ISO/IEC Information Centre keeps track of such acceptances.[64] As of February 2015, 165 SSBs from 125 Members accepted the Code; among them there are 91 central government bodies, 68 non-governmental bodies, 3 statutory bodies, 2 parastatal bodies, 3 non-governmental regional bodies (the 3 European standard-setters, i.e. CEN, CENELEC and ETSI), 1 central hybrid body, 1 central governmental/local body and 1 autonomous body.[65] In any event, the obligations under the Code relating to standards and Article 2 regarding technical regulations are for all practical purposes the same. In addition, Article 4 TBT provides that Members' obligations relating to compliance of domestic SSBs with the Code remain intact regardless of whether a domestic SSB has accepted the Code.

Even if this serves consistency, some interpretive challenges remain. For instance, paragraph D of the Code requires that standardising bodies accord equal treatment to foreign products compared with other foreign or domestic products. Nevertheless, this obligation seems to be without content, for such bodies do not deal with products directly. Requiring from SSBs the type of non-discrimination that the TBT and WTO law in general requires from States appears absurd here. Rather, one would expect the Code to focus on fair and non-discriminatory access to standardisation activities, participation, transparency, necessity and the like. For the sake of comparison and as an example of how non-discrimination could be phrased adequately, Article 5.5 of ISO/IEC Guide 59 of 1994 provides that standards shall neither be written nor adopted so as to

[64] See the Memorandum of Understanding on 'WTO Standards Information Service Operated by ISO'. Every standardisation body that accepted the Code is obliged to prepare a biannual work program in which it provides information about the standards under preparation and the standards that it adopted in the preceding period.

[65] See Committee on Technical Barriers to Trade (TBT Committee), 'List of Standardizing Bodies that Have Accepted the Code of Good Practice for the Preparation, Adoption and Application of Standards Since 1 January 1995', G/TBT/CS/2/Rev.21, 23 February 2015.

discriminate among products on the basis of the place of origin.[66] Another example of this kind is to be found in Article 5.5.3 of the ISEAL Alliance Code of Good Practice which requires that membership criteria and application procedures in standard-setting organisations be transparent and non-discriminatory.[67]

Although not explicitly referring to the relationship between the TBT and ISO, the former has numerous references to the ISO, more particularly ISO/IEC Guide 2:1991. The TBT provides that whenever terms included in this Guide are also mentioned in the TBT, then the meaning that these terms have, according to the Guide, becomes the authoritative one.[68] Thus, the ISO/IEC Guide 2:1991 constitutes important context for dispute settlement purposes by virtue of the Vienna Convention on the Law of Treaties (VCLT), directing the judiciary to have recourse to a non-WTO document to clarify certain TBT terms.[69] Together with Article 1.1 TBT, they substantiate the attempt of the TBT drafters to position the agreement within the broader standardisation community and to relate it to the existing international standard-setting processes, tilting toward the promotion of such processes. Indeed, references to non-WTO documents within the WTO treaty that allow completing the interpretation of the rights and obligations abound. In addition, the Code reproduces principles and rules known from the ISO without, however, explicitly referring to it. For instance, paragraph G of the Code provides that national SSBs shall strive to be represented in ISSBs through one delegation to ensure coherence.

IV. 'Relevant international standards'

The TBT is a harmonisation vector. The idea of favouring convergence of technical regulations globally permeates the agreement.[70] Article 2.4 requires that *relevant international standards* or *relevant parts* thereof (when they exist or are about to be adopted) be used as a basis for domestic technical regulations unless they are ineffective or inappropriate means

[66] See TBT Committee, Communication from the ISO/IEC, TBT/W/178, 15 December 1993.

[67] See ISEAL Code of Good Practice – Setting Social and Environmental Standards v5.0, 2010. For the structure and objectives of ISEAL, see Axel Marx and Jan Wouter's contribution in this volume.

[68] See introductory paragraph, Annex 1 of TBT.

[69] Similar to the Harmonised System's explanatory notes. See, for instance, Appellate Body Report, *China – Auto Parts*, paras. 149ff.

[70] Other than standards, this is also obvious in the case of conformity assessment procedures. See, for instance, Article 9.1 TBT.

for meeting the public policy objectives pursued. As per *EC – Sardines*, this obligation applies not only to technical regulations that were adopted after the entry into force of the WTO agreement but also to technical regulations that existed at that time and continued to be applicable.[71] This means that a series of international standards that were adopted before the mid-1990s and typically applied on a voluntary basis suddenly become mandatory reference points for domestic technical regulations.

In *EC – Sardines*, the concept 'as a basis for' was interpreted similarly to Article 3.1 SPS to mean that an international standard is used in this way when it is the principal constituent or fundamental principle for the purpose of enacting the technical regulation at stake, thereby revealing a strong and close relationship between the two.[72] If only parts of a given international standard are used, then those that are relevant shall be used as a basis for the technical regulation at issue. At a minimum, the international standard and the regulation cannot contradict each other. Finally, it bears noting that it is for the complaining party to prove that the international standard at issue is effective (capable of accomplishing the legitimate objective pursued) and appropriate (suitable for the fulfilment of the objective pursued) for the achievement of the objective pursued. In this analysis, it would be for the judiciary to examine and determine the legitimacy of the objectives of the measure. Effectiveness focuses on the *results* of the means used, whereas appropriateness focused on the *nature* of the means used.[73] Adjudicating bodies may be more willing to deny such qualities to international standards, for instance, when perceptions and expectations of consumers are not satisfied by it[74] or when the technical regulation at issue sets higher standards (e.g. more detailed and accurate information) for consumer protection.[75]

[71] Appellate Body Report, *EC – Sardines*, para. 205. [72] Ibid, paras. 244–245.

[73] Ibid, para. 285. [74] Ibid, para. 289.

[75] See Panel Report, *United States – COOL*, para. 7.734. This excerpt is actually one of the most striking ones in this Panel report. The Panel seems to examine indirectly, but still quite critically, the substantive content of the relevant international standard, leaving the window for the circumvention of international standards (a Codex standard in this case) wide open. In addition, the Panel appears to disregard the fact that one of the purposes of the Codex standard was indeed to protect consumers from unduly confusing and detailed information. Thus, both the international standard and the domestic labelling scheme pursued the same legitimate objective. Why, then, opt for the domestic standard when the actual objective of the TBT is convergence and harmonisation through the use of international standards? Interestingly, it would seem to suggest that the international standard may not have been 'relevant' in the first place. Unfortunately, the Panel's findings in this respect were not appealed, although the Appellate Body noted that 'a mandatory labelling scheme based on substantial transformation [i.e. the relevant CODEX standard STAN 1–1985] would, at best, contribute only *partially* to the objective of providing

Pursuant to Article 2.5, a technical regulation is presumed to comply with the TBT and more specifically Article 2.2, if it is in accordance with *relevant international standards* and provided that is prepared, adopted or applied to protect national security; prevent deceptive practices; protect human and animal health or safety; or protect the environment. The rationale behind this 'safe haven' is that voluntary international standards ostensibly incorporate international preferences and unambiguous technical superiority. Furthermore, Article 2.9 provides for additional notification requirements in case relevant international standards are not used.

Thus, other than requiring Members to use relevant international standards in a positive manner and creating a rebuttable presumption of consistency as an extra 'carrot', the agreement imposes additional burdensome conditions that Members need to comply with in case of neglecting international standards. In other words, in those areas where international standards exist, which, as noted earlier, are mainly of voluntary nature, they become the reference point and de facto mandatory normative technical material to be used by the WTO Members. As a result a mass of documents of at best uncertain legal normativity are transformed into international obligations equivalent to treaty text.[76]

Article 2 TBT mentions various times the term 'relevant international standard' but lacks any provision that would offer a definition. However, as demonstrated earlier, defining the exact scope of this term is important.[77] I mentioned earlier the TBT definition of a standard. The ISO/IEC Guide 2:2004 defines a standard as a

> document, established by *consensus* and approved by a *recognised* body, that provides for common and repeated use, rules, guidelines or characteristics for activities or their results, aimed at the achievement of the optimum degree of order in a given context. – Note: Standards should be based on the consolidated results of science technology and experience, and aimed at the promotion of optimum community benefits.[78] [emphasis added]

information to consumers on where livestock from which meat is derived were born, raised and slaughtered.' (emphasis added): Appellate Body Report, *US – COOL*, para. 486.

76 See also R. Howse, 'A New Device for Creating International Legal Normativity: The WTO Technical Barriers to Trade Agreement and "International Standards"', in C. Joerges and E.-U. Petersmann (eds), *Constitutionalism, Multilevel Trade Governance and International Economic Law* (2011), 383.

77 Cf. Appellate Body Report, *US – Tuna II*, para. 348.

78 See ISO/IEC Guide 2:2004, Art. 3.2.

Furthermore, and quite tautologically, the Guide defines as international a standard that is adopted by an international standardising/standards organisation and made available to the public.[79] Later on, the ISO/IEC Guide defines an international standardising organisation as the organisation (that is, the body that is based in the membership of other bodies or individuals and has an *established constitution and its own administration*) whose membership is open to the relevant national body from every country.[80] This seems to suggest that an international standard is not any different from a national standard in terms of *content*. Common and repeated use as well as fitness for purpose are important traits for both of them. What rather makes it distinct is the *nature* of the body preparing and adopting it – that is, the international standardising organisation.[81]

These definitions are important and relevant for the purposes of TBT interpretation, because the introductory paragraph of Annex 1 TBT provides that terms used in it that remain undefined in the TBT should have the same meaning as in the ISO/IEC Guide 2:1991.[82] Thus, in reconstructing the definition of a relevant international standard for TBT purposes, one would need to take elements from both the TBT definitions and the ISO/IEC Guide's definitions.[83]

Although not offering an explicit definition of what constitutes an international standard, the TBT definition of standard in Annex 1.2 TBT also deviates from the ISO definition in that it also considers as standards for TBT purposes those documents that are not based on consensus. This is a considerate deviation[84] because the explanatory note in Annex 1.2 TBT is clear on the TBT drafters' awareness of the fact that international standards are typically based on consensus. Thus, in *EC – Sardines*, the Panel first and the Appellate Body later rejected the EU's argument that only standards adopted by consensus can be regarded as relevant for purposes of Article 2.4 TBT.

The EU, in this case, suggested that a standard such as the Codex Stan 94, which is accepted by only eighteen countries, of which only

79 Echoed in the EU Regulation 1025/2012, Art. 2(1)(a).

80 See ISO/IEC Guide 2:2004, Art. 4.3.2. in conjunction with Art. 4.2.

81 Compare the distinction among international, European and national standards in the EU Regulation 1025/2012, Article 2(1) (a), (b) and (d).

82 Note that the relevant definitions of the ISO/IEC Guide 2:1991 have remained identical in the ISO/IEC Guide 2:2004.

83 Recall that the TBT definition of standard prevails over the ISO one as per Annex 1.2 TBT.

84 Cf. Appellate Body Report, *EC – Sardines*, para. 225.

four accepted it fully, cannot be regarded as an international standard.[85] Interestingly, to answer the EU's claim regarding the relevance of the CAC standard (i.e. an international standard), the Appellate Body had recourse to the generic definition of standard enshrined in the last two sentences of the explanatory note of Annex 1.2 TBT. This is an ambiguous interpretive technique to construct a definition by using elements from another definition and can be problematic; arguably, this has been the case here as well.

A careful reading of the explanatory note allows two interpretations that seem to be equally plausible. The Panel and Appellate Body in *EC – Sardines* found that the word 'also' in the last sentence makes clear that the word 'document' in the same sentence can only refer to international standards. However, a reading leading to the opposite outcome is equally possible: the explanatory note serves the role of *concretising* the generic term of standard enclosed in Annex 1.2 TBT and not international standards. The note refers to the two general traits of standards: their voluntary character and the method of adopting them (consensus). However, even if the one before last sentence refers to the international standardisation community, this does not change the fact that the Note is there to give flesh to the *generic* definition of standards for TBT purposes. In this respect, it is no coincidence that the Appellate Body was quite reluctant in *US – Tuna II* to revisit the analysis in *EC – Sardines*.[86] The suggested interpretation would have an important repercussion: if the last two sentences of the explanatory note are to be read independently, then international standards not adopted by consensus would not benefit from the presumption of Article 2.5 TBT.

V. Recognised international standardisation bodies

Even so, interpretive challenges remain. For instance, which bodies are 'recognised standardisation bodies'? Which among them constitute the 'international standardisation community'? The TBT refers to 'relevant international standardising bodies' in the Code with particular ease, as if it was clearly defined and identified.[87]

Contrary to the TBT, the SPS explicitly recognises the ISSBs that should be deemed as points of reference: the Codex Alimentarius Commission, the International Office of Epizootics (OIE); and the International Plant

[85] See Panel Report, *EC – Sardines*, para. 4.33.

[86] Cf. Appellate Body Report, *US – Tuna II*, para. 353.

[87] Code, paras. G and H.

Protection Convention.[88] Even if the list is not exhaustive, it does cover the most important ISSBs in the area of sanitary and phytosanitary issues. Importantly, all these bodies are intergovernmental organisations. The TBT, in turn, does not have a similar provision, although it seems that at least the ISO and IEC would be the first bodies to recognise as having such activities in the area of technical standardisation. A fundamental difference between the two areas of standardisation and indeed source of difficulties is that technical standardisation is in principle of private nature, organised within associations of bodies rather than intergovernmental organisations, as we explained earlier. Thus, any explicit reference to or incorporation of normative work done in an essentially private body would lie uncomfortably within an otherwise State-to-State international contract.[89] As Alvarez eloquently noted, 'the soft law that the WTO is hardening into an outer limit for permissible public regulation to protect health and the environment is arguably a form of self-regulation produced by industry to protect its own interests, and not government standards produced by a public agency accountable to consumers or the greater public'.[90]

The peculiar nature of technical standardisation is also implicitly acknowledged in the TBT definition of standard. A TBT standard is typically adopted by a body, be it international, regional or central government

[88] Annex A.3 SPS.

[89] In a *dicta*, the Appellate Body in *US – Tuna II* noted that by not expressly referring to particular ISSBs, the TBT aims to encourage the development of international standards by bodies that were not already engaged in standardising activities in the time of the entry into force of the TBT. This statement has a twofold meaning: first, it makes clear the Appellate Body's willingness to be open to assessing the capacity of new ISSBs to create relevant international standards within the meaning of Art. 2.4 TBT and thus benefit from the presumption of TBT consistency, which arguably would increase their legitimacy within the international standardisation community and the multilateral trading system. Second, the Appellate Body reminds us that such a recognition has a price because it may be dependant on the standardising body at issue, showing that it meets the high due process criteria of the TBT Committee Decision (see infra). This is a responsibility *shared* by ISSBs and WTO Members participating therein. In acknowledging some 'bite' to an otherwise best-endeavour TBT provision relating to special and differential treatment (S&DT), the Appellate Body clarifies its intention in the future to make WTO Members accountable for their sincere efforts to ensure that ISSBs are organised and operated in a manner that ensure representative participation, taking the particularities of the developing countries into account. See Appellate Body Report, *US – Tuna II*, para. 379 and fn. 745.

[90] J. Alvarez, *International Organizations as Law-makers* (Oxford University Press, 2005), at 221.

entity. In turn, the ISO/IEC Guide defines a body as a legal or administrative entity that has specific tasks and composition.[91]

Obviously, the many references to the ISO would seem to suggest that this non-governmental federation of national standards bodies belongs to the international standardisation community and thus that its standards are international. The same would most likely apply to IEC. Yet other ISSBs exist as well. The Annex 1.4 TBT, consistent with WTO legacy, defines international bodies in an open-ended manner; international are those bodies or systems for which membership is open to the relevant bodies of all Members. This is largely in line with the ISO/IEC Guide definition of an international standardising organisation. Thus a regional body that is open to only some of the Members would not fall within this definition. If, however, membership to this body is not a priori excluded vis-à-vis a particular Member or its relevant standardising body, then under certain circumstances, its standards may still be regarded as international standards for TBT purposes.

For this to happen, the body would need to be a recognised one. This is reminiscent of Article 4.3 of the ISO/IEC Guide 2:2004 which defines as standardising those bodies that have 'recognised activities in standardisation'.[92] However, the Guide does not specify what 'recognised activities in standardisation' would mean nor who should recognise such activities – and the same goes for the TBT. At the same time, the Guide does specify that standardisation is the activity of establishing provisions for common and repeated use, aimed at the achievement of the optimum degree of order in a given context, more particularly the activity of preparing and implementing standards.[93] It also makes clear that a recognised body in this area can be a body that, among other rules, also promulgates standards (standardising body), but also a body whose main function is, pursuant to its statutes, the preparation, approval or adoption of standards that are publicly available (standards body).[94] The former category is broader and it is probably no coincidence that the

[91] See also Panel Report, *US – Tuna II*, para. 7.679. To show the diversity of bodies that can come under this definition, the ISO/IEC Guide provides that a body can be, for instance, an organisation, an authority, a company or even a foundation. Thus, organisations can also be bodies, but the definition of organisation under the ISO/IEC Guide is narrow. See Art. 4.2 of the Guide. This, however, is not always straightforward. For instance, in *US – Tuna II*, the Appellate Body seems to have misread this. See Appellate Body Report, *US – Tuna II*, para. 356.

[92] In *US – Tuna II*, the Appellate Body found that the ISO definition of a standardising body should assist in the interpretation of the TBT term 'recognised body'.

[93] See also supra note 3.

[94] Arts. 4.3 and 4.4 of the ISO/IEC Guide 2:2004.

Code uses this broad term to also cover bodies that incidentally develop standards.

However, with respect to 'recognition', neither the Guide nor the TBT establish a quantitative benchmark. For instance, can a Member invoke as a defence against a standard that, for instance, its relevant body did not participate in the standardisation activities of a given body or that, even if it participated, it objected or voted against that standard? Taking into account that consensus does not imply unanimity in the international standardisation community, a critical mass of negative votes would be needed to raise doubts against the international character of a standard adopted within this setting.

In *US – Tuna II*, the Appellate Body dealt with the meaning of 'recognition' in this respect. It first found that recognition implies that Members know, or at least expect that the international body at stake is active in standardisation. Furthermore, recognition would be an issue to examine on a case-by-case basis, examining evidence of recognition by Members and/or standardising bodies at the regional, national or sub-national level. Thus, the Appellate Body avoided to establish a general test which would imply a *de minimis* rule for recognition, other than noting that 'the larger the number of countries that participate in the development of a standard, the more likely it can be said that the respective body's activities in standardisation are "recognised"'.[95] Thus, it was made clear that recognition by WTO Members, rather than the standardisation community (its 'peers') would cover a broad part of the scope of recognition within the meaning of Annex 1.2 TBT.

However, bodies having developed a single standard would not come outside the scope of the TBT, simply because they do not have extensive standardisation activities. Elements such as wide participation of WTO Members in the development of the standard; wide recognition of the validity and legality of the single standard; or adherence to the TBT Committee Decision of 2000 on principles for the development of international standards (the 'TBT Committee Decision')[96] would reveal a body with recognised activities in standardisation.[97]

[95] Appellate Body Report, *US – Tuna II*, para. 390.

[96] See WTO, TBT Committee, 'Decision of the Committee on Principles for the Development of International Standards, Guides and Recommendations with Relation to Articles 2, 5 and Annex 3 of the Agreement', G/TBT/9), 13 November 2000, para. 20 and Annex 4 (reproduced in WTO, TBT Committee, 'Decisions and Recommendations Adopted by the WTO Committee on Technical Barriers to Trade Since 1 January 1995', Note by the Secretariat, G/TBT/1/Rev. 10, 9 June 2011, pp. 46–48).

[97] See Appellate Body Report, *US – Tuna II*, para. 394.

In this respect, an important argument was brought forward by Canada: Canada suggested that a recognition of activities of an international standardising body cannot be assessed independently of the TBT Committed Decision of 2000 on principles for the development of international standards (the 'TBT Committee Decision').[98] Put differently, an international standardising body is 'recognised' if it develops standards or engages in standardising activities in accordance with certain recognised principles such as those developed in the TBT Committee Decision. The Appellate Body correctly agreed that evidence about adherence to the TBT Committee Decision would be relevant for determining whether the body at issue has recognised activities.[99] It is in the TBT Committee Decision that I now turn.

VI. The TBT Committee Decision

Article 2.4 TBT appears to endorse international standards and ISSBs without any meaningful qualifications or conditions. It essentially transforms voluntary standards into de facto mandatory norms. This raises concerns regarding the way that such international standards are set and whether they indeed reveal international preferences and absolute technical superiority. Practice suggests that various standards fail to achieve this. In several instances, international standards were adopted with a narrow majority of the absolute number of votes and despite conflicting scientific opinions.

Take, for instance, the case of *EC – Sardines*, as noted previously, in which a small minority of Codex Members adopted the relevant Codex standard (18 of 150 at the time). The same goes in the case of *EC – Hormones*. The GMO (genetically modified organisms) standard is of course an SPS standard, but it is indicative of the concerns that international standardisation may raise: the relevant Codex standard was adopted with a narrow majority of thirty-three votes to twenty-nine with seven abstentions. Indeed, the TBT Agreement, as interpreted by the WTO adjudicating bodies, endorses standards adopted by international SSBs without examining their representativeness, comprehensiveness or process

[98] See WTO, TBT Committee, 'Decision of the Committee on Principles for the Development of International Standards, Guides and Recommendations with Relation to Articles 2, 5 and Annex 3 of the Agreement', G/TBT/), 13 November 2000, para. 20 and Annex 4 (reproduced in WTO, TBT Committee, 'Decisions and Recommendations Adopted by the WTO Committee on Technical Barriers to Trade Since 1 January 1995', Note by the Secretariat, G/TBT/1/Rev, 10, 9 June 2011, pp. 46–48).

[99] See Appellate Body Report, *US – Tuna II*, para. 376.

of adoption. Can standards of this type be accepted as 'international' without running counter to any conceivable notion of fairness?

In 2000, the TBT Committee, which is responsible for the implementation of the TBT, adopted a decision during the TBT Second Triennial Review incorporating six principles and procedures that should be observed during the development of international standards. It was generally felt more input from a wider set of interests was needed in the international standardisation community. According to the dominant view, bodies operating with open, impartial and transparent procedures that afforded a fair opportunity for consensus among interested parties in all WTO Members would most likely prepare effective and relevant standards.[100] Clearly the TBT Committee Decision constituted a consensus observation on the side of the WTO that rules and procedures within ISSBs needed to improve and strengthen. Clearly, it was an external call for reform of the international standardisation community practices.[101]

The principles identified in the TBT Committee decision are transparency, openness, impartiality and consensus, effectiveness and relevance, coherence and addressing the concerns of developing countries. In *EC – Sardines*, the EU alleged that the CAC standard at issue was not a 'relevant international standard' because it was not adopted by consensus and thus was inconsistent with the principle of relevance laid down in this Decision. The Panel disagreed and stated that this Decision was not binding but merely a 'policy statement of preference', confirming the WTO adjudicators' unwillingness to examine with a critical eye the standards development process. The finding remained unappealed for several years.

However, in the recent *US – Tuna II* decision, the Appellate Body found that the TBT Committee Decision has not a mere hortatory meaning when interpreting standards-related TBT provisions. Rather, it constitutes a 'subsequent agreement' within the meaning of Article 31(3)(a) VCLT and thus should be read together with the TBT.[102] The Appellate Body was led to this conclusion based on various elements such as the fact that

[100] See G/TBT/1/Rev.10, p. 12.

[101] Cf. Appellate Body Report, *US – Tuna II*, para. 371 and fn. 736.

[102] Art. 31 VCLT (*General rule of interpretation*) provides:

1. A treaty shall be interpreted in good faith in accordance with the ordinary meaning to be given to the terms of the treaty in their context and in the light of its object and purpose.
2. The context for the purpose of the interpretation of a treaty shall comprise, in addition to the text, including its preamble and annexes:
   (a) any agreement relating to the treaty which was made between all the parties in connection with the conclusion of the treaty;

it was adopted by consensus; it bears specifically on the interpretation and application of a TBT provision and Members' expressed intention to (a) develop a better understanding of international standards within the TBT, (b) ensure the effective application of the TBT and (c) clarify and strengthen the concept of international standards.[103]

Indeed, agreements subsequent to the conclusion of a previous agreement aiming to specify how existing rules or obligations are to be applied (rather than to create new or extend existing obligations) can fall under Article 31(3)(a) VCLT, constituting a further authentic element of interpretation to be taken into account along with context.[104] However, considering the TBT Committee Decision as 'subsequent agreement' barely squares with the *EC – Sardines* previous finding that the last sentence of the Explanatory Note in Annex 1.2 TBT also relates to international standards. If it is so, and thus consensus should not be required for a standard to be regarded as a 'relevant *international* standard', then the TBT Committee Decision, by requiring consensus, amounts to an *amendment* of the TBT text.[105]

After juxtaposing the Committee Decision principles to the standard that the Panel found to be a 'relevant international standard' within the meaning of the TBT, the Appellate Body reversed the Panel's finding because the SSB at issue was not open to at least all Members, as participation was possible by invitation only. In a clear message towards ISSBs, the Appellate Body found that standardisation bodies must be open and transparent at every stage of standards development, thereby

(b) any instrument which was made by one or more parties in connection with the conclusion of the treaty and accepted by the other parties as an instrument related to the treaty.

3. There shall be taken into account, together with the context:

(a) *any subsequent agreement between the parties <u>regarding the interpretation of the treaty or the application of its provisions</u>*; (emphasis added).

[103] Appellate Body Report, *US – Tuna II*, paras. 371–372.

[104] Cf. Appellate Body Report, *EC – Bananas III (Article 21.5 – Ecuador II)*, para. 391; and Appellate Body Report, *US – Clove Cigarettes*, para. 265.

[105] Interestingly, this goes against the letter and spirit of the WTO treaty. We find contextual support in Article IX.2 of the WTO Agreement (WTOA) prohibiting the abuse of the provision relating to authoritative interpretation to undermine the amendment provisions in Article X WTOA. Cf. Appellate Body Report, *EC – Bananas III (Article 21.5 – Ecuador II)*, para. 383. By the same token, Decisions by WTO organs (such as the TBT Committee) cannot be used to circumvent the amendment provisions of the WTO. This is yet another indication that decoupling the last two sentences of the Explanatory Note so that its last sentence only refers to domestic and regional standard is the only reasonable interpretation of this provision.

demonstrating the potential 'bite' of the WTO as an *ex post* arbiter of the legitimacy of standards.[106]

Tuna II should be assessed within its specific context: it was confined to a discussion of practices and institutional structures of a *regional* standardising body, whereas it discussed shortly only one out of the six principles identified in the TBT Committee Decision. Even so, *Tuna II* marks a shift from previous rulings where international standards were endorsed without examining under which procedural and substantive safeguards ISSBs had elaborated them. It demonstrates the WTO's incipient determination to delve into the very essence of the international standardisation processes.

Nevertheless, it is argued that the principles set out in the TBT Committee decision are not sufficiently inclusive. From a normative point of view, for a standard to be regarded as a genuinely international standard, additional, but at the same time more concrete, criteria would need to be developed. A conceptual framework to assess the international character of international standards would need to take, inter alia, the following issues or criteria into account:

(a) ***Participation:*** Participation of stakeholders and parties directly affected would need to be ensured in a standard-setting process. In addition, participation of both governmental and non-governmental bodies can ensure dealing with a particular issue in a comprehensive manner. By the same token, a proper balance between developed and developing country participants would have to be achieved. Formulating rules ensuring proper and effective participation (e.g. through the circulation of the necessary documents well ahead of any decisions to be taken) and balancing of interests in the deliberation process may vary from SSB to SSB.
(b) ***Representativeness:*** Representativeness relates to the proper representation of interests just like participation, but it also refers to whether the mandate for the creation of a standard and the draft or final standard can be properly considered as representing an unequivocal and unambiguous majority.[107] Representativeness can also be

[106] See Appellate Body Report, *US – Tuna II*, para. 382.

[107] Ensuring adequate representation has proven very difficult even in highly integrated institutions such as the EU standard-setters. See also European Commission Communication, 'A strategic vision for European standards: Moving forward to enhance and accelerate the sustainable growth of the European economy by 2020', COM(2011) 311 final, 1 June 2011, p. 11.

assessed ex post, by identifying potential objections, criticisms and complaints.

(c) ***Voting:*** Voting rules can play a crucial role in ensuring fair representation of interests and balanced outcomes. Consensus, simple majority or qualified majority voting are to be found in the statutes of various SSBs. However, many times, abstentions are significant, leading to a handful of parties deciding on an international standard. More often than not, such a standard may be subject to criticism in a polycentric world.

(d) ***Open and transparent procedures:*** All procedural guarantees are important and of a substantive nature.[108] Ensuring openness at an early stage of the process through adequate notification and circulation of the necessary documentation to all parties to an SSB may lead to fewer objections later on and thus lead to a more expedited standards development process. Notice and comment procedures are sometimes in place but not always used effectively. Furthermore, appeal procedures and possibilities to raise objections both on technical and procedural matters shall be in place.

(e) ***Impartiality:*** Impartiality is not always easy to verify. Previous studies have questioned the sometimes disproportionate power of the secretary of a given technical committee within the major ISSBs. Excessive zeal to break a deal to the detriment of due process may ultimately lead to a bad standard.

(f) ***Scientific rigour, relevance and excellence:*** This criterion would require proper audit and ex post review of standards adopted as well as possibilities to challenge existing standards in a fair, effective and expedited manner. Before the adoption of the standard, any information contradicting the absolute superiority of a given standard may question the international character of that standard. In addition, a harder issue to solve is whether any procedural deficiencies identified throughout the standards development process can be justified through recourse to the unequivocal scientific excellence of a given standard.

(g) ***Impact assessment and sustainability:***[109] Like public-driven domestic regulations, the impact of standards on their addressees should be periodically assessed and SSBs should provide for this obligation. Notably the social and environmental impact of a given standard

[108] See Hans Lindahl's contribution to this volume (Chapter 2).

[109] Cf. For instance, see the ISEAL Impacts Code, available at: http://www.isealalliance.org/online-community/resources/iseal-impacts-code-of-good-practice.

should be periodically reviewed. Such a process should lead to the amendment of the standard, if this allows for addressing better specific non-economic interests. By the same token, when preparing a given standard, information relating to sustainability issues should feed into the preparation of a given standard. Environmental sustainability in particular should be taken into account in every stage of standards development in view of the importance of global warming. Environmental sustainability is not mentioned in the TBT Committee Decision, but the issue has come to the fore ever since. For instance, the ISO has recently published guidelines addressing sustainability in standards development. The guidelines are intended to encourage the examination of issues relating to sustainability during all stages of standards development within ISO. Importantly, the guidelines provide that the lack of considering sustainability issues in the development of a given standard can validly justify the revision of that standard.[110]

(h) ***Interinstitutional sensitivity and cooperation:*** In times of interdependence of global governance institutions, synergies of the ISSB in question with relevant IOs but also NGOs are important and many times necessary (for instance, to ensure that a given standard also is in line with pertinent labour or human rights). Allowing for such cooperation in the statutes of the SSBs is a precondition for well-crafted standards in an international arena of increasingly specialised institutions. Cooperation with similar institutions in terms of content and function such as the cooperation between ISO and IEC would not suffice; rather, complementarities and synergies among institutions should be strived for to inform the development of international standards.

D. Conclusion

With its presumption of consistency when relevant international standards are used as a basis for domestic regulations, the TBT shows undue deference to the international standardisation regimes. WTO cases in which standards were used to defeat national regulatory autonomy have led to serious questions about the legitimacy of these regimes and their standards. By initially applying narrowly the TBT and the TBT Committee Decision, the WTO adjudicating bodies have legitimised certain standards that do not reflect a critical mass of international preferences and thus

[110] ISO Guide 82:2014.

should not be used as reference points or benchmarks because of certain deficiencies identified in the process of their adoption.

In what has preceded, I showed that the erosion of State consent in international law and the transfer of power to technocrats can sometimes lead to absurd results that are in dire need of being mapped and addressed. International standard-setting is emblematic of such challenges. Anecdotal evidence suggests that international standard-setters often behave as a de facto intractable club, excluding voices that increasingly seek to be heard in such fora. There are examples demonstrating the political imbalances that create distortions and elitist standards. However, demands for more openness and representativeness that come from emerging economies and other developing countries suggest that the current formation of regulatory-making in this arena reflects anachronistic international geopolitics and can no longer hold. Power is now more equally distributed among the nations of the world. Such redistribution has to be reflected within international organisations as well, notably those that may have substantial effects on the economic emancipation and welfare of a given country.

US – Tuna II can herald a new era of more transparent and open international standard-setting. The jurisprudential shift toward more critical thinking about the activities of the international standardisation community certainly affects ISSBs such as the ISO or the CAC, which are interested in remaining relevant for the trade regulation regime. Additionally, their reputation may suffer if their standard-setting practices are found to lack legitimacy.

More empirical research is needed to develop analytical tools and heuristic devices that would objectively assess the mechanics of international standard-setting and, if needed, make the case for institutional reform within ISSBs with a view to increasing deliberation, representativeness, openness, transparency, due process and accountability without jeopardising economic and technological progress.[111] This possibility exists: arguably, the interpretation of the *COOL* Panel as to when a standard constitutes an effective and appropriate means for achieving a legitimate objective may indeed go that direction – a lock, stock and barrel interpretation of the message conveyed by the Appellate Body in *US – Tuna II*.

[111] For a first, very tentative attempt, see M. Morikawa and J. Morrison, 'Who Develops ISO Standards? A Survey of Participation in ISO's International Standards Development Processes' (Pacific Institute, 2004).

6

Technical standard-setting in the financial sector

JUAN A. MARCHETTI

A. Introduction

In the case of financial services, the term 'standards' has been used in two different ways, focusing on two distinct perspectives – a technical one and a regulatory one. The technical perspective is exemplified by the definition of a standard utilised by the International Organisation for Standardisation (ISO): 'a document that provides requirements, specifications, guidelines or characteristics that can be used consistently to ensure that materials, products, processes and services are fit for their purpose'.[1] The concept of 'fitness for purpose' is, in turn, defined as the ability of a product, process, or service to serve a defined purpose under specific conditions.[2] These standards generally go unnoticed. As a study from the US Congress acknowledged more than two decades ago, '[t]hey are mostly quiet, unseen forces, such as specifications, regulations, and protocols, that ensure that things work properly, interactively, and responsibly. How standards come about is a mystery to most people should they even ponder the question'.[3] We may call these 'technical standards'.

On the other hand, regulatory authorities in the financial services sector often use the term 'standards' and 'international standards' to designate

I thank Stijn Claessens, Panagiotis Delimatsis and Petros C. Mavroidis for helpful comments on a previous draft of this chapter. Any errors remain my own. The views expressed in this chapter are personal and should not be attributed to WTO Members or the WTO Secretariat.

[1] ISO Website, answer to question, 'What is a standard?' Available at: http://www.iso.org/iso/home/standards.htm.

[2] ISO, 'ISO/IEC Guide 2:2004 – Standardization and related activities – General vocabulary' (2004). WTO (2012), 'Technical Standards in Services,' Note by the Secretariat, S/WPDR/W/49, 13 September 2012.

[3] US Congress, Office of Technology Assessment (1992), 'Global Standards: Building Blocks for the Future', TCT-512 (Washington, DC: US Government Printing Office, March 1992), http://www.fas.org/ota/reports/9220.pdf.

principles and rules of financial regulation and supervision. For example, the Financial Stability Board (FSB), established to coordinate at the international level the work of national financial authorities and international standard-setting bodies and to develop and promote the implementation of effective regulatory, supervisory, and other financial sector policies, has identified twelve key standards that are broadly accepted as representing minimum requirements for good practice that countries are encourages to meet or exceed. Chief among these are the Core Principles for Effective Banking Supervision, established by the Basel Committee on Banking Supervision; the Insurance Core Principles, set by the International Association of Insurance Supervisors (IAIS); and the Objectives and Principles of Securities Regulation, agreed by the International Organization of Securities Commissions (IOSCO).[4] This is the 'regulatory perspective'. We may call these 'regulatory standards' or, simply, 'regulations'.

This chapter focuses on the first perspective, which remains somewhat underdeveloped in the trade literature. Indeed, although there has been a lot of talking and writing on domestic regulation in services and the prospects for disciplines in this area, that literature has hardly focused on 'technical standards', and much less so on these standards as applied to financial services.[5]

The chapter is organised as follows. Section B provides an overview of the General Agreement on Trade in Services (GATS) provisions dealing

[4] The actual 'standard-setting' bodies include, apart from these three agencies and the FSB itself, the Committee on the Global Financial System (CGFS), the Committee on Payment and Settlement Systems (CPSS), the Financial Action Task Force on Money Laundering (FATF), the International Association of Deposit Insurers (IADI), the International Accounting Standards Board (IASB), and the International Auditing and Assurance Standards Board (IAASB), the International Monetary Fund (IMF), the Organisation for Economic Co-operation and Development (OECD), and the World Bank. The IMF and the World Bank assess compliance with all the core international financial sector standards through their Financial Sector Assessment Program and Reports on the Observance of Standards and Codes program.

[5] Although negotiations on disciplines for domestic regulation have been ongoing since the inception of the WTO in 1995, WTO Members only focused in earnest on 'technical standards' more recently. Indeed, although a proposal for disciplines on technical standards in services, in the context of negotiations under Article VI:4 of the GATS, was submitted by Mexico and Switzerland in 2005 (document S/WPDR/W/32/Rev.1, 28 October 2005), only in 2011 did the Working Party on Domestic Regulation (the body in charge of the negotiations just referred to) requested the WTO Secretariat to provide background information on technical standards in services. See WTO (2005) 'Proposal for Disciplines on Technical Standards in Services', communication from Mexico and Switzerland, S/WPDR/W/32/Rev.1, 28 October; and WTO (2012), 'Technical Standards in Services', Note by the Secretariat, S/WPDR/W/49, 13 September 2012.

with technical standards, raising some issues for further discussion. Section C, the longest in the chapter, discusses technical standards in financial services. The section looks into the economics of technical standards, provides an overview of the main technical standards in the financial services industry, discusses the international standard-setting process (including relevant political economy issues), and, finally, provides some examples of trade obstacles arising from inadequate (or the lack of) technical standards. The final section provides some concluding observations based on the foregoing analysis.

B. Technical standards in the GATS

The GATS addresses technical standards as part of Domestic Regulation, that is, as part of measures that are neither limitations on market access nor contrary to the national treatment principle.[6] More precisely, technical standards are referred to in the context of the negotiating mandate contained in Article VI:4 of the GATS to establish any necessary disciplines with a view to ensuring that 'measures relating to qualification requirements and procedures, *technical standards* and licensing requirements do not constitute unnecessary barriers to trade in services'. Such disciplines shall aim to ensure that such requirements are, inter alia, (a) based on objective and transparent criteria, such as competence and the ability to supply the service; (b) not more burdensome than necessary to ensure the quality of the service; and (c) in the case of licensing procedures, not in themselves a restriction on the supply of the service. Therefore, although it does not impose a direct 'necessity test' on Members, the mandate in Article VI:4 of the GATS points in that direction.[7]

The term 'technical standards' is not defined in the GATS. Neither is it defined by the first disciplines developed by WTO Members under the Article VI:4 mandate – the Disciplines on Domestic Regulation in the Accountancy Sector, adopted in 1998.[8] Nevertheless, discussions among

[6] Juan A. Marchetti and Petros C. Mavroidis, 'What Are the Main Challenges for the GATS Framework? Don't Talk about Revolution' (2004) *European Business Organization Law Review*, Vol. 5, 546–547.

[7] WTO (2001), 'Report on the Meeting Held on 11 May 2001.' Note by the Secretariat, document S/WPDR/M/11, 7 June 2001, at para. 29. For a discussion on 'necessity tests', see WTO (2003), 'Necessity Tests in the WTO', Note by the Secretariat, S/WPDR/W/27, 2 December 2003.

[8] WTO (1998), 'Disciplines on Domestic Regulation in the Accountancy Sector', adopted by the Council for Trade in Services on 14 December 1998, document S/L/64.

WTO Members in the context of the negotiations of such disciplines have tended to suggest that 'technical standards' could be understood as criteria or rules specifying the characteristics of the service (e.g. the format of financial reports), as well as the manner in which it should be performed (e.g. the way in which an audit must be performed, including the checks that must be performed, the way work should be documented, and so on).[9] Technical standards have also been described as 'requirements which may apply both to the characteristics or definition of the service itself and to the manner in which it is performed'.[10] It has been argued that the distinction made between the 'characteristics of a service' and the 'manner in which it is performed' is akin to that made in the Technical Barriers to Trade (TBT) Agreement between 'product characteristics' and 'related processes and production methods'.[11]

Pending the entry into force of the disciplines to be developed under Article VI:4, Article VI:5 of the GATS seeks to ensure that specific commitments are not nullified or impaired through licensing and qualification requirements and technical standards that (a) are not based on objective and transparent criteria or are more burdensome than necessary to ensure quality and (b) could not reasonably have been expected at the time the specific commitments in the relevant sectors were made. In addition, Article VI:5 provides that in determining whether a WTO Member's domestic regulations are in conformity with these obligations, 'account shall be taken of international standards of relevant international organisations applied by that Member'. A footnote to Article VI:5 clarifies that the term 'relevant international organisations' refers to 'international bodies whose membership is open to the relevant bodies of at least all Members of the WTO'. The current provisions in the GATS do not go as far as the TBT agreement in laying down a general obligation on Members to use international standards when they are available, thereby establishing a rebuttable presumption that any measure that is consistent with international standards would be considered not to create an unnecessary obstacle to trade. Nevertheless, the GATS provisions in this area do seem to point in a similar direction.

[9] WTO (2012), supra note 5, p. 6.

[10] WTO (1996), 'The Relevance of the Disciplines of the Agreements on Technical Barriers to Trade (TBT) and on Import Licensing Procedures to Article VI.4 of the General Agreement on Trade in Services', Note by the Secretariat, document S/WPPS/W/9, 11 September 1996, p. 3.

[11] WTO (1996), supra note 10, p. 6.

C. Technical standards in financial services

I. The basic economics of technical standards

A technical standard is a document that provides requirements, specifications, guidelines, or characteristics that can be used consistently to ensure that materials, products, processes, and services are fit for their purpose. These standards provide the basic infrastructure that allows the international financial system to function on a day-to-day basis.[12] Like 'plumbing' they are not often the subject of intellectual conversation. We simply assume the plumbing will work.

Standards serve to reduce risks, transaction costs, and incompatibility, thus enabling trade to take place. Standards are usually classified for analytical purposes according to four objectives they are meant to attain: compatibility or interoperability, minimum quality, information and product description, and variety reduction.[13] In real life, however, standards do not fall neatly and exclusively into a single category; indeed, many standards fulfil several purposes. But this analytical distinction is useful because each purpose of standardisation produces different economic effects, and the analytical models used to understand these effects are different.

Developments in information and communications technologies have demonstrated the huge economic importance of standards for *interoperability*. Two particular economic phenomena influence producer and customer decisions. First, producers and customers face *switching costs*. Before they are committed to a particular standard, they are relatively free to choose between different configurations. But once they have invested in a particular system or standard, they will find it increasingly expensive to switch to another. Second, producer and customer choices are influenced by *network effects*, also called *network externalities*. The basic idea of these network effects is that it is desirable to choose a system that is widely used by others.

When both these phenomena exist, there is a risk that markets can get locked into inferior designs because both sides are reluctant to switch to something better unless they can be sure that all others will too, and

[12] William R. White, 'International Agreements in the Area of Banking and Finance: Accomplishments and Outstanding Issues' (1996), BIS Working Paper No. 38, October.

[13] Knut Blind, 'The Economics of Standards: Theory, Evidence and Policy' (2004), Edward Elgar Publishing Limited, Cheltenham. Peter G. M. Swan, 'The Economics of Standardization' (2000), Final Report for Standards and Technical Regulations Directorate, UK Department of Trade and Industry, 11 December.

in any case, switching is costly. This has been described as a problem of technological lock-in. Prompt definition of high-quality standards would help overcome these problems.

Interoperability standards have had broad implications for the worldwide expansion of financial services activities. In recent years, the financial services industry has been moving toward electronic solutions, migrating from paper-based systems to electronically stored information, or from e-business to e-records to e-evidence. The increased use of electronic communication improves the efficiency of commercial activities, enhances trade connections, and allows new access opportunities for previously remote parties and markets. It is in that context that financial services suppliers have been striving for commonality of operations through interoperable standards, so as to increase productivity, reduce costs, achieve economies of scale, and ultimately develop global operations.

Take the example of the payment processing industry. Similar to the production of a network switch, data processing requires capital investment, such as computer equipment and complex telecommunication infrastructure that connects ATM owners, merchants, banks, and networks. Hence, as more transactions are routed through a processor's data processing platform, the cost per transaction drops. The standardisation of payment instruments and processes allows participants in the payment industry (i.e. banks and payment processors) to benefit from economies of scale. If standardisation takes place at a global level, it opens access to new markets, both for increasing the revenue base for existing payment instruments and for launching innovations on a broader scale.

Standards also reassure consumers as to the *quality of products and services*. Consumers face a bewildering variety of products and usually find it hard to assess which is best suited to their purposes. When consumers cannot assess the quality of a product or service before buying – a problem usually known as 'information asymmetry' – markets fail.[14] If buyers cannot distinguish high quality from low quality before purchase, then it is hard for the high-quality seller to sustain a price premium. In the absence of this premium, and if the high-quality seller's costs exceed those of the low-quality seller, then the former may not be able to survive. Bad sellers (who sell low-quality produce) drive out good-quality sellers by undercutting them.

[14] G. Akerlof, 'The Market for Lemons: Quality Uncertainty and the Market Mechanism' (1970), *Quarterly Journal of Economics*, Vol. 89, 488–500.

Minimum quality standards can reduce what economists call *transaction costs* and *search costs*. If the standard defines the product in a way that reduces buyer uncertainty, then first the risk to the buyer is reduced, and second there is less need for the buyer to spend time and money evaluating the product before purchase. Consider a commodity market, for example: how could it exist in the absence of standards? Traders must be able to buy and sell large volumes without even viewing their trades. This is only possible if there is complete confidence about what it is that is being traded. That presumes a clearly defined standard grade and certification that all produce traded meets that grade.

Variety reduction performs two different functions. First, it seeks to exploit economies of scale by minimising the wasteful proliferation of minimally differentiated models. The trade-off operating here is between choice and price. Variety reduction standards do not need to be defined publicly: the economies of scale can in principle be obtained with an idiosyncratic model range. This is the best-known function of variety-reduction standards.

There is, however, an even more important role for variety reduction, and this operates to the benefit of the producer as much as for the customer. Standards can also reduce the risks faced by suppliers – even if this also means they face more competition. The availability and use of standards often shape future technological trajectories, and these are instrumental in the development and growth of new markets. In the formative stages of a market for a new technology, standards can play an important role in achieving focus and cohesion among the pioneers.

Standards of information and product description are usually treated as a distinct category from the foregoing three categories, but for many purposes, it is sufficient to treat these as a hybrid of the other three. Take the example of different grades of petrol: four-star, unleaded, and super-unleaded. These are standards of product description that also offer the other three features. Most motorists are confident that one type of four-star is *compatible* with another, and so can fill up at a Shell station one week and BP the next. Equally, these grades satisfy certain quality standards. And of course, there are major economies of scale in distribution from the limited range of petrol grades.

In addition to the arguments and theories in the preceding paragraphs, two concepts may need further explanation: regulatory capture and raising rivals' costs. The concept of 'regulatory capture' is the idea that some producers may lobby so skilfully that they persuade the regulator to define regulations in the interest of the producers rather than in the interest of the customer (as originally intended). In that context the concept of 'raising rivals' costs' may be highly relevant.[15] Some high-cost and high-quality producers may find it in their interest to lobby for an unnecessarily high minimum quality standard because that will in effect exclude their lower-cost, lower-quality rivals from the market.

A major question is who should take responsibility for setting standards. Health and safety standards are usually the responsibility of government or regulators (thus financial regulators, through CPSS/IOSCO, have taken the lead in setting standards for safety in payment and settlement systems). Public sector bodies, such as telecommunications regulators, may also play a coordinating role in standard setting. At the opposite extreme from these standards set by public bodies are those set by a coalition of voluntary participants, most notably the open standards supporting the Internet (TCP/IP data transfer protocols, HTML display, XML data storage standard). Such standards are the outcome of innovation and cooperation between various users, often in public sector defence and research institutions. The 'open-source' software movement reflects a continued commitment to this mechanism of creating and developing standards. Finally standards may be set by industry participants, either by individual firms, by consortia of firms, or by industry established standard-setting organisations (or SSOs). This is the usual practice for many engineering, product compatibility, and communication standards. The FIX consortium and SWIFT, to which I turn in the next section, are examples of such industry-based standard-setting organisations in securities trading and settlement. The standard-setting process is likely to favor such industry-based mechanism in cases in which public sector bodies lack the necessary technical knowledge and in which existing incumbents in the industry have a strong interest in influencing the development of standards.

The academic literature has paid particular attention to cases of 'de facto' standards for complementary products with 'adoption' or 'installed base' externalities (also often referred to as 'network externalities'). The term 'de facto' refers to standards established through competition in the marketplace, after being introduced by a single producer or group of producers. Payment card networks are a case in point. The utility from joining the network as a merchant is directly proportional to the number of customers also joining, and vice versa. Standards are essential because

[15] Steven C. Salop and David T. Scheffman, 'Raising Rivals' Costs' (1983), *American Economic Review*, Vol. 73, No. 2, 267–271.

the two products (the point-of-sale [POS] reading device and the payment card) will not work together without them. In these cases, there is then an 'adoption' externality associated with the number of ATMs and POS devices – the wider the availability of these installations, as well as the acceptance of a particular payment card, the more attractive a specific payment card is to the consumer.

There is an ongoing and somewhat inconclusive debate in the network economics literature about the relationship among standards, innovation, and efficiency.[16] In some circumstances standards may create a lock-in into inefficient technology. In other cases, competition between standards can actually create excessively rapid change, with customers or firms forced to upgrade to new standards more frequently than can be justified on pure dynamic efficiency grounds, because of the 'network' benefit of staying on the same standard as others. There is also considerable variation in the extent and operation of these network effects in different industries. Although in some services industries, such as mobile telephony, there is a substantial 'first-mover' advantage, this is not necessarily the case in financial services.[17] Typically such first-mover advantage is relatively small and, in the case of services supporting the process of exchange – for example, payments instruments or clearing and settlement services – there is typically no first-mover advantage at all.[18] In these cases, it is necessary for an innovation to be adopted simultaneously by the industry as a whole, and this means no individual firm can benefit from being the first to market. Thus financial services standards, although potentially supporting improvements in customer service, are not a key driver of profit and innovation. If the adoption of a new standard is costly, then it is unlikely to take place, even when from a customer perspective the benefits are large enough to outweigh the costs. An open standard may support considerable cost reductions for customers but still be resisted by providers.[19]

16 Alistair Milne, Discussion of Foresight Report Section 7 (Computers and Complexity) and supporting Driver Review DR31 'Standards in Computer Based Trading: A Review' authored by Kevin Houstoun, The Foresight Report on the Future of Computer Trading in Financial Markets: Conference at London School of Economics, 11 January 2013.

17 Economists speak about first-mover advantage in cases in which the fact of being the first to introduce an innovation supports higher margins and profits, until the time when other competitors are able to replicate the innovation.

18 Milne (2013), supra note 16, p. 5.

19 Alistair Milne, 'Standards and Competition in Post-trade Securities Processing' (2007), *Journal of Financial Transformation* Vol. 20, The Capco Institute, September 2007.

II. Technical standards in financial services: an overview

As the reader may have already guessed, many of these standards – or at least their development and expansion in recent decades – are closely linked to the application of information and communications technologies (ICT) to financial services. The financial industry relies on ICT to transmit messages relaying financial transactions and other similar financial messages between banks, firms, and institutions located around the world. Standards have thus become critical for financial services where networks prevail or the degree of automation is high.

Technical standards have, for example, enabled the development of the card transaction business, as the entire card-processing infrastructure is built on ISO standards, including the personal identification number (PIN). One single ISO standard is used to transact billions of dollars in the retail card business annually: *ISO 8583 (Financial transaction card originated messages – Interchange message specifications – Part 1: Messages, data elements and code values).* This standard allows all of the various parties to a retail transaction to complete the transaction within seconds. Additionally, the well-known PIN standard (ISO 9564) is used for initiation of card operations and customer security of card transactions. The PIN is used millions of times globally each day. In the securities side of the financial services business, standards form the basis of the communication of information, known as electronic banking, including those for coding and numbering the securities (ISO 6166) and those that transact securities movement (ISO 15022).

ISO 15022 is currently the predominant securities standard in cross-border settlement, reconciliation, and corporate action processing. It was introduced around 1998 to replace ISO 7775, which was much less structured and often omitted crucial settlement information. The adoption of ISO 15022, mandated in 2003, has led to a dramatic increase in straight-through-processing (STP) rates. In settlement messages, for example, it is common to come across STP rates of more than 95 percent. About half of the 15 million messages that are exchanged on the SWIFT network every day are ISO 15022. A new universal financial industry message standard – ISO 20022 – was released in 2004 as a replacement for ISO 15022. ISO 20022 provides a methodology intended to create consistent financial message standards, helping make the transfer of this data more reliable. The latest version of ISO 20022, issued in 2013, is designed to allow financial message developers more flexibility in the development of these communications, potentially improving the timeliness and accuracy of financial transactions, among other potential benefits.

Other standards predominantly used in the financial services industry include FIX, FpML, SWIFT, and XBRL. FIX is used for routing orders from buy-side to the sell-side of securities transactions, for distributing quotes and market data, and for some post-trade, pre-settlement activities. FpML (Financial Products Mark-up Language) is used to represent complex derivatives contracts in a computer readable format. It is widely used between broker-dealers and other securities industry players to exchange information on, for example, Swaps and CDOs. SWIFT proprietary standards, also known as MT messages, are the standard for messaging in correspondent banking, foreign exchange, and documentary credits. More than 9,000 financial institutions around the world use this standard to exchange millions of messages per day over the SWIFT network. XBRL (eXtensible, Business Reporting language) is the machine-readable format for encoding financial statements. In addition, there are other standards used in computer trading that are either codifications of reference information, such as the ISIN standard for instrument identification, which are peripheral to the trading process.

Finally, there are proprietary standards in which a single platform producer has implemented a single standard sufficiently widely that it has become a de facto standard. Examples of such proprietary standards include ITCH and OUCH. ITCH is the direct data feed protocol technically owned by NASDAQ but also used by other exchanges such as BATS/Chi-X, which modify it to suit their own requirements. OUCH is the simple order and execution protocol often used as the basis for exchanges proprietary interfaces. Despite many interfaces building on ITCH or OUCH to connect to multiple exchange platforms, one needs to use multiple implementations rather than a single connector. Although not officially sanctioned, these proprietary de facto standards have become widely adopted.[20]

Table 6.1 provides some examples of standards in use in the financial services sector according to the four purposes outlined in the preceding section.

[20] The preceding discussion in this subsection draws, inter alia, from Kevin Jon Houstoun, 'Standards in Computer-based Trading: A Review' (2012), Foresight Report Section 7 (Computers and Complexity) and supporting Driver Review DR31, UK Government Office for Science, 19 September 2012.

Table 6.1. *Examples of standards in use in the financial services sector*

Category	Example
Compatibility (or interface)	ISDA documentation standards for OTC derivatives ISO 20022 – universal financial industry message schemes SEC's Nationally Recognized Statistical Rating Organization
Quality	AS 3806 – compliance systems BS 10500 – anti-bribery management system BS 8453 – compliance framework for regulated financial services firm ISO 22222 – personal financial planners ISO 22301 – business continuity management ISO 27001 – information security systems ISO 31000 – international risk management standard Fairbanking Hedge Fund Standards Board
Variety reduction	BS 8477 – customer service IFRS – accounting standards ISO 10002 – guidelines for complaints handling ISO 14001 – environmental management systems ISO 9001 – quality management systems Professional bodies – lawyers, accountants, actuaries, securities professionals SAS 70 – auditing of financial controls
Information standards	ACORD standards for insurance document exchange numerous ICT standards covering security, messaging, magnetic cards, etc. SWIFT – numerous information transmission standards

Source: Michael Mainelli and Chiara von Gunten, Z/Yen Group[21]

[21] 'Backing Market Forces: How to Make Voluntary Standards Markets Work for Financial Services Regulation' (2013), Report commissioned by the BSI (British Standard Institution), CISI (Chartered Institute for Securities & Investment) and Long Finance, November. http://www.longfinance.net/images/BSICISILF_Backing%20Market%20Forces_November2013.pdf (accessed on 15 January 2014).

III. The international standard-setting process

The previous section described some of the standards used in the financial services industry. But who sets these standards? There is in fact a multitude of national private standard-setting associations. At the international level, the ISO is the world's largest developer of technical standards. Founded in 1947, the ISO counts 161 member countries (as of February 2014) represented by their national standard-setting bodies. National standard-setting bodies contribute to standard development at the ISO level and represent the ISO in their respective countries. Since its creation, the ISO has published more than 19,500 standards, of which standards on financial services represent only a modest fraction (51 as of February 2014) compared with other sectors.

ISO standards are developed by groups of experts from all over the world, within technical committees (TCs). TCs are made up of representatives of the relevant industry, consumer associations, academia, non-governmental organisations, and government. ISO's full members (member bodies) can decide whether they would like to be a participating member (P-member) of a particular TC or an observing member (O-member). P-members participate actively in the work and have an obligation to vote on all questions submitted to vote within the TC. O-members follow the work as an observer but cannot make any comments about the development process or vote. Correspondent members can become O-members of a particular TC.

International Standards are developed by ISO technical committees (TC) and subcommittees (SC). ISO does not decide when to develop a new standard. Instead, it responds to a request from industry or other stakeholders such as consumer groups. Typically, an industry sector or group communicates the need for a standard to its national member who then contacts ISO. A new work item proposal (NP) is submitted for vote by the members of the relevant TC or SC to determine the inclusion of the work item in the program of work (proposal stage). The proposal is accepted if a majority of the P-members of the TC/SC votes in favor and if at least five P-members declare their commitment to participate actively in the project. At this stage, a project leader responsible for the work item is normally appointed. The relevant TC/SC sets up a working group of experts (preparatory stage). Successive working drafts may be considered until the working group is satisfied that it has developed the best technical solution to the problem being addressed. At this stage, the draft is forwarded to the working group's parent committee for the consensus-building phase.

As soon as a first committee draft is available (committee stage), it is registered by the ISO Central Secretariat. It is distributed for comment and, if required, voting, by the P-members of the TC/SC. Successive committee drafts may be considered until consensus is reached on the technical content. Once consensus has been attained, the text is finalised for submission as a Draft International Standard (DIS).

The DIS is circulated to all ISO member bodies by the ISO Central Secretariat for voting and comment within a period of three months (enquiry and approval stages). It is approved for submission as a Final Draft International Standard (FDIS) if a two-thirds majority of the P-members of the TC/SC are in favour and not more than one-quarter of the total number of votes cast are negative. If the approval criteria are not met, the text is returned to the originating TC/SC for further study and a revised document will again be circulated for voting and comment as a draft International Standard. In such cases, the final draft International Standard (FDIS) is circulated to all ISO member bodies by the ISO Central Secretariat for a final yes/no vote within a period of two months. If technical comments are received during this period, they are no longer considered at this stage but registered for consideration during a future revision of the International Standard. The text is approved as an International Standard if a two-thirds majority of the P-members of the TC/SC is in favor and not more than one-quarter of the total number of votes cast are negative. If these approval criteria are not met, the standard is referred back to the originating TC/SC for reconsideration in light of the technical reasons submitted in support of the negative votes received.

Once a final draft International Standard has been approved, only minor editorial changes, if and where necessary, are introduced into the final text. The final text is sent to the ISO Central Secretariat, which publishes the International Standard (publication stage). All International Standards are reviewed at least every five years by all the ISO member bodies. A majority of the P-members of the TC/SC decides whether an International Standard should be confirmed, revised, or withdrawn.

A fast-track procedure is also available. If a document with a certain degree of maturity is available at the start of a standardisation project – for example, a standard developed by another organisation – it is possible to omit certain stages. In this fast-track procedure, a document is submitted directly for approval as a draft International Standard (DIS) to the ISO member bodies (enquiry stage) or, if the document has been developed by an international standardising body recognised by the ISO Council, as a final draft International Standard (FDIS, approval stage), without passing through the previous stages.

The ISO currently comprises 236 TCs including the TC 68, the technical committee designated to develop technical standards for the financial services industry. Created in 1948, the TC 68 counts thirty participating countries and forty-eight observing countries.[22] In addition to countries' delegations, seven organisations are in liaison with ISO TC 68: the Association of National Numbering Agencies, the European Central Bank, the European Payment Council, MasterCard, the Society for Worldwide Interbank Financial Telecommunication (SWIFT), the UN Centre for Trade Facilitation and E-business (UN/CEFACT), and VISA. The TC 68 comprises three subcommittees (SCs): SC 2 – Financial services, security; SC 4 – Securities and related financial instruments; and SC 7 – Core banking. The United States plays a leading role in ISO TC 68, with the American National Standards Institute (ANSI) holding the committee's secretariat, which has been delegated to Accredited Standards Committee X9 (ASC X9).

Although technical standards are usually developed at the request of the industry, sometimes they arise as a result of regulators' initiatives. The recent development of 'legal entity identifiers' (LEIs), promoted at the global level by the G-20, the Committee on Payment and Settlement Systems, the Technical Committee of the IOSCO, and the FSB, is a case in point. The 2008/2009 global financial crisis brought heightened awareness of the need for transparency of financial transactions to accurately assess exposure to counterparty risk. The idea behind LEIs is to overcome the current fragmented system of firm identifiers by creating a common identifier for financial institutions. A unique ID associated with a single legal entity, LEIs allow for consistent identification of parties to financial transactions, facilitating a consistent

[22] The participating countries are Algeria, Australia, Austria, Barbados, Brazil, Bulgaria, Canada, China, Denmark, Finland, France, Germany, India, Italy, Japan, Kenya, Korea (Rep. of), Luxembourg, the Netherlands, Norway, Portugal, Russia, Singapore, South Africa, Sweden, Switzerland, Tunisia, Turkey, the United Kingdom, and the United States, which holds the Secretariat. The observing countries are Angola, Argentina, Armenia, Armenia, Belarus, Belgium, Botswana, Chile, Colombia, Congo (The Democratic Republic of the), Croatia, Cuba, Cyprus, Czech Republic, Côte d'Ivoire, Egypt, Greece, Hong Kong China, Hungary, Iceland, Iran, Ireland, Kazakhstan, Lithuania, Malawi, Malaysia, Malta, Mauritania, Mauritius, Moldova, Mongolia, Montenegro, Mozambique, Namibia, Pakistan, Philippines, Poland, Romania, Serbia, Seychelles, Slovak Republic, Spain, Sri Lanka, Swazilandia, Tanzania, Thailand, Ukraine, Zambia, and Zimbabwe.

and integrated view of exposures. The LEI is a unique 20-character alphanumeric code assigned to any organisation or firm involved in a financial transaction. Such an identifier for each legal entity allows regulators to conduct more accurate analysis of global, systemically important financial institutions and their transactions with all counterparties across markets, products, and regions, allowing regulators to better identify concentrations and emerging risks. It will also allow local regulators to realise these same benefits. For risk managers in financial institutions, the LEI will increase the effectiveness of tools aggregating their exposures to counterparties.

Financial regulators began discussing the need for a global standard for identifying financial contracts and their participants in 2010. Late in 2011, the G-20 directed the Financial Stability Board (FSB) to begin developing a framework for a global LEI standard.[23] In May 2012, a specific LEI standard, ISO 17442, was developed by ISO. In June of that year, the G-20 endorsed an FSB report that contained the blueprint for the LEI system. The report outlined a three-tiered public-private governance framework that would protect the public interest while meeting private sector needs. Overseeing the LEI system is a Regulatory Oversight Committee (ROC). The ROC has established committees to set up the rest of the governance framework for the global LEI system. The centre tier to the framework is the Global Legal Entity Identifier Foundation, organised in Switzerland, which is entrusted with building the technology infrastructure of the LEI system and ensure adherence to governing principles and standards, including reliability, quality, and uniqueness. The foundation's Board of Directors is the second tier. Its role is to direct the organisation. The third tier of the LEI system is an international network of Local Operating

[23] The US Securities and Exchange Commission (SEC) and the Commodity Futures Trading Commission (CFTC), helped to spur adoption by proposing swaps rules that required use of an LEI. The CFTC was the first regulator to mandate the use of an identifier in regulatory reporting. CFTC Rule 17 CFR Part 45 – Swap Data Recordkeeping and Reporting Requirements, which went into effect in October 2012, require swap dealers and other major swap participants executing OTC derivatives trades to report those transactions subject to CFTC oversight to a Trade Repository, and to identify themselves, their counterparties and any underlying reference entities of the contracts, with standard identifiers. Because the CFTC will require such identifiers before the availability of the global LEI as designated by the FSB, the CFTC has mandated use of a 'CFTC Interim Compliant Identifier' or 'CICI'. To allow the industry to comply with this requirement, the CFTC named SWIFT and DTCC as the solution providers for CICIs, until the global LEI system is defined via the process led by the Financial Stability Board. Globally, the FSB has established an Implementation Group of global regulators, which has sought advice from the private sector, including the financial services industry, via a Private Sector Participatory Group.

Units to register entities, assign LEIs, validate and maintain the reference data associated with each LEI, and make the data continuously available to the public and regulators, free of charge. As of February 2014, a dozen early-stage registrars have issued about 189,000 LEIs in 169 countries.[24]

IV. Trade obstacles arising from inadequate technical standards (or the lack thereof)

That technical standards could restrict the cross-border supply of financial services became apparent in the first report produced by the Giovannini Group.[25] This report, produced at the request of the European Commission, identified fifteen barriers to the integration of clearing and settlement services within the European Union. Talk about these has resurfaced more recently in the context of the regulatory reform of the over-the-counter derivatives market, in particular with regard to the interoperability of central counterparties (CCPs) and trading venues.

The first category of barriers identified by the Giovannini Group concerned 'barriers related to technical requirements/market practice', such as national differences in information technology and interfaces, which add to the cost of cross-border clearing and settlement by requiring a higher level of manual input. The recommendation here was to standardise communication between the various clearing and settlement systems and the adoption of an European Union (EU)-wide protocol defining message formats between systems and their members.

Another technical barrier arose from differences in national rules governing corporate actions (e.g. the offering of share options, rights issues, etc.). Because corporate actions often require a response from the securities owner, national differences in how they are managed may require specialised local knowledge and/or the lodgement of physical documents locally and so inhibit the centralisation of securities settlement and custody. The recommendation here was to implement ISO 15022 message standards for communicating corporate actions between central securities

[24] The information on the LEI development process in the preceding paragraphs draws from remarks of Linda Powell, Chief Data Officer, Office of Financial Research, U.S. Department of the Treasury at the GS1 Global Forum 2014, Brussels, Belgium, 18 February 2014. Available at: http://www.treasury.gov/initiatives/ofr/news/Documents/Powell_speech_GS1_02-18-2014_FINAL.pdf.

[25] The Giovannini Group, 'Cross-Border Clearing and Settlement Arrangements in the European Union' (2001), Brussels, November. Available at: http://ec.europa.eu/internal_market/financial-markets/docs/clearing/first_giovannini_report_en.pdf (accessed on 1 September 2013).

depositories (CSDs), so as to speed up information dissemination across systems.

The need for market participants to interface with multiple post-trading systems in the context of cross-border transactions was also found to constitute a significant barrier because it led to a duplication of costs. National differences in settlement periods (which could be considered a technical standard governing the process or the actual production of a service) was also found to affect cross-border clearing and settlement, particularly in cases where local market practices differed from international practice (e.g. Germany settling on T+2 while the international convention is T+3). Finally, national differences in securities issuance practice that arose due to the lack of an efficient same-day distribution mechanism. Among the more important shortcomings in this area was an uneven capability across the securities markets in Europe to allocate ISIN numbers to securities issues in real-time. Standardised electronic links among issuing agents, dealers, and settlement systems was recommended to enable the speedy exchange of issuance information and ISIN codes, facilitating same-day issuance.

Before that, in 1999, a report by the European Central Bank (ECB) had found that bilateral money transfers were costly and inefficient due to the use of nonstandard customer interfaces, incompatible formats between domestic and foreign banks, and the low degree of automation in banks' internal systems.[26] More recently, a Green Paper by the European Commission explains that fragmented standardisation in the field of card payments has a threefold effect. First, the lack of common standards limits the range of potential service providers to domestic acquiring payment service providers, thus hampering the achievement of a competitive single market for payment services. Second, merchants have to maintain different systems and protocols for managing data exchange in the acquiring process – at least one for each country in which they operate but in many cases even more – thereby reducing the opportunities to centralise operations and limiting efficiency gains. Third, the lack of common standards for the data exchange between the merchant and the acquiring bank often prevents debit cards from being accepted abroad.[27]

[26] European Central Bank (ECB9), 'Improving Cross-Border Retail Payment Services: The Eurosystem's View' (1999), September 1999, p. 10. Available at http://www.ecb.europa.eu/pub/pdf/other/retailpsen.pdf (accessed on 2 September 2013).

[27] European Commission, 'Towards an Integrated European Market for Card, Internet and Mobile Payments' (2012), COM(2011)941 FINAL, Brussels, 11 January.

In the WTO context, the recent case on *China – Certain Measures Affecting Electronic Payment Services* (hereinafter *China – EPS*) showed that technical standards, either by design or purpose, could become a trade barrier. Indeed, several of the legal instruments considered by the Panel contained 'business specifications' and 'technical standards' as the Panel called them.[28] As stated by China in the course of the proceedings, these 'establish a national inter-bank network for clearing and settling RMB payment card transactions and otherwise create uniform technical and commercial standards that allow this inter-bank payment card network to function'.[29] Without getting into the details of this dispute, suffice to say that some of these requirements, such as the requirement that all bank cards issued in China bear the Yin Lian/UnionPay logo (i.e. the logo of China UnionPay's [CUP's] network) and be interoperable with that network and that all terminal equipment in China be capable of accepting Yin Lian/UnionPay logo cards, were found by the Panel to be inconsistent with China's mode 1 and mode 3 national treatment obligations under Article XVII of the GATS because, through these requirements, China modified the conditions of competition in favor of CUP, failing therefore to grant national treatment to EPS suppliers of other Members, contrary to China's commitments.

A recent report by the Asian Development Bank (ADB) identified several barriers to trading and settlement of bonds in ASEAN+3 countries (comprising the ten ASEAN members plus China, Japan, and Korea).[30] Many of these barriers consist of standards – for example, the non-use of international standards for securities messaging in a local market. The international standard message formats such as ISO 20022 are regarded as necessary for enabling straight-through processing (STP) in securities post-trade processing.[31] This involves not only settlement but also pre-matching. If local, proprietary standards are used, there is a need for interface and translation either at the global custodian or local custodian, with associated costs of development and maintenance, and the risk of

[28] Panel Report, *China – Measures Affecting Electronic Payment Services*, WT/DS413/R, adopted 31 August 2012, para 2.5.

[29] Panel Report, *China – Measures Affecting Electronic Payment Services*, para. 7.211.

[30] Asian Development Bank, 'Barriers to Cross-border Investment and Settlement in the ASEAN+3 Bond Markets, Final Report, Part III' (2010), ABMI Group of Experts, July 2010.

[31] Straight-through processing (STP) enables the entire trade process for capital markets and payment transactions to be conducted electronically without the need for re-keying or manual intervention.

error is greater. In some cases, the local CSD does not use ISO messaging standards.

The non-use of international standards for securities numbering in a local market may also be perceived as a barrier to trade. The relevant international standard is ISIN (International Securities Identification Number), in accordance with ISO 6166, which provides a globally unique identification to all securities. As with securities messaging, lack of ISIN use makes STP more difficult and increases the risk of error. Most ASEAN+3 markets now have established local agencies for issuing and administering ISIN numbers for locally issued securities. The limitations may be the following: ISINs are not available on the issue date of the bonds, making trading and settlement more difficult; ISINs are not widely used by local market participants; or ISINs are not used by the local CSD.

Standards governing the settlement cycle (i.e. the number of days between trade date and settlement date) may also constitute barriers, according to the ADB report. Most markets operate a standard settlement cycle. Typically this is T+1 for government bonds and T+2 or T+3 for corporate bonds (and equities). A short settlement cycle is better for local market participants because it reduces counterparty risk. However, non-resident investors may find it difficult to settle on T+1 if they or their global custodian are located in a different time zone. For this reason, such investors look for the ability to negotiate a longer settlement period (T+2 or T+3 is the favored cycle).

Standards on 'trade and settlement matching' (i.e. the matching of trade details between counterparties) may also present problems. Matching may take two forms: trade matching and pre-settlement matching. In the case of trade matching, details of the trade are compared between the counterparties to ensure that there is no misunderstanding of the terms of the trade. This should of course be performed as soon as possible after the trade is executed, and in any event before the end of the business day. With automated trading systems (e.g. exchange systems or electronic OTC systems) matching is done at the time of trade, so there is no need for subsequent trade matching. However, many bond trades are done by telephone. As to pre-settlement matching, details of the agreed trade are compared between the counterparties' settlement agents (e.g. local custodian and local broker) to ensure that all information needed for settlement is in place. Most markets in ASEAN+3 operate some form of matching systems but some do not, and this can create uncertainty. The absence of automated matching is likely to lead to increased settlement fails and make it more difficult to shorten the settlement cycle.

Physical certificates may also raise obstacles. Most bonds today are in dematerialised form, held in book entry at the local securities depository or central bank system or (in the case of most international bonds) on the books of the ICSDs. Some bonds are still in paper certificate form. The disadvantage of physical certificates are obvious – the need for manual examination, risk of loss, damage or forgery, and cost of storage. Typically, these remaining physical bonds are not of great interest to cross-border investors, and indeed are unlikely to be traded at all. The ideal situation, clearly, is dematerialisation. An intermediate step would be to hold physical certificates, where they exist, in the vaults of the local securities depository (immobilisation).

D. Concluding observations

The purpose of this chapter has been to explore issues related to technical standards in financial services. As such, it has ventured – to the best of my knowledge – into relatively uncharted territory, at least from the trade policy perspective. Given the purpose of the chapter, I would not seek to draw definitive conclusions with regard to the development of disciplines in the context of the GATS. Rather, I limit myself to some additional observations that seem pertinent and require further consideration and research:

(a) Technical standards, such as those developed by ISO TC68 and other related organisations, may be described as 'product' and 'process' standards, their purpose being to ensure interoperability and compatibility of systems, variety reduction, minimum quality, and information. These types of standards are basically developed for use by financial services firms and the industry as a whole. However, even though technical standards are generally developed to cater for specific industry needs, they may also arise at the initiative of regulators, as the experience with the LEI standard shows.

(b) The majority of technical standards are designed for voluntary use, in the sense that they are developed by groups of experts, with a predominant presence of sectoral interests, without being mandated by law. It is only when observance of a particular technical standard is required by law – for example, the LEI standard – that they become mandatory. In cases in which the risks faced by society – such as counterparty risk, in the case of the LEI standard – are considered to be significantly high, one might expect the incorporation of technical standards into regulation.

(c) The relationship between technical standards and specific policy objectives (e.g. mitigation of financial risk) deserves further attention. Take the example of settlement cycles – that is, the number of days between trade date (T) and settlement date, which have become 'standardised' in most markets. Typically this is T+1 for government bonds and T+2 or T+3 for corporate bonds and equities. A short settlement cycle is better for local market participants, as it reduces counterparty risk, which most regulators would consider a legitimate policy objective. However, foreign (non-resident) investors may find it difficult to settle on T+1 if they or their global custodians are located in a different time zone. For this reason, such investors look for the ability to negotiate a longer settlement period (T+2 or T+3 is the favored cycle), even if that does not contribute to the control of counterparty risk.

(d) In addition, the legitimacy of the international standard-setting process becomes crucial if future multilateral disciplines on technical standards in services emulate the TBT Agreement and establish a presumption that any measure that is consistent with international standards should be considered not to create an unnecessary barrier to trade. It has been argued in that regard that sometimes technical standards in financial services have addressed short-term private interests rather than public interest.[32] In some cases, any public benefits have been by-products rather than the original intent of the standardisation initiative. For example, when introduced in 1992, the FIX protocol was initially seen by Salomon Brothers as an opportunity to lock in Fidelity's order flows to its own firm (and to the few other brokers who adopted FIX). However, being an open standard, its widespread adoption – today in equity markets any buy-side order management system is able to communicate with any sell-side order management system because of FIX – had the opposite impact: it increased competition and resulted in reduced commissions all around. More generally, as FPL, SWIFT, and FpML are all effectively

[32] Kevin Jon Houstoun, 'Standards in Computer-Based Trading: A Review' (2012), Foresight Report Section 7 (Computers and Complexity) and supporting Driver Review DR31, UK Government Office for Science, 19 September 2012.

consortia of banks and vendors, the standards they have developed, and continue to promote, originally addressed the private interests of the banks and vendors in those consortia.

(e) Lastly, the consensual and highly technical nature of the standardisation process is meant to reduce and mitigate conflicts, but it can also serve as an implicit barrier to the participation of developing countries.[33] Active membership in technical committees is a crucial condition for influencing standardisation. The secretariat and chairmanship of ISO technical committees (TC), critical positions in the standard-setting process, are usually held by those members that have strong interest in the standards being developed there. Moreover, even though the standardisation process is formally open to every member of ISO, not every country has the same infrastructure and technical expertise to participate effectively.[34] Developing countries do not participate in many committees at all – they remain 'standard takers' instead of 'standard makers'. A study conducted in 2004 found that Western European countries represented only 12 percent of ISO's total members but made up for nearly half of the voting P-members in the TCs and held 47 percent of the chairs and 58 percent of the secretariats. African countries, by contrast, constituted nearly 30 percent of ISO's members but participated in only 4 percent of the TCs.[35]

[33] Marcel Heires, 'The International Organization for Standardization (ISO)' (2008), *New Political Economy*, Vol. 13, No. 3. Based on interviews with developing country delegates to ISO General Assembly, this author explains that opposition to proposals for new standards without presenting an elaborated alternative can be difficult. Moreover, the experts in the respective committees are usually specialists in their fields and have all been working together for some time. This makes it harder for new members without the experience and expertise to enter into their circle.

[34] Ibid.

[35] Pacific Institute, 'Who Develops ISO Standards? A Survey of Participation in ISO's International Standards Development Processes' (2004). Available at http://www.pacinst.org/wp-content/uploads/sites/21/2013/02/iso_participation_study3.pdf (accessed on 15 January 2014).

7

Standardising trade in services?

PETROS C. MAVROIDIS

A. Introductory remarks

National standards and regulatory measures may act as barriers to trade, either deliberately or inadvertently. This is so because although standard-setting often reflects a 'genuine' need for regulation to address a market failure of some kind, it can also be influenced by political economy forces, and, consequently, the risk for capture of the process should not be underestimated. The political economy literature on product standards concludes that these often will be beneficial for economic actors but that they can also be used for protectionist purposes. This is also the case for domestic regulation, which can be captured by incumbents to 'raise rivals costs' or used as an instrument to discriminate against foreign suppliers.[1] Allegations of protectionist abuse have been the basis of numerous trade disputes over the years. These motivated the negotiation and inclusion of specific disciplines on product standards for goods in the General Agreement on Tariffs and Trade (GATT)/World Trade Organization (WTO) and the building of bridges between the trade and international standard-setting community. The key WTO agreements in this area are the Agreement on Technical Barriers to Trade (TBT) and the Agreement on Sanitary and Phytosanitary (SPS) measures. The TBT Agreement addresses technical requirements (mandatory standards) imposed by governments for goods; the SPS agreement deals with health and safety-related norms for agricultural products (foodstuffs, plant and animal health). The two

I am grateful to Jagdish Bhagwati, Bernard M. Hoekman, and Juan-Alberto Marchetti for numerous discussions on the subject of this chapter.

[1] There are many examples of product standards and regulatory norms that were used for protectionist purposes. A classic example was a Japanese product standard adopted in 1986 by the Consumer Product Safety Association for skis sold in Japan. The Association argued that American or European standards for skis were not appropriate for Japan because snow in Japan was 'wetter' and that Japan had unique geothermal activity (Vogel, 1992).

standards agreements provide a means for WTO Members to 'in-source' the results of international cooperation on product safety-related norms.

Empirical research has shown that the costs for firms associated with differences in services regulation across countries are significant (see e.g. Kox and Nordas, 2007; Francois and Hoekman, 2010). However, there are no equivalent mechanisms to SPS and TBT in the WTO for standardising services. This raises the questions of why this is so and whether similar types of agreement should be pursued in the context of services trade. In Section B, we discuss why the coverage of the TBT Agreement is limited to goods only and summarise the main elements of this agreement.[2] Section C summarises the relevant disciplines and the approach that was taken in the General Agreement on Trade in Services (GATS) toward services regulation and standards. Section D discusses whether there is a case for a TBT analogue for services and what WTO Members might consider doing in this area given the high degree of uncertainty as to how international standardisation of services will evolve in the years to come. It also includes the conclusions of this chapter.

B. Why a discrepancy?

I. Trade in goods

A critical feature of the TBT Agreement is that it encourages the use of harmonisation as a way of reducing TBT-related trade costs (Art. 2.4). This provision presupposes the existence of 'international standards', and requires from WTO Members to use them, unless if they are 'ineffective' or 'inappropriate'. The international standards that emerge from efforts by standard-setting institutions will reflect a common view of how best to address a specific need to regulate through the adoption of a technical measure. Under the TBT Agreement, there is a presumption that such international standards are least trade restrictive in that the norms are considered to satisfy the necessity test. The presumption is that by having many countries involved in the norm development process, whatever is agreed is regarded as being non-discriminatory in intent, no matter the actual effect on trade. As a result, the TBT Agreement makes a step toward 'rationalisation' of domestic policies.

[2] I will not discuss the SPS Agreement because extending this agreement to services was, by the nature of this agreement, impossible; it is limited to very specific types of tangible products.

Table 7.1. *ISO standards for financial services*

Years	Number
1990–1994	2
1995–1999	8
2000–2004	6
2005–2009	17
2010–2014	18
Cumulative total	51
Memo: ISO total (all sectors/areas)	19,977

Source: http://www.iso.org/iso/home/about/iso-in-figures.htm

II. Trade in services

International standards are elaborated for services as well as for goods. Thus the question arises why the TBT Agreement only covers trade in goods. One very basic reason is that when the first TBT Agreement was enacted in the late 1970s, the GATT coverage did not extend to trade in services.[3] To our knowledge, services were never raised as a matter of concern in the deliberations of the TBT Committee.

Another reason is that, compared with goods, there is much less activity around the development of international standards for services. The homepage of the International Organization for Standardisation (ISO) is suggestive in this respect: services standards represent less than 0.5 percent of its total activity. This is not to say that there is no activity on services. For some sectors, like accountancy and some financial services products, there have been a number of standards issued (Table 7.1).

Other international regulatory/standard-setting bodies that deal with services include the International Air Transport Association (IATA) (air transport), the International Accounting Standards Board (IASB) (accounting), the International Maritime Organization (IMO) (maritime transport), the UN Economic Commission or Europe (UNECE) (road transport), the International Telecommunications Union (ITU), the Bank for International Settlements (BIS), the International Organization of Securities Commissions (IOSCO) and the International Association of Insurance Supervisors (IAIS). These bodies have established international regulatory norms and standards in their respective areas. For example,

[3] On the genesis of GATS, see Marchetti and Mavroidis (2011).

the IASB has developed the International Financial Reporting Standards (IFRS), and the BIS has produced the Basel Core Principles for Effective Banking Supervision. Many of the norms become mandatory for services suppliers that are active in the sectors concerned and that operate in their respective jurisdictions. For example, some eighty-five countries require the use of IFRS by companies listed on domestic stock exchanges.

The action when it comes to internationalisation of standards for services is centered on the characteristics of services suppliers or on common definitions of concepts. International standardisation of what the required characteristics are, for example, capital adequacy requirements for financial service providers, or how to define, measure and report specific concepts to ensure international comparability, for example, definitions of the various elements of the financial statements of companies for accounting purposes – is pursued through associations of national regulators. Although international standardisation of regulations pertaining to services providers is actively pursued in some areas, most notably financial services (see Claessens and Kodres, 2014, for a recent assessment and discussion), in many sectors, regulation is purely national. Indeed, for some service sectors, the regulatory regime is sub-national, especially in countries with a federal system (Arnold, 2005; Suddaby, Cooper and Greenwood, 2007).

In light of the factors noted, it seems clear that the 'payoff' of emulating the TBT Agreement is likely to be limited. A key feature of the TBT Agreement is the reference to international standards. In the services context, these are either already operating, in that where these are being developed they are adopted by the participating nations (some areas of financial services) so that it is unclear what is 'the problem' that requires use of the WTO, or they do not exist. We are then left with the question whether principles such as least-trade restrictive and necessity can be applied to instances of national regulation of services and service providers. This is something that, not surprisingly, arose during the negotiation of the GATS.

C. The GATS

I. The obligation to accord national treatment

In theory, if a country has standards for the supply of accountants, it need not treat suppliers from foreign nations as it does domestic suppliers even if a foreign country has adopted identical standards for accountants.

The fact that national treatment is not a 'general obligation' provides a potential rationale to consider negotiation of a GATS TBT insofar as that provides a route for WTO Members to make national treatment the norm with respect to services for which international standards exist.

II. Domestic regulation in GATS

The GATS includes disciplines regarding domestic regulation. These are in part reminiscent of obligations in the TBT Agreement.[4] Art. VI.4, VI.5 GATS prove this point because they reflect the necessity requirement that must be respected when issuing qualification and licensing requirements, as well as technical standards, and request from WTO Members to avoid undoing the value of their commitments when doing so. The GATS thus includes disciplines on 'standards', albeit the terminology used in the body of Art. VI GATS might be prone to confusion ('technical standards' are distinguished from 'licensing and qualification requirements', which could be standardised as well). However, this provision addresses only 'national' standards. Art. VI GATS leaves it open to WTO members to use standards that they themselves have unilaterally defined or that have been elaborated in international standard-setting institutions such as the ISO. A Working Party on Domestic Regulation was mandated to develop disciplines called for by Art. VI.4 to ensure that licensing and qualification requirements and related standards are not unnecessary barriers to trade in services.

III. Transparency in trade in services

Art. III GATS covers standards, and requests that they are used in transparent manner. Transparency does not amount to participation in the elaboration of standard, the duty to explain the rationale (for example) for standardisation and all other obligations that the TBT imposes. The TBT Agreement has stronger transparency disciplines. Each WTO member must notify the WTO when it plans to adopt technical requirements that do not conform to an international standard, allow reasonable time for other members to comment, as well as a reasonable period of time for exporters to adapt to new requirements. Moreover, Members must establish a national enquiry point where traders may obtain documents and answers regarding: technical regulations adopted or proposed by bodies

[4] Delimatsis (2007) offers a comprehensive treatment of this provision.

which have legal power to enforce them; standards adopted or proposed by central or local government bodies or by regional standardising bodies; and conformity assessment procedures, existing or proposed, applied by enforcing bodies.

IV. Preliminary conclusion

Under the current regime, WTO Members must simply be transparent and act in nondiscriminatory manner toward foreign suppliers. They can, but do not have to, go beyond that and afford foreign suppliers with national treatment. The distinction between product and process standards noted earlier and the fact that services are generally nontangible credence or experience goods means that the focus of regulation and standards is frequently on the producers as opposed to the service product. This does much to explain the lack of a TBT analogue under the GATS because the TBT mostly deals with product regulation as opposed to the firms and production processes used to generate a product. Moreover, the absence of international standardisation of services is another factor that differentiates services from goods because this is a core feature of the TBT Agreement.

D. A TBT for GATS?

In this chapter, I asked whether we need a TBT analogue for the GATS. The response is negative. First, there is little in the way of product standardisation, as the data I produced supra shows, and thus also little in the way of internationalisation of services standards. Nor should we try to agree on substantive internationalisation of services at the WTO level because such endeavours are best left to international regulatory bodies and all ensuing cooperation should be channelled accordingly. In this realm, one should not underestimate the potential for a 'cooperation freeze': adding in the threat of the Understanding on Rules and Procedures Governing the Settlement of Disputes, and countermeasures, may have a chilling effect on the willingness, or even the ability, of regulators to cooperate on internationalisation of standards for services providers. The GATS negotiators did address the matter and came up with the agreement that reflects a tipping point. What would be achieved by including a reference to international standards – the missing bit in GATS Art. VI? Arguably, not much. Why? Governments and regulators either agree to do it, as it happens in the context of financial services, or they do not, as is the case in many services. An added complication is that such norms may be public or private. Few would be on board, and the only viable perspective here seems to us to be that offered by a plurilateral agreement. But even here it is clearly not a panacea given the difficulty in agreeing to adopt common disciplines in the GATS on professional services (accountancy).

References

Adlung, Rolf. 2006. Services Negotiations in the Doha Round: Lost in Flexibility?, *Journal of International Economic Law*, 9(4): 865–893.

Arnold, Patricia. 2005. Disciplining domestic regulation: the World Trade Organization and the market for professional services, *Accounting, Organizations and Society*, 30: 299–330.

Claessens, Stijn, and Laura Kodres. 2014. The Regulatory Responses to the Global Financial Crisis: Some Uncomfortable Questions, IMF Working Paper WP/14/46.

Delimatsis, Panagiotis. 2007. *International Trade in Services and Domestic Regulations: Necessity, Transparency and Regulatory Diversity*. Oxford: Oxford University Press.

Francois, Joseph, and Bernard Hoekman. 2010. Services Trade and Policy, *Journal of Economic Literature*, 48(3): 642–692.

Kox, Henk, and Hildegunn Nordas. 2007. Services trade and domestic regulation, MPRA Paper No. 2116. At: http://mpra.ub.uni-muenchen.de/2116/.

Marchetti, Juan A., and Petros C. Mavroidis. 2011. The Genesis of the GATS, *European Journal of International Law*, 22(3): 689–721.

Suddaby, Roy, David J. Cooper and Royston Greenwood. 2007. Transnational regulation of professional services: Governance dynamics of World level organizational change, *Accounting, Organizations and Society*, 32: 333–362.

Vogel, David. 1992. Consumer Protection and Protectionism in Japan, *Journal of Japanese Studies*, 18(1): 119–154.

PART III

The politics and dynamics of international and regional standardisation

8

The power of the penholder

The missing politics in global regulatory governance analysis

WALTER MATTLI AND JACK SEDDON

A. Introduction

Traditional understandings of international governance, organised around the idea of sovereign states vesting authority in centralised intergovernmental organisations (IGOs) to govern the conduct of states, have been found increasingly wanting. By now it is widely accepted that the structure and arrangement of international governance has been transformed in response to what Robert Keohane has called the 'governance dilemma' – the mismatch between the demand and supply for global governance arising in a world of increased interdependence and sovereign states – and the reconstitution of state-society relations associated with the rise of the regulatory state.[1] Strong linkages between actors from different institutional levels – sub-national, national and supranational – and sites of authority – public and private – have been shifting authority away from its traditional vestiges and changing its form, away from hierarchy and direct regulation toward softer forms of law and indirect market-based instruments.

A bewildering range of concepts, including 'delegated governance', 'experimental governance', 'multi-level governance', 'multi-stakeholder governance', 'hybrid governance', 'joint governance' and 'network governance', 'incorporated trans-governmentalism', 'institutional twining',

[1] G Majone, 'The New European Agencies: Regulation by Information' (1997) 4 (2) *Journal of European Public Policy* 262, at 262–275; D Levi-Faur, 'The Global Diffusion of Regulatory Capitalism' (2005) 598 (1) *Annals of the American Academy of Political and Social Science* 12, at 12–32; and RO Keohane, 'Governance in a Partially Globalized World' (2001) 95 (1) *American Political Science Review* 1, at 1–13.

'multi-actor or collaborative schemes', 'multi-tiered regulation', 'orchestration arrangements', 'regime complexes' and 'supranational regulatory state', have been deployed to capture these institutional linkages and wider transformations in international governance.[2] In the interest of parsimony, we interpret all such relations as governance partnerships. Each governance partnership can be specified analytically as a formal or informal relational arrangement between a *founder partner* (any actor that institutes a partnership) and one or more *enlisted partners* (frequently transnational actors) that, having certain common goals and objectives, while remaining organisationally independent, decide to share in the exercise of regulatory authority over a common set of target actors or issues. Regulatory authority is broadly defined as responsibilities in the organisation and control of economic, political and social activities by means of making, implementing, monitoring and enforcing rules.

Public-public, public-private and private-private partnerships today exist in virtually every arena of global governance. At the risk of overstatement, where there is international governance, there increasingly are governance partnerships. For instance, beyond those in trade already addressed in this volume, the global financial architecture comprises various disparate partnerships established among political networks (e.g. the G-20), inter-governmental organisations (IGOs; e.g. the International Monetary Fund [IMF]) and a plethora of standard-setting bodies (e.g. the International Organisation of Securities Commissions [IOSCO]). Equally, the United Nations Environment Programme (UNEP) constitutes a disparate regime of partnerships.

[2] E.g. R Dehousse, 'Regulation by Networks in the European Community' (1997) 4 (2) *Journal of European Public Policy* 246, at 246–261; L Hooghe and G Marks, *Multi-Level Governance and European Integration* (2001); M Thatcher, 'Delegation to Independent Regulatory Agencies' (2002) 25 (1) *West European Politics* 125, at 125–145; K Raustiala 'The Architecture of International Cooperation: Transgovernmental Networks and the Future of International Law' (2002) 43 (1) *Virginia Journal of International Law* 1, at 1–92; A-M Slaughter, *A New World Order* (2004); A Héritier and D Lehmkuhl, 'The Shadow of Hierarchy and New Modes of Governance: Sectoral Governance and Democratic Government' (2008) 28 (1) *Journal of Public Policy* 1, at 1–17; and D Coen and M Thatcher, 'Network Governance and Multi-level Delegation' (2008) 28 (1) *Journal of Public Policy* 49, at 49–71; B Eberlein and A Newman, 'Escaping the International Governance Dilemma? Incorporated Transgovernmental Networks in the European Union' (2008) 21 (1) *Governance* 25, at 28; K Abbott and D Snidal, 'The Governance Triangle: Regulatory Standards Institutions and the Shadow of the State' in W Mattli and N Woods (eds), *The Politics of Global Regulation* (2009), at 44–88; K Abbott and D Snidal, 'Strengthening International Regulation Through Transnational New Governance: Overcoming the Orchestration Deficit' (2009) 42 *Vanderbilt Journal of Transnational Law* 501, at 501–578.

Partnerships have attained a particularly prominent status in European Union (EU) governance.[3] Pioneered in the 1980s through the New Approach to product standard setting, against a background of legal constraints and bureaucratic conservatism, there now exist more than twenty pan-European specialised agencies that regulate the common market, working outside the structure of the EU's central institutions. The EU Commission also retains hundreds more alliances with non-governmental organisations (NGOs), trade associations and other actors.[4] The frequency with which the Commission seeks to partner with others should not surprise: its mandate is sweeping; its resources, however, are few. Its own administrative staff is small – 'much smaller than that of the local government of the city of Rotterdam', for example.[5] The inevitable consequence, as put by Bowen, is that 'because of the under-resourced nature of the European Commission, the institution depends heavily on external resources.'[6] It is important to emphasise that regional governance partnerships are not unique to the EU: the Association of Southeast Asian Nations and the North American Free Trade Agreement, for instance, have also witnessed the sharing of regulatory authority with an increasing number of specialised agencies and networks.[7]

Governance partnerships can be understood, in part, in functionalist terms: by sharing policy responsibilities and authority, partnerships improve efficiencies in the allocation of collective institutional goods; they also permit greater specialisation and stratification over time as the environmental systems they regulate become more complex. Partnerships are therefore said to be 'synergistic', helping to coordinate and empower decentralised actors and resolving international cooperation problems.[8]

We accept the functionalist basis of these studies but deem the picture incomplete. Lacking a clearly articulated theory of power, and accepting at face value formal partnership contracts, functionalist frameworks seldom recognise – still less satisfactorily explain – the distributional effects of partnerships. Power and conflict may skew the workings of partnerships, favouring some partners more than others – sometimes in unintended or

[3] Dehousse, supra note 2.

[4] B Kohler-Koch and R Eising, *The Transformation of Governance in the European Union* (1999).

[5] P Bouwen, 'The European Commission', in D Coen and J Richardson (eds), *Lobbying the European Union: Institutions, Actors, and Issues* (2009), at 20.

[6] Ibid, 20. [7] Slaughter, supra note 2.

[8] Eberlein and Newman, supra note 2, 28; and Raustiala, supra note 2, 6.

unexpected ways. We thus propose an extension to shed light on the politics of governance partnerships. We ask: Who are the winners and losers in these governance arrangements? What endows governance partners with power? How and why do partnerships evolve the way they do? What are the long-term distributional effects of varying power endowments among partners?

We are reminded here of Krasner's famous critique over two decades ago of neo-liberal institutionalism: 'indeed, institutional arrangements . . . [are] necessary to resolve coordination problems. . . . Without [these arrangements] all parties would . . . [be] worse off. There are, however, many points along the Pareto frontier'.[9] Different points have different distributional implications; that is, some points favour certain parties more than other parties. Krasner argued that relative power among states plays a crucial role in understanding the choice of points on the Pareto frontier and thus the nature of institutional arrangements. Krasner's general argument about the central role of power remains valuable and pertinent. Sharing regulatory authority is bound to give rise to distributional squabbles – that is, disagreements over which point along the Pareto frontier should be chosen. However, the world Krasner described is in many respects different from the world of today – as evidenced by the rise in governance partnerships. In particular, state power – central to Krasner's analysis – is less relevant to an analysis of the politics of regulatory partnerships where rival organisational interests are as important as state interests and arrangements frequently emerge in spite of states.

We argue that although power is central to an analysis of the distributional implications of partnerships, the sources of power have seen subtle but critical changes. In a growing number of issue areas in global governance, the decisive source is not traditional state power based on military might or the size of the economy, but rule-making power, or what we call 'the power of the pen'. Such power has been migrating during the last decade or two in several key areas of the global economy from the state to the non-state domain.[10]

The thesis of this study is that governance partners – whether founder-partners or enlisted partners – often assume one of two complementary

[9] S Krasner, 'Global Communications and National Power: Life on the Pareto Frontier' (1991) 43 (3) *World Politics* 336, at 337.

[10] W Mattli and T Büthe, 'Setting International Standards: Technological Rationality or Primacy of Power?' (2003) 56 (1) *World Politics* 1, at 1–42; and T Büthe and W Mattli, *The New Global Rulers: The Privatization of Regulation in the World Economy* (2011).

roles: rule-making or rule-supporting. Rule-makers tend to 'win' – in the sense of succeeding in selecting their preferred Pareto points – for three reasons: first, their technical knowledge generates information asymmetries, protecting their interests over those of prospective 'losers'. Second, a rule-making role is naturally more constitutive than a rule-supporting role. Rule-makers are in the driver's seat; rule-supporting is more discreet and backstage. And third, rule-makers benefit from network effects. As a result, a rule, once set and implemented, is hard to change. The ease with which rule-makers obtain their preferred Pareto points is not constant, however. The theory holds that the price of 'victory' also depends on the number of (actual or potential) partners. For example, a single/focal rule-maker interacting with multiple rule-supporters extracts most concessions from its partners and thus pays the lowest price.

The study is organised as follows. Section B develops the theoretical framework of the distributional dimension of governance partnerships. Section C examines the framework's conjectures through a set of brief case studies. Section D concludes the chapter.

B. An analysis of the distributional dimension of governance partnerships

Governance partnerships are cooperative arrangements. Actors can secure mutual benefits from cooperation because they have different comparative advantages – that is, complementary skills and capabilities. In general terms, these skills and capabilities include regulatory competencies, institutional legitimacy, as well as operational capacities (financial and material, human and intellectual, informational and knowledge-based). It is self-evident, especially in the international arena, that few institutions are endowed with all these resources and capacities. By pooling resources and joining forces, institutions can 'fill gaps' in their regulatory capacities and so position themselves to better tackle global challenges.

However, complementary comparative advantages do not imply perfect harmonies of interest. The participants in arrangements – IGOs, NGOs, business organisations, public-private groups and trans-governmental networks – will inevitably exercise distinct forms of authority, operate within unique institutional and legal contexts, and be subject to varying societal pressures and organisational pathologies.

The skill sets and governance capabilities associated with particular partners, and their preferences over the structure and arrangement of partnership schemes can all vary enormously. Some partners may be large and well-resourced IGOs, or they may be small and weak regulatory networks seeking to leverage the power of IGOs or wealthy business organisations. Equally, the partners that they enlist may be highly dependent on the partnership or free to exit at will.

This section seeks to identify the key factors likely to shape and determine the nature of the relationship between governance partners – in particular, their relative dependencies and resulting distributional implications.

I. The analytical framework

Our analytical framework makes two fundamental assumptions. First, the nature of the strategic relationship between founder partners and those they enlist is rarely perfectly harmonious but best captured by the Battle of the Sexes Coordination game. In such games, all parties have an overriding preference to come to a common solution, but they disagree about the details of the common solution. Each party prefers a solution over none; in addition, each party prefers a solution that conforms to its own specific preference than that of the partner. Governance partnerships thus may favour one party more than the other. In this sense, they can be said to have distributional implications.

Second, the framework assumes that actors are boundedly rational rather than synoptically rational. The assumption of bounded rationality refers to cognitive and informational limitations that decision-makers experience when faced with complex or urgent problems.[11] Bounded rationality understands the informational problem differently from the standard rendition of information problems. Traditional rational choice models typically characterise the lack of information in terms of a probabilistic distribution over the possible outcomes (i.e. risk rather than uncertainty).[12] Bounded rationality emphasises that – in addition to being unable to do the requisite calculations – actors may not know this distribution. Applied to this study, bounded rationality simply means that parties cannot fully anticipate the distributional implications of their

[11] H Simon, *Models of Bounded Rationality* (1997).

[12] Risk refers to probabilities over known possibilities, and choice under risk is readily resolved in terms of maximising expected utility. This can be distinguished from uncertainty, which entails unknown possibilities and/or probabilities and is not amenable to the same rational calculus. For an extended discussion of how the bounded-rationality assumption can improve institutional theories in the field of International Relations, see J Jupille, W Mattli and D Snidal, *Institutional Choice and Global Commerce* (2013).

partnership agreement. Unintended consequences are therefore significant, perhaps especially so in areas that are highly technical and involve specialised expertise. In other words, it is necessary to look well beyond formal partnership contracts to understand the distributional implications of cooperation. Eventual outcomes can differ dramatically from formally sanctioned goals.

The theory presented in this section posits that two key factors explain the distributional consequences of partnership agreements: (1) rule-making versus rule-supporting role or functionality and (2) single/focal versus multiple (actual or potential) partners. We first offer an elaboration of these factors and then conclude with the presentation of the main conjectures. These conjectures are put to test in the subsequent section.

1. First explanatory factor: rule-making versus rule-supporting functionality

Rule-supporting partners, which can be founder partners or enlisted partners, provide a wide range of services, including provision of information on local conditions or compliance levels, endorsement or appointment of rule-makers, ongoing legitimation and oversight and various forms of logistic and material support to rule-makers. Rule-makers, which, again, can be either founder partners or enlisted partners, define the general parameters and/or the specific details of a solution to a cooperation challenge. In areas where partners diverge on how best to approach and solve a cooperation problem, agents with rule-making capabilities will have a disproportionate impact on the ultimate solution. They are in a position to select their preferred Pareto points – that is, they are likely to set regulatory parameters that reflect their substantive policy preferences and ensure that their organisational and economic interests are maximised subject to a constraint of minimum acceptability to rule-supporting partners.

In what ways does rule-making confer more power than a rule-supporting role in a partnership? That is, what are the sources of rule-making power or 'power of the pen'? At least three sources are worthy of mention. First, as amply illustrated in this study, technical or expert knowledge generates great information asymmetry between partners, keeping the rule-supporter potentially in the dark or at bay. The latter partner may trust the rule-maker but not fully understand the regulatory solutions proposed or grasp the long-term implications of these solutions. Lack of information and knowledge makes it difficult to effectively oppose a regulatory proposal that may have unfavourable distributional implications. In other words, asymmetry reduces the constraints under which the rule-maker operates, enabling the forging of solutions that will favour the interests of the rule-maker more so than those of the rule-supporter.

Second, rule-making is more constitutive than rule-supporting. A rule-maker is in the driver's seat, charting the course of action and continuously deciding which turn to take or not. Rule-makers can use their power to directly enhance their own capacity for political action. By contrast, rule-supporting activities – although critically important to effective partnership – are more discrete and backstage. A decision by an IGO, for example, to formally endorse a rule-making partner can tremendously boost the legitimacy and prestige of the rule-maker. However, the rule-supporting act of endorsement is a single event. Endorsement can be revoked, of course, as can logistic or financial support. Nevertheless, rule-supporting events have less direct impact on the shaping of cooperation outcomes than the more continuous activity of rule-making.

Finally, rules, once set and adopted, may be hard to change because of network effects. The power to institute rules inevitably attracts actors outside the partnership with an interest in the form and content of rules, gradually concentrating and reinforcing the locus of relevant information and deliberation, decision and authority. The benefits of adopting and complying with new rules also increase as more targets convert to these rules. The logic of these increasing returns is to spawn large networks of vested interests in and defenders of the rules and rule-makers, increasing the cost to opponents seeking to dislodge their status. Network effects entrench and magnify the power of the rule-maker over time and, ceteris paribus, render it more difficult for the governance partner to successfully rectify the rules' distributional biases.[13]

Rule-making power is often said to manifest itself at the agenda-setting and negotiation stages of a regulatory process and is thought to derive from formal rule-making authority. This picture is seriously incomplete and thus can be highly misleading. First, rule-making is not confined to agenda-setting and negotiations. Significant rule-making may occur at the implementation and enforcement stages. The implementation of general

[13] WB Arthur, 'Competing Technologies, Increasing Returns, and Lock-in by Historical Events' (1989) 99 (394) *The Economic Journal* 116, at 116–131; PA David, 'Clio and the Economics of QWERTY' (1985) 75 (2) *American Economic Review* 332, at 332–337; D North, *Institutions, Institutional Change and Economic Performance* (1990); and P Pierson, *Politics in Time: History, Institutions, and Social Analysis* (2004).

legislation and rules, for example, may necessitate further regulatory clarification, inviting the implementing partner to engage in more detailed rule-making. In regulation, the devil is frequently in the details, implying that the implementing partner can possess tremendous (de jure or de facto) rule-making power. As discussed later in the chapter in detail, the European Commission drafted general securities market legislation in the 1990s and enlisted an independent expert group called the Committee of European Securities Regulators (CESR) to produce detailed guidelines on the meaning of these rules to ensure consistent implementation and regulatory convergence across Europe. CESR seized on its seemingly narrow rule-making mandate at the implementation stage to expand its jurisdiction incrementally and shape financial market legislation in ways that the Commission had not intended and found difficult to stop.

Similarly, enforcement is not necessarily simply a judicial-technical support operation. Rule-making at the enforcement stage – judicial rule-making or 'activism' – is a well-known phenomenon extensively covered in the academic literature. The European Court of Justice is just one example of an enforcement body that has successfully engaged in judicial activism and shaped the regulatory core of European integration in ways no head of state would have thought possible at the time of the signing of the Treaty of Rome.[14]

Second, with the growing complexity of regulatory issues in the global economy, it no longer can be taken for granted that all those endowed with formal rule-making authority possess actual rule-making capability and power. A striking illustration is provided by the United Nations Commission on International Trade Law (UNCITRAL).[15] The Commission was established in 1966 to develop a unified body of international trade law. Formally, decisions are taken by state delegations representing the sixty members of the Commission making for apparently broad governmental representation. Non-state delegations – mainly industry and expert groups with special interest in global trade issues – are invited at the discretion of the Secretariat. In reality, however, only a tiny group has the requisite financial and technical resources to attend consistently; they are the true rule-makers. These delegates come from advanced economies

[14] A-M Burley and W Mattli, 'Europe before the Court: A Political Theory of Legal Integration' (1993) 47 (1) *International Organization* 41, at 41–76; W Mattli and A-M Slaughter, 'Revisiting the European Court of Justice' (1997) 52 (1) *International Organization* 177, at 177–209; and A Stone Sweet, *The Judicial Construction of Europe* (2004).

[15] T Halliday, J Pacewicz and S Bloc-Lieb, 'Who Governs? Delegation and Delegates in Global Trade Lawmaking' (2013) 7 (3) *Regulation & Governance* 279, at 293.

and disproportionately from non-state groups. The latter not only possess extensive technical knowledge but also 'have a powerful interest in the establishment of laws and regimes consistent with their expertise and competitive advantage'.[16] UNICTRAL is only one of a growing number of cases where formal rule-making authority in global affairs no longer neatly coincides with effective rule-making power.

In conclusion, the distinction between rule-making and rule-supporting functionality seems clear and simple in the abstract. In reality, it can be challenging to identify the true rule-makers. They may have no formal rule-making authority and/or operate in relative secrecy. Only careful empirical investigation will reveal the site of real power. Further, in some partnerships, we do not have simply one rule-maker and one rule-supporter; each partner may possess rule-making *and* rule-supporting capabilities – but to varying degrees. In these circumstances, the jurisdictional boundaries of rule-making authority may sometimes be the object of contestation. However, this complexity generally is not a problem for the analysis. For example, the founder partner may define the broad contours of a regulatory regime and then charge a partner with working out the regulatory details and giving precise meaning to broad framework provisions. To the extent that the latter activity requires much greater technical expertise and is more continuous than the defining of broad contours, effective rule-making power can be said to reside with the enlisted implementing partner.

2. Second explanatory factor: single/focal versus multiple (actual or potential) partners

The distribution of power in governance partnerships is also shaped by the number of potential partners in a given domain because this will alter the relative price payable to secure or maintain a partner's cooperation. A single or focal partner facing a steady supply of (actual or potential) partners incurs a lower opportunity cost (suspended gains if a partnership breaks down) than an easily substitutable partner. The former will find it easy to arrange an alternative partnership with another 'supplier', whereas the latter may be excluded from such an arrangement for some time and thus incur a high opportunity cost. As a result, an actor is expected to accrue fewer benefits or pay a higher price to institute and/or maintain a partnership with an organisation in a monopoly position than if the

[16] Ibid, 281.

actor faces many (actual or potential) partners. Only a monopoly player can set a monopoly price. However, as our cases demonstrate, the price will also vary depending of the contestability of that monopoly. That is, a partner may be able to 'manufacture' a more contestable environment and so weaken the position of a monopolist by creating alternative partners or building the capacity to internalise functions that would otherwise be shared through the partnership. In theory, both founder and enlisted partners may be single/focal in their respective issue areas, or they may face competition. A single founder partner, displeased with the price or quality of the service rendered by a selected partner and facing a large supply of alternative providers, will shop around for a more suitable and pliant organisation. Similarly, an enlisted partner not satisfied in a partnership may shop around for a more favourable deal if more than one potential partner operates in the issue area.

In our analysis, it does not matter whether a founder partner or enlisted partner is 'single' or 'focal' – both have the same effect on distributional outcomes. The two terms, however, are not necessarily identical. Single is a quantitative expression – it means 'just one'. 'Focality' emphasises a qualitative aspect: a focal organisation may not be the only player, but it is the obvious one for a given cooperation challenge. It may have a good track record built on technical expertise, financial resources and perhaps even legitimacy. A single organisation may or may not be richly endowed and possess great legitimacy. A single organisation that is well endowed and widely accepted as legitimate may be said to also be focal. Some single organisations lack focality (as just defined) but remain key partners because the costs and risks of creating new and competing organisations may be high.

3. The conjectures

The combination of the two central variables of our framework – functionality (i.e. rule-making versus rule-supporting) and contestability (i.e. single/focal versus many) – generates four conjectures listed in Figure 8.1. The figure plausibly assumes that in an governance partnership only one founder partner is involved; and this actor faces either a single/focal partner or multiple partners.

At one extreme is the case of a focal partner assuming the task of principal rule-maker. This enlisted partner interacts with a founder partner whose main contribution to the partnership may be one of encouragement and endorsement. In this arrangement, we would expect the enlisted partner to be the primary beneficiary. It defines the rules (or policy) that

		Enlisted Partner	
		Rule-Making Role	Rule-Supporting Role
Enlisted Partner	Single/Focal	Relative Power of Enlisted Partner strongest Founder Partner few options but to accept Enlisted Partner's terms and proposals *Distributional Implications Strongly Favouring Enlisted Partner* (Conjecture 1)	Founder Partner has strong say in shaping solution but may need to find compromise with Enlisted Partner *Distributional Implications Favouring Founder Partner* (Conjecture 3)
	Multiple (Menu of options)	Enlisted Partner has strong say in shaping solution but may need to compromise with Founder Partner (It is constrained by possibility of forum shopping by Founder Partner) *Distributional Implications Favouring Enlisted Partner* (Conjecture 2)	Relative Power of Enlisted Partner weakest Enlisted Partner largely serviant *Distributional Implications Strongly Favouring Founder Partner* (Conjecture 4)

Figure 8.1. Types of enlisted partner and distributional implications

will govern the behaviour of states or private actors. In doing so, it will first and foremost seek to promote or safeguard its own regulatory preferences and organisational interests. If displeased with (some of) these rules, the founder will have few means at its disposal to put pressure on the enlisted partner to reconsider or revise its rules (Conjecture 1). At the other extreme is the case of an enlisted partner confined to a rule-supporting role and easily substitutable. Such an enlisted partner faces competition from other bodies to which the founder could turn if dissatisfied with its services. In this case, the distributional implications strongly favour the founder partner (Conjecture 4).

The model generates two further conjectures. In cases where the enlisted partner – selected from a menu of possible partners – assumes a rule-making role, it will be able to set the agenda and/or define the regulatory details, subject to minimal acceptability by the founder. If displeased, the founder may shop for a more compromising or pliant rule-making partner. The distributional consequences of the governance partnership here will only weakly favour the enlisted partner (Conjecture 2). Finally, in cases where the founder is in the rule-making driver's seat, the outcome will favour the founder partner. However, its rule-supporting single partner has a way of ensuring that the founder respects its basic interests. If unhappy with the arrangement, the single partner can walk away from the partnership, jeopardising an effective solution. The founder will want to avoid such an outcome by offering enough concessions to keep the enlisted partner in the partnership (Conjecture 3).

In conclusion, it is worth noting again that in a context of bounded rationality, the various distributional consequences of agreements may emerge only over time; they are learned through the experience of working together. This is particularly true of partnership arrangements in highly technical and complex issue areas.

C. Illustrating and testing the conjectures

The main focus of this section is to examine the conjectures regarding the downstream distributional implications of governance partnerships using short case studies (the cases are listed in bold in Figure 8.2). We highlight two cases with long histories in which the European Commission has acted as the initiator of a partnership; this introduces an important control in the study, as the formal mandate and power of the Commission is broadly constant across the examined issue areas. The Commission plays a central role in the EU legislative process by its right of legislative initiative. It is also the guardian of the legal framework; in this capacity, it is involved in the supervision and implementation of EU policies and in checking that EU law is respected in all member states. The selected cases vary in the functionality assigned to the Commission's partners and the number of actual or potential partners available to the Commission. If the theory presented in the preceding section is correct, we should observe a wide range of distributional outcomes across our cases, with the Commission gaining more or less vis-à-vis its partners over time as a

	Rule-Making Role Enlisted Partner	Rule-Supporting Role Enlisted Partner
Single/Focal Enlisted Partners	**Financial Stability Forum (FSF) – *Recognised Standard Setting Bodies (including BCBS/IOSCO/IAIS/ CPPS /FATF/IASB)*** World Trade Organization (WTO) – *Recognised Standard-Setting Bodies (ISO/IEC/Codex Alimentarius Commission)* G-20 – *Financial Stability Board (FSB)* **Commission – *International Accounting Standards Board (IASB)*** **Commission – *Committee of European Securities Regulators (Lamfalussy Level 3)*** Commission – *European Committees for Standardization (e.g. CEN/CENELEC)* Commission – *European Banking Authority* (Level 3) Commission – *European Insurance and Occupational Pensions Authority (EIOPA)* (Level 3)	Recognised Standard Setting Bodies (BCBS/IOSCO etc.) – *International Finance Institutions (IMF and World Bank)* **International Federation of Accountants (IFAC) – *Public Interest Oversight Board (PIOB)*** Basel Committee on Banking Supervision (BCBS) – *Credit Rating Agencies* NAFTA Commission for Environmental Cooperation – *Joint Public Advisory Committee* Commission – *Body of European Regulators for Electronic Communications (BEREC)* Commission – *European Competition Network (ECN)* Commission – *The European Environment Agency (EEA)* Commission – *European Regulators Group for Electricity and Gas (ERGEG)*
Multiple (actual/potential) Enlisted Partners (contestable environment)	**Financial Stability Board (FSB) – *Recognised Standard Setting Bodies (including BCBS/IOSCO etc.)* (Increased FSB capacity has made the RSSBs' rule-making monopoly marginally more contestable since 2007)** International Telecommunication Union (ITU) – *IT Consortia* Extractive Industries Transparency Initiative (EITI) – *Business Organisations and NGOs* United Nations Kimberly Process Certification Scheme (KPCS) – *Business Organisations and NGOs* OECD – *Intergovernmental Agencies (including the Bank for International Settlement, IOSCO and BCBS etc.)* (Principles of Corporate Governance) Commission – *Environmental NGOs* (1980s, early 1990s) Commission – *Working Party on the protection of individuals with regard to the Processing of Personal Data (Article 29 Working Party)*	Global Environmental Facility (GEF) – *The United Nations Development Programme, the United Nations Environment Program, the World Bank, Business Organisations and Environmental NGOs* International Labour Organisation – *Labour Rights NGOs* Global Reporting Initiative (GRI) – *Environmental NGOs, Business Organisations, Academic Institutions, and Intergovernmental Agencies* **Commission – *Committee of European Securities Regulators (Lamfalussy Level 2)*** Commission – *Environmental NGOs (since late 1990s)* Commission – *European Banking Authority (Level 2)* Commission – *Human Rights NGOs*

Figure 8.2. Examples of governance partnerships – founder–enlisted partner pairs

result of this variation in functionality and contestability. Our empirical research extends well beyond these cases. We conclude this empirical section with a summary of some of the additional evidence.

I. Commission-IASB Partnership: Strongest-Enlisted Partner Case (Conjecture 1)

Our first case involves the Commission and the International Accounting Standards Board (IASB); the Commission is the founder, and the IASB, a focal rule-maker, is the enlisted partner. As an independent expert body with a history going back to the early 1970s, the rule-maker has long enjoyed considerable authority in global accounting. After several futile attempts to establish a single set of European accounting rules, the Commission turned to the IASB, endorsing this private-sector rule-maker and committing to using IASB rules within the EU. This public endorsement conferred much-needed legitimacy upon the IASB, boosting its prestige as a global regulator and triggering a sharp increase in the demand of its rules and services by governments around the world. Enlisting the IASB enabled the Commission to attain its objective of European convergence on a single set of accounting rules. Lacking effective means to control the IASB, however, the Commission paid a steep price for this convergence. The Commission has often been displeased with the IASB's rules, but caught over a barrel by a non-substitutable partner, it has had to take them with few (if any) concessions.

We first set the context leading up to EU partnering with the IASB, then introduce the enlisted partner, and conclude with illustrations of the distributional implications of the partnership.

1. The context

The European Union's stated main goal has been the establishment of a truly single market in goods, services, labour and capital. The harmonisation of financial reporting rules and practices is one key step toward the realisation of this objective; it enhances the comparability of financial statements, which, in turn, contributes to financial integration and a more efficient allocation of capital in Europe. The member states charged the EU Commission as early as the 1960s with the task of harmonising accounting rules as well as company laws across Europe.

The cooperation challenge in this domain, however, was daunting due to the sharply contrasting styles and orientations of financial reporting practices among the various member states, reflecting deeply rooted divergences in philosophy and purpose of accounting. In most European continental countries, for example, accounts are used for tax purposes; the accounting and valuation methods thus are strongly influenced by legal and taxation requirements. An accountant's goal is to comply with strict legal rules and reduce the tax liability of a company. In the United Kingdom, by contrast, financial reports to shareholders are distinct from tax calculations, and accounting rules are more flexible.[17]

Unsurprisingly, the Commission-led convergence effort proved highly protracted; it resulted in two directives more notable for their failures and *lacunae* than their achievements. They concerned themselves with broad issues rather than detailed guidance, leaving great latitude to national legislatures when implementing the directives – latitude that compromised the objective of cross-country comparability of financial statements.[18] Further, many important issues remained untouched by the directives, including leases, pensions, taxation and currency transactions.

2. The Commission-IASB partnership

Disenchanted by slow progress and repeated setbacks, the Commission realised it needed an entirely new way of approaching the harmonisation problem. It decided to take the bold step of entering into a governance partnership with the IASB, the focal private-sector rule-maker in global accounting. The partnership seemed perfectly sensible. Europe needed a single set of financial reporting rules, and the IASB was the sole international producer of these rules. In February 2002, the Commission endorsed the IASB by unveiling a regulation requiring listed EU companies to prepare consolidated accounts in accordance with international accounting standards by 2005. This measure benefitted both parties. For Europe, it constituted a significant step forward in the realisation of a fully integrated European capital market. The IASB derived benefits as well. Europe was the first region to confer binding nature on its rules, thereby immensely boosting the legitimacy and prestige of the IASB.

[17] M Haskins, K Ferris and T Selling, *International Financial Reporting and Analysis: A Contextual Emphasis,* 2nd edn (2000), at 53; D Ordelheide, 'The True and Fair View – A European and German Perspective' (1993) 2 (1) *European Accounting Review* 81, at 81–90; and D Alexander 'A European True and Fair View?' (1993) 2 (1) *European Accounting Review* 59, at 59–80.

[18] C Nobes and R Parker, *Comparative International Accounting* (2004); and E Emenyou and S Gray, 'EC Accounting Harmonization: An Empirical Study of Measurement Practices in France, Germany and the UK' (1992) 22 (89) *Accounting and Business Research* 49, at 49–58; and P Walton, 'Harmonization of Accounting in France and Britain: Some Evidence' (1992) 28 (2) *Abacus* 187, at 187–199.

The IASB developed from the International Accounting Standards Committee (IASC) founded in 1973 by 16 accountancy bodies from nine countries (United States, United Kingdom, Canada, Australia, the Netherlands, Germany, France, Japan and Mexico). Its stated mission was to 'formulate and publish in the public interest accounting standards to be observed in the presentation of financial statements[,] . . . promote their worldwide acceptance, and . . . work generally for the improvement and harmonization of regulations, accounting standards, and procedures relating to the presentation of financial statements'.[19]

Over the years, the IASC incrementally broadened its membership and deepened its institutional structure.[20] In 1982, for example, the IASC entered into an agreement with the International Federation of Accountants (IFAC) under which IFAC recognised the IASC as the sole official source of international accounting standards and promised to promote the use of international standards in all IFAC member countries.[21] In return, the IASC recognised the IFAC as the authorised representative of the accounting profession worldwide, accepted all members of the IFAC as members of the IASC, and gave the IFAC full authority to nominate candidates for membership on the governing Board of the IASC.

The Board was IASC's key rule-making body. Initially it comprised nine representatives from accounting firms of the founding countries – all volunteers. Participation increased sevenfold by the late 1990s. Board members met four times a year for about a week each to discuss and issue international accounting standards (IASs), exposure drafts and related documents. They were assisted by steering committees, one for each proposed new or revised standard. A two-thirds majority was required for publication of an exposure draft and a three-quarters majority for the adoption of a standard. Board activities were funded largely by Board members and supported by a small secretariat located in London.[22] In 2002, the IASC morphed into the IASB.[23] Much of the institutional structure remained in place, but some important changes were implemented to improve organisational effectiveness.

[19] IASC (original) Constitution, Article 2.

[20] D Cairns, *The IASC: Evolution, Achievements and Prospects* (1999).

[21] The IFAC was formed in 1977 and comprises today some 140 professional organisations from more than 100 countries.

[22] K Camfferman and S Zeff, *Financial Reporting and Global Markets: A History of the International Accounting Standards Committee, 1973–2000* (2007).

[23] Jupille, Mattli and Snidal, supra note 12, 165–196.

3. The distributional dimension of EU-IASB partnership

The analytical framework of this study suggests that a founder partner enlisting a focal independent rule-maker as a partner is likely to pay a high price for the arrangement. If displeased with (some of) the rules, the founder will have few effective options to bring about change. What evidence is there to suggest that the IASB is the clearer winner in the partnership with the Commission, in other words, that distributional net gains have been accruing disproportionately to the IASB?

First, it is important to reiterate that both partners have been benefitting from their arrangement. The Commission has finally achieved its objective of having a single set of financial reporting standards across Europe. The IASB, in turn, received a big boost to its legitimacy as a global regulator from EU endorsement, prompting major countries such as China, Japan, Brazil, India, Canada and South Korea to set out timetables for adoption or convergence to IASB standards. By 2010, about one hundred countries required compliance with IFRS, up from less than twenty countries ten years earlier. In short, IASB's organisational interests have been well served: demand for its rules and services has surged and is likely to keep growing for years to come. Remarkably, however, the IASB has been reaping these gains largely cost-free – that is, without concessions to the Commission. By contrast, the Commission has been paying dearly for convergence, as illustrated by the following episode.

In 1999, plans were drawn up to overhaul the IASC to increase its rule-making efficiency. In particular, the Board was cut down to only fourteen members – now full-time and highly paid professionals from the accounting profession. A majority of eight members was needed to adopt a new standard. When the reformed organisation resumed operations – now as the IASB, five Board members were American accounting professionals. The United States, the United Kingdom, Canada, South Africa and Australia together had no fewer than ten members. The Commission was outraged. The *Financial Times* ran stories titled 'Brussels lost voice' and 'American coup that shut out (continental) European influence from the reformed IASC'.[24] The Commission possessed no real powers to change course. Financially, the IASB is independent. It is privately funded – largely through voluntary contributions from the 'Big Five' accounting firms and multinational corporations as well as the sale of standards and related documents. Further, the Commission has no formal role in

[24] Jupille, Mattli and Snidal, supra note 12, 184–193.

IASB's governance structure and cannot intervene directly in the private rule-making process of its enlisted partner.

The Commission has repeatedly pressed the IASB to overhaul its governance system on the grounds that, as the first region to adopt IFRS, Europe deserved much wider representation in IASB rule-making. This argument cuts no ice with the IASB. It has countered that as a private-sector expert body, membership selection must be guided not by political or regional criteria but by technical expertise. With no alternative partner in sight, the Commission is in no position to extract significant concessions.

II. The Commission–CESR partnership: two contrasting cases (Conjectures 4 and 1)

Another instance of a Commission-led governance partnership occurred in the context of building an integrated EU securities market. This example is particularly felicitous because the enlisted partner – the independent Committee of European Securities Regulators (CESR) – has a double functionality, rule-making and rule-supporting in separate but related domains. This allows for hypotheses testing in conditions close to those found in controlled experiments. The theoretical expectations are that the partnership will (strongly) favour the enlisted partner in relative terms where it acts as (focal) rule-maker and disfavour it where it acts as rule-supporter. The example also offers a nice illustration of rule-making power at the implementation stage in contrast to the more familiar rule-making at the agenda-setting stage, the potential to be misled by formal partnership contracts and the constraints of bounded rationality on the ability of actors' to fully anticipate the downstream distributional effects of collaboration.

After sketching the context that gave rise to a governance partnership in EU securities market regulation, we describe the two partnership arrangements and then analyse their distributional dimensions.

1. The context

The creation of a single financial market in Europe has been a longstanding but elusive goal of the Commission. One essential component of such a market is a single set of financial reporting rules. As discussed earlier, the adoption of such rules was an immensely protracted and arduous process. It is only when the Commission started to experiment with partnerships, enlisting the IASB, that the pace of progress quickened.

Another foundational element of an integrated financial market is a set of common rules for securities markets.

The Commission's powerful DG for the Internal Market and Services (DG Market) has exclusive responsibility for proposing new directives in this area. Assisted by the Economic and Financial Affairs Council (ECOFIN) composed of finance and economics ministers of the EU member states, DG Market attempted in the 1970s and 1980s to harmonise securities laws.[25] These efforts largely failed; they produced a small number of securities directives peppered with exemptions and generally deemed of poor technical quality.[26] Significantly, independent domestic securities regulators played almost no role in these early attempts – an anomaly attributable to jurisdictional turf battles and jealousies between finance ministries and securities regulators in most EU member states.[27]

2. The Commission-CESR partnership

Pressure to integrate financial markets grew intense again with the adoption of a single European currency in 1995.[28] The Commission responded by announcing the 'Financial Services Action Plan' (FSAP) but recognised that 'if we are to successfully implement the FASP, we will need to overhaul the way we develop financial services legislation'.[29] It also established a committee of 'wise men' chaired by Alexander Lamfalussy, a former director of the Bank for International Settlements (BIS), with a mandate to chart the best way forward. After two years of consultations and deliberations, the committee proposed a new system, the most innovative central component of which was a partnership between the Commission and the previously excluded securities regulators – then organised in an

[25] T Tridimas, 'EU Financial Regulation: Federalisation, Crisis Management and Law Reform', in P Craig and G de Burca (eds), *The Evolution of EU Law*, 2nd edn (2011), at 783–804.

[26] E Ferran, *Building an EU Securities Market* (2004); N Moloney, *EC Securities Regulation*, 2nd edn (2008).

[27] Eddy Wymeersch (Chairman of CESR 2009–2012), Interview, 27 January 2012, Brussels, Belgium.

[28] Fabrice Demarign (Secretary General of FESCO 1997–2001, Member of CESR 2001–2006), Interview, 19 January 2011, Paris, France. Nigel Wicks (Member of the Lamfalussy Committee), Interview, 13 February 2012, London, United Kingdom.

[29] Commission, Financial Services: Implementing the Framework for Financial Markets: Action Plan (Brussels, 232 final, 1999), 16. Available at http://ec.europa.eu/internal_market/finances/docs/actionplan/index/action_en.pdf (accessed 4 August 2012).

independent club, the Committee of European Securities Regulators.[30] The Commission decision enlisting CESR describes the organisation simply as 'an independent advisory group' on implementing provisions.[31] CESR was not designed to formally legislate, nor was it intended that CESR would intervene directly in the day-to-day supervision of financial markets.[32] CESR held four plenary sessions a year and conducted most of its technical work through standing committees and task forces supported by a small secretariat located in Paris.

CESR had two distinct roles – one rule-supporting, the other rule-making – in Lamfalussy's multi-level governance system.[33] This system can be summarised as follows. At level 1 (L1), the Commission proposed 'framework' legislation that was then adopted by the Council of Ministers and the European Parliament under the co-decision procedure. The Commission next developed level-2 (L2) rules to clarify and supplement the adopted framework legislation. Much of this technical work took place in the new European Securities Committee (ESC) run by the Commission with involvement from finance ministries. In this L2 process, the Commission enlisted CESR in an advisory capacity; CESR was asked to comment on draft standards and assist with technical and market information. Rule-making authority remained with the Commission, Council of Ministers and European Parliament; CESR's first role simply was to support the L2 process.

CESR played a second role – this one much more consequential – at level 3 (L3). This level was designed to support convergence and consistency in the implementation and application of L1 and L2 rules. CESR's L3 task was to 'foster and review common uniform day-to-day implementation and application of Community legislation, issuing guidelines, recommendations, and standards to be adopted by CESR members in

30 CESR replaced the Forum of European Securities Commissions (FESCO) established in 1997.

31 Commission Decision 2001/527/EC of 6 June 2001 establishing the Committee of European Securities Regulators (CESR) OJ 2001, L 191/43 Art. 1 and 2.

32 Nigel Wicks, Interview, 13 February 2012. David Wright (Deputy Director-General for the Internal Market and Services of the Commission and Rapporteur of Lamfalussy Committee), Interview, 21 December 2011, Brussels, Belgium.

33 Lamfalussy, *Final Report of the Committee of Wise Men in the Regulation of European Securities Markets* (15 February 2001) (Brussels). Available at http://ec.europe.eu/internal_market/securities/doc/lamfalussy/wisemen/final-report-wise-men_en.pdf (accessed 4 August 2012).

their regulatory practices on a voluntary basis'.[34] Seemingly technical and narrow, this provision laid down the principle that CESR had rule-making power through the issuance of 'guidelines', 'recommendations' and 'standards'. As discussed subsequently, CESR quickly seized on this new power to expand its jurisdiction and shape financial market legislation in ways the Commission had not intended and found difficult to stop.[35] In the words of CESR's influential first chairman, Arthur Docters van Leeuwen: '[We saw] a once in a lifetime opportunity to create something and gain a bigger voice in Europe and at home'.[36]

3. The distributional dimension of EU-CESR partnership

What is the evidence that CESR's rule-supporting L2 role disproportionately benefitted the founding partner, whereas its rule-making L3 role disproportionately favoured the enlisted partner?

CESR entered the L2 partnership arrangement with high hopes but was soon disappointed. The reasons are twofold. First, the Commission frequently rejected or ignored CESR advice when such advice failed to fit with its general regulatory preferences. For example, it rejected CESR's view in relation to the L2 market abuse rules that investment recommendations by journalists be subject to a specific regime or CESR's recommendation that credit-rating agencies be subject to the investment research rules of the market abuse regime.[37] Further, the Commission sometimes marginalised CESR in discussions over certain sensitive issues such as bond market transparency as well as hedge-fund and private-equity regulation, seeking advice from other expert groups, including groups it established in 2004 partly to diminish CESR's monopoly status – namely, the European Securities Markets Expert Group (ESME) and the Forum of User Experts in the Area of Financial Services (FIN-USE).[38] In principle, CESR was the Commission's first port of call for L2 assistance. In practice, the Commission created and then turned to alternative providers whenever it so pleased. Consistent with the theory's conjecture, the presence

[34] Charter of the Committee of European Securities Regulators (11 September 2001), Article 4.3.

[35] David Wright, Interview, 21 December 2011.

[36] Docters van Leeuwen (chairman of CESR 2001–2009), Interview, 15 February 2012, The Hague, Netherlands.

[37] Moloney, supra note 26, chapter 13.

[38] David Wright, Interview, 21 December 2011. Docters van Leeuwen, Interview, 15 February 2012.

of alternative rule-supporting groups furthered weakened the position of CESR in L2 matters.

CESR seemed to have been 'bought by the Commission' – as aptly put by a frustrated securities regulator.[39] Indeed, the Commission emerged as clear winner in the L2 arrangement. It retained all of its legislative powers, and obtained from its partner valuable technical assistance and market information for the 'cheap price' of CESR inclusion in the Lamfalussy system. With this support, the Commission successfully tabled a large number of major pieces of financial legislation in a remarkably brief period.[40] This outcome contrasts sharply with the pre-partnership period of protracted and largely ineffectual securities market regulation led by the Commission.

Although its gains were modest in L2 issues, CESR did astonishingly well in L3 matters.[41] At L3, CESR was required to draft guidelines on the precise meaning of L1 and L2 rules to secure their consistent implementation and regulatory convergence across the EU. CESR took an increasingly expansive view of the measures needed to achieve this objective, thus solidifying its institutional power and overcoming the limited formal powers ceded to it in its formal partnership agreement.[42] As summarised by Moloney: 'Through level 3 activities, CESR . . . acquired a degree of influence over the financial markets which is remarkable given its establishment in 2001. . . . This has occurred through a range of quasi-regulatory [and] supervisory mechanisms . . . no[t] . . . specified in CESR's founding Decision'.[43] CESR used its L3 powers to effectively amend EU legislation or introduce rules in areas not yet covered by the Commission. However daring its actions, CESR routinely justified them in terms of the need for supervisory convergence and clarity in implementation.[44] Its guidelines and standards, although formally not binding, were quickly 'regarded by market [actors] as tantamount to regulatory fiat'.[45] CESR exploited blurred boundaries between securities market regulation and

[39] Fabrice Demarign, Interview, 19 January 2011.

[40] Examples include the Market Abuse Directive (2003), Prospectus Directive (2004), Markets in Financial Instruments Directive (2006) and the Transparency Directive (2007).

[41] Docters van Leeuwen, Interview, 15 February 2012.

[42] Docters van Leeuwen, Interview, 15 February 2012.

[43] N Moloney, 'The Committee of European Securities Regulators and Level 3 of the Lamfalussy Process' in M Tison, H de Wulf, C van der Elst and R Steennot (eds), *Perspectives in Company Law and Financial Regulation* (2009), 451–452.

[44] Fabrice Demarign, Interview, 19 January 2011.

[45] Moloney, supra note 43, at 455.

corporate governance to expand its rule-making domain and brought issues for which it clearly had no formal legal authority, such as cross-border takeovers and audit services, into its orbit.[46] Finally, CESR took steps to position itself as the focal interlocutor for operational negotiations with the US Securities and Exchange Commission (SEC), the International Organization of Securities Commissions (IOSCO) as well as a range of societal groups, including the financial services industry.[47]

In sum, CESR succeeded in less than a decade to leverage a seemingly narrow and technical rule-making role in the L3 arrangement into a wide-ranging quasi-regulatory undertaking, leaping from local obscurity to global prominence. The Commission watched CESR's growing influence with concern and reprimanded its power-grabbing partner, occasionally even in public. But it could do little else. It depended on CESR for proper implementation of EU securities markets legislation, and no substitute was in sight.[48]

CESR morphed into the European Securities and Markets Authority (ESMA) in the aftermath of the recent global financial crisis. Some observers have emphasised the restrictions imposed by EU institutions on ESMA and doubt its ability to become an effective European Supervisory Authority.[49] Our theory, however, suggests that notwithstanding how conservative and distrustful EU member states and the Commission were in its design, ESMA inevitably will grow in regulatory stature where it holds the pen and monopolises political action. This space may well prove large enough to realise the dream held by some securities regulator at the birth of CESR of one day seeing their organisation grow into a fully-fledged 'Euro-SEC'.[50]

III. A summary of additional cases

The preceding partnership cases were selected, in part, because of their relatively long history; they offer sufficient evidence to examine distributional implications that unfold over time. Many more European cases, although more recent, also fit the theory (see Figure 8.2). Beyond Europe, evidence we have collected on the dense complex of partnerships that

46 Moloney, supra note 43, at 457.
47 Docters van Leeuwen, Interview, 15 February 2012.
48 David Wright, Interview, 21 December 2011.
49 E Wymeersch, 'The Reforms of the European Financial Supervisory System: An Overview' (2010) 7 (2) *European Company & Financial Law Review* 240, at 240–265.
50 Michel Prada (Member of CESR 2001–2011), Interview, 19 January 2012, Paris, France.

together make up the global financial architecture also provides consistent support for our framework. Space permits three examples.

The first partnership, conceived initially in the G-22 working groups established in the aftermath of the Asian financial crisis and eventually created at the direction of the G-7, saw a number of standard setting bodies (SSBs), such as the BCBS, IOSCO and International Association of Insurance Supervisors (IAIS), fill a major gap in their institutional capacities by partnering with the International Finance Institutions (IFIs; i.e. the IMF and WB). Before the establishment of this partnership in 1999, the SSBs active in global finance had limited means to encourage the uptake and consistent application of their regulatory standards. The IFIs were the logical solution to this resource gap: in contrast to the SSBs, the IFIs had near-universal membership and direct access to the developing country targets of standards. In addition, the IFIs had the operational resources necessary to perform these labour-intensive tasks and the potential to bring instruments of 'hard law' into the sector, although this potential was not realised until after the recent financial crisis. In this partnership, the IFIs were reduced to a rule-supporting function, monitoring the implementation of standards through the Financial Sector Assessment Program (FSAP), which produces Reports on the Observance of Standards and Codes (ROSCs) and Financial Sector Stability Assessments (FSSAs).

The distributional effects of this deal were dramatic and immediately felt. Overnight the SSBs experienced a 'huge uplift' in the reach and force of their standards, and they paid very little for this gain.[51] By contrast, the IFIs were left ruing the need to perform what they considered a Cinderella task. They also saw the arrangement as crystallising a massive shift in power toward the Basel-based regulatory groupings because the latter were anointed as the primary rule-makers for matters of financial stability.[52]

However, consistent with Conjecture 3, the SSBs have felt the constraints of non-substitutable partnership on occasion. For example, when the IFIs required a methodology by which to benchmark national compliance with IOSCO's Core Principles of Securities Regulation (the Principles), key members of IOSCO expressed strong reservations and pledged

[51] Michel Prada, Interview, 19 January 2012.

[52] P Blustein, 'How Global Watchdogs Missed a World of Trouble' (2012) 5 *CIGI Papers*. Available at http://www.cigionline.org/publications/2012/7/how-global-watchdogs-missed-world-of-trouble (accessed 29 April 2013).

resistance. They feared that the methodology would undermine a long-standing commitment to flexibility in the interpretation and application of the Principles designed to accommodate countries with differing regulatory systems. However, IOSCO had little practical choice in the end but to develop the methodology to sustain the partnership.[53]

The second partnership, instituted in the same period by the G-7, established relations between the newly created Financial Stability Forum (FSF), a group of major national financial authorities such as finance ministries and central bankers, and the same set of SSBs. When the FSF enlisted the SSBs, it did little more than endorse their standard-setting activities.[54] The technocratic work of the regulators gained the impetus of a quasi-political authority at a cheap price.[55] The FSF, by contrast, lacking in a clear direction and effective leadership, drifted into stasis having delivered this endorsement. As one regulator recalled: 'The FSF was set up to avoid another crisis like Asia. In this, it failed miserably. It simply did not do what it was set up to do'.[56]

To function effectively, the political masters of the FSF recognised that they had to somehow shake up the status quo and position the organisation to 'get under the noses' of the SSBs.[57] The Financial Stability Board (FSB), established by the G-20 to replace the FSF in 2009, set out to recalibrate the partnership accordingly.[58] The FSB has had some success in this regard, but not through major constitutional reforms or the acquisition of formal powers over the SSBs as principal-agent theories might lead us to expect. All such proposals – of which there were many – found themselves derailed by a lack of political support from the FSB's state masters and strong resistance from the SSBs.[59]

[53] Tony Neoh (Chair of IOSCO Technical Committee 2003–2006), Interview, 22 January 2013, Beijing, China.

[54] Andrew Crockett (Chairman of FSF 1999–2003), Interview, 10 January 2012, San Francisco, USA.

[55] Confirmed in interviews from regulators across the regulatory spectrum: in insurance, Alessandro Iuppa (Chair of the IAIS Executive Committee 1995–2000), Interview, 9 January 2012, Washington, DC, USA; in securities, Jane Diplock (Chair of IOSCO Executive Committee), Interview,20 November 2012, Wellington, New Zealand; and in banking, Andrew Crockett, Interview, 10 January 2012.

[56] Svein Andresen (Secretary General, Financial Stability Board since 2007), Interview, 5 April 2012, Basel, Switzerland.

[57] Anonymous (Deputy Director US Treasury Department), Interview, 14 September 2011, Washington, DC, USA.

[58] Andrew Crockett, Interview, 10 January 2012.

[59] Ethiopis Tafara (Director of the Office of International Affairs at the United States Securities Exchange Commission), Interview, 17 January 2012, Washington, DC, USA.

The FSB has instead achieved some modest gains in less formal ways. In particular, through a process analogous to creation of alternative partners, the FSB has been seeking to develop a rule-making capacity that can potentially contest the SSBs' monopoly, albeit only at its margins and in a limited sense. For instance, in November 2011, the FSB announced policy measures to address systemically important financial institutions, and it issued an international standard for resolution regimes, immediately triggering complaints and accusations from the SSBs of power grabbing. Members of the FSB, especially the finance ministries that are excluded from the SSBs, retorted that rule-making gives them much needed leverage over their partners.

It would be imprudent and premature to claim a shift in the locus of the power of the pen, however. As our theory anticipates, powerful network effects are acting to hold it in place. As Peter Sands, the CEO of Standard Chartered, put it with reference to global banking regulations, 'I do wonder if all this concern [about the FSB] is a bit misplaced. . . . If you really want to focus on where the key standards are being set[, it is] in the Basel Committee'.[60]

The final, less well-known, partnership is found in the realm of international auditing standards. In response to the Enron and WorldCom scandals, in 2003 the International Federation of Accountants (IFAC) began to look around for partners that might be enlisted to deflect the difficult questions being asked about the audit profession setting its own rules. Reflecting a large pool of actors with a long-standing interest in international audit standards, seven international actors responded to this call including the BCBS, IOSCO, European Commission and the World Bank. They pooled their resources in an elaborately conceived structure comprising a new Monitoring Group and an independent public international body called the Public Interest Oversight Board (PIOB) that was tasked with ensuring that the IFAC's independent standard-setting boards serve the public interest in their rule-making activities.

The effect of this partnership has been to provide the IFAC boards with an invaluable source of ongoing public legitimacy without anything like a commensurate change in their normal operating procedures. The PIOB has no influence over the substantive content of IFAC standards, nor any power to declare a standard contrary to the public interest, it checks only that 'due process' has been followed and that societal stakeholders have been consulted and listened to in the production of standards and work

[60] Quoted in Brookings Institute Transcript, 2011.

plans. However, consistent with Conjecture 3, the fact that the Monitoring Group members created a single entity means they can effectively jeopardise the IFAC's solution to its legitimacy deficit. This has encouraged the IFAC to make some concessions following the Monitoring Group's first five-year review of the partnership.

There are reasons for some scepticism about how meaningful these concessions are, however. The PIOB faces enormous challenges in performing even its narrow tasks. These difficulties compellingly illustrate the manner in which information asymmetries can reduce the constraints under which rule-makers operate.

Members of the PIOB have to be accustomed with the IFAC board procedures, they have to grasp the scope and impact of their standard-setting work, and they have in some way to assess how complex regulations should define and accommodate the public interest among many other vested and powerful stakeholders. In the highly technical world of audit standards, the manifold complexity of this task encourages box-ticking and rubber-stamping. Staffers sent to perform firsthand oversight concede that they often find it impossible to assess whether the public interest has been taken into account, and it is still more unlikely that a regulatory proposal could be opposed on such a basis.[61]

D. Conclusion

This study offers a simple yet powerful account of the distributional implications of governance partnerships – a mode of international governance enjoying rapidly growing popularity because it enables participants to realise common objectives through collaborative arrangements. The relationship between partners, however, rarely is perfectly harmonious. Disagreements over the specifics of a solution – the precise points on the Pareto frontier – are bound to erupt. Which party is likely to prevail and at what price? We have argued that in many cases, one party will assume a role that is predominantly rule-making in nature (broadly defined), whereas the other partner specialises in rule-supporting tasks. Over time, greater benefits from the partnership will accrue to the rule-maker rather than the rule-supporting partner. Ruler-makers derive their relative power from three sources: their technical knowledge generates information asymmetries, rendering it difficult for their partners to immediately comprehend

[61] Teresa Rodríguez (IOSCO Policy Advisor), Interview, 1 August 2012, Madrid, Spain.

the distributional implications of proposed rules; ruler-makers are in the driver's seat, whereas rule-supporters are one step removed from the action; and network effects make the reversal of rules by opponents difficult. The theory also explains why the price paid by rule-makers for attaining their preferred Pareto-points is not constant but varies. The critical explanatory factor here is the number of available partners. Single/focal rule-makers interacting with multiple rule-supporters pay the lowest price and are likely to be highly successful in pushing their regulatory preferences and maximising their organisational interests.

The empirical part of this study examined our theory's main conjectures. The focus was on the EU Commission – one of the pioneers of partnership arrangements. The Commission's mandate is vast; however, its resources are few, and political support of its state masters, at times, is in doubt. The case studies show how partnerships with a wide range of actors across the full spectrum of the regulatory process have enabled the Commission to overcome these handicaps and deliver effective policies. At the same time, and consistent with the theory, such partnerships have had unexpected and unintended distributional consequences, and the price paid for successful common policies has depended both on the number of partners and the roles assumed by them. A summary of additional cases suggests that our theory travels well beyond Europe and applies to a wide range of global governance partnerships.

We see our study as an overdue corrective to the prevailing analytical tenor in much of the rapidly growing literature on new transnational governance. This literature highlights the significance of the shift from 'government' to 'governance'. Its acceptance of functionalist premise, however, at times has led to rather sanguine depictions of institutional efficiency, ignoring questions of who gets what, when and why in new modes of global governance. Functionalist accounts need not be wrong or irrelevant – indeed, we have built our theory on functionalist foundation – although they sometimes seem based more on faith or convenience rather than critical investigation. They tend to fail, however, to systematically ponder the distributional dimension of the new governance arrangements.

Where distributional issues have been explored, principal-agent models dominate. These models naturally focus on the ways in which principals control agents to minimise agency losses. For instance, principal–agent analyses of European governance partnerships have highlighted the limited powers of the EU's partner organisations, their inability to initiate legislation and other constraints, thus assigning these organisations a distinctly subordinated role in pan-European regulation.[62] However, as we have shown in European accounting and securities regulation, partners enlisted to help the European Commission draft and implement rules have had an unexpectedly dramatic impact on regulatory outcomes. The reduction of the EU's partners to 'information', 'coordination' or 'implementation' devices by theoretical fiat can be highly misleading.

The distributional dimensions of partnerships are overlooked by principal-agent models largely because they assume hierarchical relations and therefore exclude phenomena such as two-way information asymmetries and mutual dependencies. The general emphasis remains on aligned interests and complementary resources. However, in the real world, the gains from partnerships are seldom equally or equitably distributed, partners are recalcitrant, they will demand a price for their commitment and they will fight bitterly to amend structures that marginalise their interests. The need to capture these causes for conflict among partners and the distributional implications of their varying power endowments motivated this study.

We have developed a simple yet powerful analytical framework that systematically addresses these neglected distributional aspects of governance partnerships. These partnerships are revealed as a relational, dynamic and irreducibly political phenomenon that sometimes dramatically changes the locus of power and authority in global governance arrangements. The surprising, even disturbing, outcomes we have found underscore the enormous stakes involved. They also serve to caution against new partnerships being cast in terms of unclouded optimism. An increased prescription of partnership deals may help to ameliorate the governance dilemma we face, but it will also often cause the *pen*dulum of power to swing in ways that will prove difficult to reverse and we may live to regret.

[62] Héritier and Lehmkuhl, supra note 2, 1–17; and Coen and Thatcher, supra note 2, 49–71.

9

Between standards and regulation

On the concept of 'de facto mandatory standards' after *Tuna II* and *Fra.bo*

HARM SCHEPEL

A. Introduction

In 1985, the European Commission launched the New Approach to technical harmonisation and standards, a comprehensive overhaul of its program of internal market legislation relying to a large extent on private European standards.[1] In 1995, the US Congress passed the National Technology Transfer and Advancement Act, instructing federal agencies to use, where possible, standards developed by private organisations.[2] Under the influence of a newly dominant regulatory philosophy weary of command-and-control legislation, private standards have rapidly become an important part of regulatory arrangements around the world.[3] This, in turn, has opened up space for the increased regulatory use and adoption of

[1] Council Resolution on a new approach to technical harmonisation and standards [1985] OJ C 136/1. See e.g. U DiFabio, *Produktharmonisierung durch Normung und Selbstverwältung* (1996), and M Egan, *Constructing a European Market – Standards, Regulation and Governance* (2001).

[2] Pub Law 104–13. The Office of Management and Budget has had a circular in place since 1982 encouraging federal participation and use of voluntary standards. For early enthusiasm, see RW Hamilton, 'Prospects for the Nongovernmental Development of Regulatory Standards', (1983) American University Law Review 455. The current version of OMB Circular A-119 is available at http://www.nist.gov/standardsgov/omba119.cfm. Samuel Krislov has aptly described the system as one 'at cross purposes with itself, which 'minimizes government interference in the formative stages and casts a legal shadow over everything else.' S Krislov, *How Nations Choose Product Standards and Standards Change Nations* (1997), 133.

[3] See e.g. H Schepel, *The Constitution of Private Governance – Product Standards in the Regulation of Integrating Markets* (2005), and T Büthe and W Mattli, *The New Global Rulers – The Privatization of Regulation in the World Economy* (2011).

international private standards, a space eagerly exploited by regional and international trade law.[4]

The use of private standards in public regulation creates, naturally, a lot of unease in liberal democracies under the rule of law.[5] The default answer of legal systems to concerns of constitutional propriety and legitimacy is to relegate standards either to the realm of purely private, voluntary 'guidelines' and 'recommendations' or to adopt them wholesale as binding public law, re-enacting standardisation as regulatory process. In either, problems of privatised law-making simply disappear from sight. But regulatory practice is rather more complex than either of these solutions would allow for.

Under the New Approach, product legislation limits itself to formulating 'essential legal requirements'. The idea is then that compliance with harmonised European standards endorsed by the Commission lends but a presumption of conformity to these requirements. The standards themselves, supposedly, remain 'private' and 'voluntary'. This edifice depends, of course, on the possibility of alternative means of meeting the 'essential requirements', which, for that purpose, are to be worded 'precisely enough in order to create, on transposition into national law, legally binding obligations which can be enforced'.[6] That this is largely fiction became abundantly clear in 2002, when the General Product Safety

[4] See e.g. A Sykes, *Product Standards for Internationally Integrated Goods Markets* (1995), and E Wijkström and D McDaniels, 'Improving Regulatory Governance: International Standards and the WTO TBT Agreement', (2013) 47 *Journal of World Trade* 1013.

[5] There is an enormous, predominantly German, literature on these issues in Europe. For discussion and references, see H Schepel, *The Constitution of Private Governance*, supra note 3. For the US context, see, e.g., J Freeman, 'Private Parties, Public Functions and the New Administrative Law', (2000) 52 *Administrative Law Journal* 813; PR Verkuil, 'Public Law Limitations on Privatization of Government Functions', (2006) 84 *North Carolina Law Review* 397, and TRT Wolf, 'Existing in a Legal Limbo: The Precarious Legal Position of Standards-Development Organizations', (2008) 65 *Washington and Lee Law Review* 807. There has also been a fairly recent upsurge of criticism of the idea of 'regulation by commodity': private standards are protected by copyright and not freely accessible. See e.g. ES Bremer, 'Incorporation by Reference in an Open-Government Age', (2013) 36 *Harvard Journal of Public Law and Policy* 131; PL Strauss, 'Private Standards Organizations and Public Law', (2013) 22 *William & Mary Bill of Rights Journal* 497, and NA Mendelson, 'Private Control over Access to the Law: The Perplexing Federal Regulatory Use of Private Standards', (2014) 112 *Michigan Law Review* 737. For discussion of these matters in the European Union, see R. van Gestel and H-W Micklitz, 'European Integration Through Standardization: How Judicial Review Is Breaking Down the Club House of Private Standardization Bodies', (2013) 50 *Common Market Law Review* 145.

[6] Section B (III), Annex II (the 'Model Directive'), Council Resolution on a new approach to technical harmonization and standards, [1985] OJ C 136/1.

Directive turned the requirement that products be 'safe' into a New Approach 'essential requirement'.[7]

Comparable arrangements abound in the United States, even if federal policy is less systematic. A few fairly random examples from the US Code of Federal Regulations may help to illustrate the point. The US Consumer Product Safety Commission 'strongly encourages all firms to comply with voluntary consumer product safety standards and considers, where appropriate, compliance or non-compliance with such standards in exercising its authorities under the Consumer Product Safety Act and other federal statutes'.[8] The US Occupational Safety and Health Administration (OSHA) states that all protective eye and face protection devices *must* comply with certain ANSI standards, but adds:

> Protective eye and face protection devices that the employer demonstrates are at least as effective as protective eye and face protection devices that are constructed in accordance with one of the above consensus standards will be deemed to be in compliance with the requirements of this section.[9]

OSHA also requires that atmospheric tanks '*shall* be built in accordance with acceptable good standards of design', only to add that they *may* be built in accordance with certain standards from Underwriters' Laboratories and the American Petroleum Institute.[10]

In none of these examples is compliance with the product requirements laid down in the referenced standards formally 'mandatory', but in all of them it is obviously a sensible idea to comply with them anyway, because non-compliance will be 'considered' by enforcement authorities, because the burden of demonstrating that devices are at least as safe as devices built in accordance with standards may be cumbersome and costly, or because the procedures and hassle of proving that alternative methods meet 'acceptable good standards of design' lead to irrational expenses

[7] Directive 2001/95/EC on General Product Safety [2002] OJ L 11/4. Article 3 (2) presumes products manufactured in compliance with harmonized standards to be 'safe'. Article 2 (b) defines a 'safe product' as 'any product which, under normal or reasonably foreseeable conditions of use including duration and, where applicable, putting into service, installation and maintenance requirements, does not present any risk or only the minimum risks compatible with the product's use, considered to be acceptable and consistent with a high level of protection for the safety and health of persons.'

[8] US Code of Federal Regulations, Title 16, Chapter II (B), § 1115.8 Compliance with product safety standards.

[9] US Code of Federal Regulations, Title 29 (B), Chapter XVII, § 1910.133 Eye and Face Protection.

[10] US Code of Federal Regulations, Title 29 (B), Chapter XVII, § 1910.106 Flammable liquids.

of time and money. When compliance with a particular standard is not 'mandatory' in a legal sense, but in practice the only feasible way for products to be marketed, access to markets may be barred and courts will seek to discipline these 'de facto mandatory standards'. This chapter discusses two recent instances of this: the *Tuna II* litigation in the World Trade Organization (WTO)[11] and, first, the judgment of the Court of Justice of the European Union in *Fra.bo.*[12]

B. Marketing copper fittings in Germany: *Fra.bo*

Standards in the German water industry are set by the *Deutsche Vereinigung des Gas – und Wasserfaches* (DVGW), a nonprofit body governed by private law. In 2005, the DVGW amended its standard for copper fittings to introduce a new requirement: the elastomeric waterproof joints of fittings were now to withstand a test of immersion in boiling water for 3,000 hours. When Fra.bo, an Italian company, failed to obtain a certificate to the amended standard,[13] it brought an action in German courts against the DVGW for contravention of its rights under Article 34 TFEU on the free movement of goods. The *Oberlandesgericht* in Düsseldorf duly referred the question to the Court of Justice.

On the face of it, *Fra.bo*'s action was a long shot. The DVGW is a private body over which the government has no influence whatsoever. From a legal standpoint, its standards are about as 'voluntary' as can be. Certification by the DVWG lends but a presumption of conformity with the general requirement that only products that follow 'the acknowledged rules of technology' may be used in the industry. That requirement, in turn, is found not in binding legislation but in the Regulation on General Conditions for Water Supply (the *AVBWasserV*), which lays down general sales conditions for water supply undertakings and their customers, from which parties are free to depart. If the Court was to come to the aid of

[11] Panel Report, United States – Measures Concerning the Importation, Marketing and Sale of Tuna and Tuna Products, WT/DS381/R, adopted on 15 September 2011, and Appellate Body Report, United States – Measures Concerning the Importation, Marketing and Sale of Tuna and Tuna Products WT/DS381/AB/R, adopted on 16 May 2012.

[12] Case C-171/11 *Fra.bo*, judgment of 12 July 2012, ECLI:EU:C:2012:453.

[13] There was an issue in the case about the refusal by the DVWG to recognize as equivalent the inspection reports issued by an Italian laboratory. Although potentially important in the larger scheme of things, the matter was of no consequence for the issues discussed here: as far as is evident from the judgment, Fra.bo never put its fittings to the 3,000 hours test.

Fra.bo in the company's efforts to market its fittings in Germany, it was going to have to stretch the scope of Article 34 TFEU in either (or both) of two ways. It could either extend the substantive scope of 'measures having equivalent effect to quantitative restrictions' and hold Germany liable for impeding market access to products not certified by the DVGW. This solution is more plausible than it may seem. In *Commission* v. *Belgium*, the Court had already found a regulatory arrangement virtually identical to the one at issue in *Fra.bo* to fall foul of Article 34 TFEU.[14] The crucial difference with *Fra.bo* lies in the legal status of the standards body: the *Institut Belge de Normalisation* being a public body, the Court had no trouble classifying its standards as 'state measures'.[15] As a consequence, the Court had no need to distinguish between the regulatory provisions and the standards concerned to identify the source of the trade restriction: it treated the whole arrangement as one 'State measure' capable of 'dissuading' operators to market their products in Belgium by the mere fact of 'encouraging' them to obtain national marks of conformity from the IBN for products lawfully marketed in other Member States. Such an easy escape was not available in *Fra.bo*, as the Court explicitly acknowledged.[16]

The alternative would be to extend the personal scope of Article 34 TFEU and hold the DVGW itself liable for limiting the exercise of *Fra.bo*'s free movement rights. This solution, too, is rather more plausible than it may seem at first glance. The Court has long endowed the provisions on the free movement of workers, the freedom to provide services and the freedom of establishment with 'horizontal direct effect',[17] objecting to the activities of sporting federations,[18] bar associations[19] and trade unions.[20] But until *Fra.bo*, the Court had refused to do the same for the free movement of goods.[21]

14 Case C-227/06 *Commission* v. *Belgium*, Judgment of 13 March 2008. Summary publication in [2008] ECR I-46. The full text of the judgment is available (in French only) from the Court's website: http://www.curia.europa.eu.

15 Ibid, para 37. 16 Case C-171/11 *Fra.bo*, supra note 12, para 24.

17 See generally on the doctrine, H Schepel, 'Constitutionalising the Market, Marketising the Constitution, and to Tell the Difference: On the Horizontal Application of the Free Movement Provisions in EU Law', (2012) 18 *European Law Journal* 177.

18 See e.g. Case 36/74 *Walrave and Koch* [1974] ECR 1405, and Case C-415/93 *Bosman* [1995] ECR I-4921.

19 Case C-309/99 *Wouters* [2002] ECR I-1577.

20 Case 438/05 *Viking* [2007] ECR I-10779, and Case C-341/05 *Laval* [2007] ECR I-11767.

21 The most explicit rejection of horizontal direct effect of Article 34 TFEU is to be found in Case C-159/00 *Sapod Audic* [2002] ECR I-5031, para 74.

To many commentators, the Court appeared to choose this second option in its judgment.[22] From the way it cast the question before it and the way it answered it, this seems indeed to be the case. The Court framed the question it faced as follows:

> It must therefore be determined whether, in the light of inter alia the legislative and regulatory context in which it operates, the activities of a private-law body such as the DVGW has the effect of giving rise to restrictions on the free movement of goods in the same manner as do measures imposed by the State.[23]

And the Court formulated the operative part like this:

> Article [34 FFEU] must be interpreted as meaning that it applies to standardisation and certification activities of a private-law body, where the national legislation considers the products certified by that body to be compliant with national law and that has the effect of restricting the marketing of products which are not certified by that body.[24]

The problem is, however, that in arriving at this answer the Court concentrated exclusively on the 'legislative and regulatory context' and seemed not be interested at all in the activities of the DVGW itself. It first dismissed objections about the status of the Regulation as an optional instrument:

> Although the AVBWasserV merely lays down the general sales conditions as between water supply undertakings and their customers, from which the parties are free to depart, it is apparent from the case-file that, in practice, almost all German consumers purchase copper fittings certified by the DVGW.[25]

It then dismissed the idea of viable alternatives to certification by the DVGW as a means of demonstrating compliance with the 'acknowledge rules of technology', holding that 'administrative difficulties' and 'additional costs' make such alternative procedures 'of little or no *practical* use'.[26] On the basis of these considerations, then, the Court concluded that

[22] See e.g. the exuberant editorial by N Nic Shuibhne, 'The Treaty Is Going to Get You . . . .', (2012) 37 *European Law Review* 367; W-H Roth, 'Die "horizontale" Anwendbarkeit der Warenverkehrsfreiheit (Art. 34 AEUV)', (2013) *Europäisches Wirtschafts- und Steuerrecht* 16, and H van Harten and T Nauta, 'Towards direct horizontal effect for the free movement of goods?', (2013) 38 *European Law Review* 677.

[23] Case C-171/11 *Fra.bo*, supra note 12, para 26.

[24] Ibid, para 32.

[25] Ibid, para 30. Emphasis added.

[26] Ibid, para 29. Emphasis added.

> it is clear that a body such as the DVGW, by virtue of its authority to certify the products, *in reality* holds the power to regulate the entry into the German market of products such as the copper fittings at issue in the main proceedings.[27]

The key to understanding this schizophrenic reasoning lies, arguably,[28] in the blind sport of the preliminary reference procedure.[29] It obviously hadn't occurred to the *Oberlandesgericht* to even ask whether 'the legislative and regulatory context' itself was compatible with the regime on the free movement of goods; it only asked whether Article 34 TFEU could be applicable to the DVGW. Although it cast its judgment as an answer to the latter question, the Court really answered the first. In the light of *Commission* v. *Belgium*, the Court *must* hold the *AVBWasserV* itself to be a restriction of the free movement of goods by 'encouraging' operators to obtain national certificates for products lawfully marketed in other Member States. In that light, the Court's indifference toward the 'activities' of the standards body itself is perfectly logical: whether the DVGW demands copper fittings to resist 3,000 hours or one day of immersion in boiling water is not the issue when the restriction lies in the mere fact of having to obtain the certificate. What's more, there is nothing much the DVGW could possibly do to address the objections raised by the Court: that, *in practice*, almost all German consumers only buy DVGW-certified products may either be due to governmental sponsorship or to the fact that the DVGW issues solid and dependable standards that consumers have faith in, or both. In the former case, it is hard to see why the standards body should be held responsible. In the latter, it seems perverse policy to increase scrutiny of a standards body in proportion to how well it serves its purpose. That alternative ways of demonstrating compliance with 'the acknowledged rules of technology' are costly and cumbersome is good news for the standards body but hardly something it can be blamed for.

The result is sobering and depressing: in this way, the pertinent and interesting questions raised by the facts in *Fra.bo* all remain unanswered and even beyond scrutiny. Can it be said that the 3,000-hour test reflects the 'acknowledged rules of technology'? How could we tell? Is it reasonable and legitimate to set standards not for reasons of health and safety but, as in this case, with a view to increasing the durability of construction

[27] Ibid, para 31. Emphasis added.

[28] See the note by H Schepel, (2013) 9 *European Review of Contract Law* 186.

[29] Article 267 TFEU.

products?[30] How, and through which procedures, did the DVGW come to the decision to introduce the new test? In sum, the Court of Justice launches the concept of a 'de facto mandatory standard' either to hold private bodies responsible for restrictive effects they cannot reasonably be blamed for, or for holding States liable for something they do not impose and have no control over.

It seems certain that *Fra.bo* will have consequences not just for the control of national standards but of harmonised standards under the New Approach,[31] and even of international standards. After all, the Court considers it settled case law that Article 34 TFEU applies to the institutions of the European Union in the same way as it applies to Member States:[32] in other words, any measure taken by the Commission that 'dissuades' operators from marketing their products in the internal market will be subjected to the discipline of the free movement of goods. The similarities between the regulatory arrangements at issue in these two cases and the New Approach will not be lost on anyone: under the New Approach, too, certification to harmonised standards is formally voluntary, but de facto it is prohibitively expensive and complicated to prove conformity with the essential requirements by alternative means. Under the New Approach, too, consumers purchase, *in practice*, only products bearing CE marks. The European standards body, CEN, too, 'in reality' holds the power to regulate entry into the internal market of products covered by the New Approach. And so it is hard to think of a good reason why harmonised standards and the regulatory arrangement of the New Approach itself should not be held to fall within the reach of Article 34 TFEU. And so it would be hard to think of a plausible reason why any disaffected producer whose products are excluded by a harmonised standard – and *every* standard excludes certain products and favours others – could not,

30 This question especially is arguably better dealt with under Article 101 (3) TFEU than it is under the free movement of goods. The *Oberlandesgericht* did actually refer a question on the liability of the DVGW under the competition rules, but, regrettably, the Court saw 'no need' to answer it in light of its answer on the question on Article 34 TFEU.

31 See generally H Schepel, 'A New Approach to the New Approach: The Juridification of Harmonised Standards in EU Law', (2013) 20 *Maastricht Journal of Comparative and European Law* 521.

32 See e.g. Joined Cases C-154/04 and C-155/04 *Alliance for Natural Health* [2005] ECR I-6451, para 49. See e.g. K Mortelmans, 'The Relationship between the Treaty Rules and Community Measures for the Establishment and Functioning of the Internal Market-Towards a Concordance Rule', (2002) 39 *Common Market Law Review* 1303, and K. Sørensen, 'Reconciling Secondary Legislation with the Treaty Rights of Free Movement', (2011) 36 *European Law Review* 339.

now, seek judicial review in any court in the European Union of private standards under New Approach Directives. And that makes it easy to see to which levels of absurdity *Fra.bo* may lead: if we are to take the Court seriously, national private standards bodies may now be held liable under Article 34 TFEU for the restrictive effects of its – obligatory and blind – transposition as a national standard of a European harmonised standard, which could well be, in turn, the adoption of an ISO standard.

C. Marketing tuna in the United States: *Tuna II*

The Technical Barriers to Trade (TBT) Agreement operates on two relevant distinctions. First, as regards the instruments involved, it defines a standard as a document providing 'rules, guidelines and characteristics for products' with which compliance is not mandatory. A 'technical regulation', on the other hand, is a document laying down product characteristics with which compliance is mandatory.[33] As to the source of the document at issue, the TBT Agreement is agnostic: it leaves open the possibility of public authorities issuing a non-binding 'standard'[34] and also explicitly envisages the situation in which a 'non-governmental body' enacts a technical regulation.[35] The disciplines for 'standards' and 'regulations' are different: standards are regulated by the annexed Code of Good Practice,[36] whereas technical regulations are subjected to Article 2 of the Agreement itself. The substantive obligations in both sections are, however, very similar.[37] Article 2 and the Code are obviously binding on central governmental regulators and standards bodies;[38] for private regulators and standards bodies, Members are to take all 'reasonable measures as may be available to them' to ensure that they comply with the respective obligations.[39]

[33] Paragraphs 1 and 2, Annex 1, TBT Agreement.

[34] This follows from Article 4, TBT Agreement.

[35] Article 3, TBT Agreement is titled '*Preparation, Adoption and Application of Technical Regulations by Local Government Bodies and Non-Governmental Bodies*'.

[36] Annex 3, TBT Agreement.

[37] See e.g. A Arcuri, 'The TBT Agreement and Private Standards,' in T Epps and MJ Trebilcock (eds.), *Research Handbook on the WTO and Technical Barriers to Trade* (2014), 485.

[38] Article 4 (1) TBT Agreement ('Members shall ensure that their central government standardizing bodies accept and comply with the Code of Good Practice for the Preparation, Adoption and Application of Standards in Annex 3 to this Agreement [referred to in this Agreement as the 'Code of Good Practice']).

[39] Articles 3 (1) and 4 (1), TBT Agreement.

It is against this background that the WTO judiciary had to deal with a challenge by Mexico to measures of the United States designed to prevent death and suffering of dolphins during the catch of tuna. At issue was a State measure that made it unlawful to label tuna as 'dolphin-safe' or to carry related designations if the product did not comply with a set of specific requirements, largely concerned with the conditions in which it was caught, promulgated by the Secretary of Commerce. To the consternation of commentators[40] and a dissenting voice on the Panel, both the majority on the Panel and the Appellate Body chose to classify the measure as a 'technical regulation', holding it to be of 'mandatory compliance'. This seems an odd decision because meeting the requirements was 'mandatory' only to carry the label, not to market tuna in the United States: nothing in the regulations involved made the right of access to the US market conditional on the label. It also seems odd, at first sight, to make such a fuss – including the dissent, a rare occurrence in WTO law – over a point on which, *in casu*, nothing turned: even if designated a 'standard', the measure – promulgated by the US Secretary of Commerce – would still have had to comply with the Code of Good Practice which does not seem to fail to impose any of the obligations contained in Article 2 that were of relevance in the case.[41] One has to assume, then, that the panellists and appellate judges were well aware of the implications of their holding for quite another constellation of facts, one in which the use of *private* standards – issued by private bodies not automatically bound to the Code of Good Practice – is furthered, endorsed or enforced by government action.[42]

It makes sense, then, to read *Tuna II* in this light. For present purposes, it seems best to start with the new intellectual hero of WTO punditry,[43] the

40 See e.g. R Howse and PI Levy, 'The TBT Panels- US-Cloves, US-Tuna, US-COOL', (2013) 12 *World Trade Review* 327; P Mavroidis, 'Driftin' Too Far from Shore – Why the Test for Compliance with the TBT Agreement Developed by the WTO Appellate Body Is Wrong, and What Should the AB Have Done Instead', (2013) 12 *World Trade Review* 509; KT Mull, 'Making Sense of "Mandatory" Measures in the TBT Agreement: Why the Majority Panel's Determination in US-*Tuna II* Rendered the Distinction between Technical Regulations and Standards to Be Meaningless', (2013) 25 *Georgetown International Environmental Law Review* 367; A Davies, 'Technical Regulations and Standards under the WTO Agreement on Technical Barriers to Trade', (2014) 41 *Legal Issues of Economic Integration* 37, and Delimatsis, this volume (Chapter 5).

41 The point is made forcefully by A Davies, supra note 40.

42 The point is ignored by A Davies, supra note 40.

43 See also the discussion at: http://worldtradelaw.typepad.com/ielpblog/2011/09/technical-regulations-vs-standards-in-the-tuna-panel-report.html

(anonymous) author of the separate opinion in the panel's report. To him (or her), it is, rightly, rather obvious that any label certifying conformity to any standard makes it 'mandatory' that the requirements laid down in the standard are actually met – otherwise, the label would be meaningless.[44] (S)he also, rightly, finds nothing strange in the fact that a State would throw its weight behind a label 'to compel producers to fulfil the promises they make in relation to compliance with voluntary standards'.[45] This happens all the time all around the world, through unfair trading laws, prohibitions of misleading advertising and other such devices. But as long as the label itself is not mandatory to market a product, the standard to which that label certifies conformity cannot be held to be 'mandatory' for purposes of the definition of a 'technical regulation'. As (s)he puts it,

> it is the *voluntary* decision of those producers to label their products in a certain way what triggers the obligation to comply with such standards. The fact that operators may be legally accountable for any misleading or false declarations in the event that they choose to advertise compliance with a standard without in fact meeting its requirements does not modify the essentially voluntary nature of such standard.[46]

The reasoning so far is obvious, and obviously right. And it is not as though the Panel majority and the Appellate Body were not aware of the distinction between labelling requirements and market access requirements.[47] Their point was quite another one, however poorly made. The flaw in the dissent – and in much of the commentary on the cases – is that it treats the 'dolphin-safe' label in much the same way as it would treat a CE mark, or a certificate attesting conformity to the ASME Boiler Vessel Code, or a DVGW certificate for copper fittings. But it is one thing for a label to convey that tuna has been caught according to certain requirements that will positively affect the life of dolphins and quite another for a label to be the only possible way to make *any* claim about tuna being caught in a way that does not hurt dolphins. As the panel put it:

[44] Panel Report, *Tuna II*, above note 11, para 7.150.

[45] Ibid, para 7.157 [46] Ibid, para 7.156. Emphasis in original.

[47] The Panel made this clear explicitly. Ibid, para 7.141. ('In this respect, we find it useful to highlight that we are not suggesting that *any* situation in which access to a label reflecting compliance with a particular standard is reserved for products that comply with the specific requirements of that standard would amount to a situation in which a mandatory technical regulation exists. In our view, the measures at issue in the present case go significantly beyond that.')

> The measures leave no discretion to resort to any other standard to inform consumers about the 'dolphin-safety' of tuna than to meet the specific requirements of the measure. Effectively, the 'dolphin-safe' standard reflected in the measures at issue is, by virtue of these measures, the *only* standard available to address the issue. Through access to the label, the measures thus effectively regulate the 'dolphin-safe' status of tuna products in a binding and exclusive manner and prescribe, both in a positive and in a negative manner, the requirements for 'dolphin-safe' claims to be made. This distinguishes this situation from one in which, for example, various competing standards may co-exist in relation to the same issue, with different but related claims, each of which may be protected in its own right.[48]

From this angle, the approach seems compelling enough. It is hard to see how it would spell the end of voluntary labelling schemes such as the ones under the aegis of the ISEAL Alliance.[49] A government endorsing and protecting the certification scheme of Forest Stewardship Council would not be enacting a 'technical regulation' as long as the FSC label is legally held to communicate nothing more than the fact that timber has been produced in a way that meets the FSC's standards. This will only change the moment a government outlaws all other, competing schemes and declares the FSC label the only possible way of claiming to sell timber that is 'rainforest-safe'.

But what of regulatory arrangements such as the one in *Fra.bo*, or in the examples from the United States given in the Introduction? The Panel majority attached a footnote to the paragraph just quoted referencing the definitions of 'exclusive reference' and 'mandatory standard' in ISO/IEC Guide 2, which reads as follows:

> 'exclusive reference (to standards)': 'reference to standards that states that *the only way* to meet the relevant requirements of a technical regulation is to comply with the standards referred to' [emphasis added].

48 Ibid, para 7.144. The dissent's opinion descends into tautology. Ibid, para 7.161. ('This conclusion is not modified by the limitations placed by the measures on the use of alternative "dolphin-safe" marks or labels and on the use of certain terms. These aspects of the measures should be understood in connection with the overall architecture of the measure, which aims to ensure that any claims made in relation to the impact on dolphins of the catching of the tuna contained in tuna products are consistent with the conditions set out in the measures, and do not mislead consumers.') See also the Appellate Body Report, supra note 11, para 195.

49 For such anxiety, see e.g. LJ Ankersmit and JC Lawrence, 'The Future of Environmental Labeling: Tuna II and the Scope of the TBT,' (2012) 39 *Legal Issues of Economic Integration* 127, and R Howse and PI Levy, supra note 40, at 357.

> 'mandatory standard': 'standard the application of which is made compulsory by virtue of a general law *or exclusive reference* in a regulation' [emphasis added].

Allowance made for the systematic differences between the definitions employed in the ISO/IEC Guide and those in the TBT Agreement, the logic seems clear.[50] Regardless of whether the requirement of construction according to the 'acknowledged rules if technology' or to 'acceptable good standards of design' itself is 'mandatory', the WTO judiciary will classify as a 'technical regulation' any exclusive reference to a standard that makes that standard the only way to meet that requirement. Conversely, one would assume, *any* legislation that provides that a requirement – even a mandatory requirement – may be met by alternative means to compliance with the referenced standards, will be classified as a 'standard' for purposes of the TBT Agreement. This result is, undeniably, at odds with the logic of Annex 1. But is has its advantages. First, it constitutes an all out attack on the use of exclusive references – and exclusive references to private standards are an unqualified evil that cannot be condemned enough. Second, it limits itself to scrutinising regulatory technique and does not delve into whether alternative means are *practically* useless, whether, *in practice*, consumers only buy products certified to the standards at issue or whether the standards body concerned, as a consequence, *in reality* regulates entry into the market. In other words, it stays way from the concept of 'de facto mandatory standards'.

The dissent, on the other hand, pays the price for its interpretative rigor maintaining the *effet utile* of Annex 1 when it recognises that the neat distinction between labelling requirements and requirements for market access may collapse when it becomes impossible, *in practice*, to access the market without the label. And so here we go:

> The 'dolphin-safe' label may be considered *de facto* mandatory in order to market tuna products in the United States, if doing otherwise becomes impossible, not because it would contradict a mandatory provision in the measures, but because it would be prevented by a factual situation that is sufficiently connected to the actions of the United States. Thus, this analysis is two-fold. First, the impossibility of marketing tuna products

[50] Under the TBT Agreement, of course, a 'mandatory standard' *is* a technical regulation. ISO/IEC Guide 2 distinguishes between standards and regulations on the nature of the document, not on whether the requirements contained in them are 'mandatory' as in the TBT Agreement. See also the *Explanatory note* to Annex 1.2 ('Standards as defined by ISO/IEC Guide 2 may be mandatory or voluntary. For the purpose of this Agreement standards are defined as voluntary and technical regulations as mandatory documents.')

> in the United States without the 'dolphin-safe' label must be established. Second, such impossibility must arise from facts sufficiently connected to the US dolphin-safe provisions or to another governmental action of the United States.[51]

This is, frankly, a hopeless test. Paradoxically, it is even more hopeless in the circumstances of *Tuna II* itself than it would be in most other imaginable scenarios. The problem lies, again, in the fact that the dissent equates the 'dolphin-safe' label with any other label simply certifying compliance with product requirements. The 'factual situation' making it *practically* impossible to market tuna without the label that the dissent envisages would consist, presumably, of widespread consumer and retailer preferences. But it is hard to imagine that this collective yearning would be for tuna caught according to the precise prescriptions of the Secretary of Commerce rather than, generally, for 'dolphin-safe' tuna.[52] In that sense, the lone panellist misses the point completely: the problem in *Tuna II* lies not in the way the 'dolphin-safe' label *shapes* (discriminatory) consumer preferences but in the (discriminatory) way it allows operators to *respond* to consumer preferences.

But even in circumstances such as those in *Fra.bo*, the test soon runs into trouble. Where, *in practice*, consumers buy only DVGW-certified products, the scheme would pass the first prong of the test. In the second prong, then, the question is whether 'what consumers want' is causally related to the regulatory scheme at issue or to 'another government action'. This is virtually impossible to determine in the best of circumstances. Standards are deeply embedded in the social and legal fabric of markets: they are used in tort law and general product safety law, in public works contracts and supply contracts, in insurance policies and even in other standards. Any attempt to disentangle a causal relationship between a standard's adoption 'in the marketplace' on the one hand and regulatory or judicial action on the other is inevitably going to be arbitrary. It surely helps for a standard to be recognised by a piece of regulation. But it is equally clear that regulators are not going to recognise a particular standard from a particular standards body unless they are widely recognised and used in the marketplace. Standards are bound to be attractive to manufacturers when they are recognised by courts in product liability cases as meeting

[51] Panel Report, *Tuna II*, above n 11, para 1.175.

[52] It is this point that distinguishes the 'Dolphin-safe' label from a 'self-justifying trade barrier'. See Panel Report, *EC – Sardines*, WT/DS231/R, adopted as modified by the Appellate Body 23 October 2002, para 7.127.

appropriate legal standards of 'good engineering practice' or 'the state of the art'. But it is equally clear that courts would not recognise standards in these cases unless they are widely used and respected.

Even if all of this could be accomplished, the dissent's approach would still ask the wrong questions of the wrong people, ultimately in much the same way as the Court of Justice does. The government is held responsible if and insofar as it can be said to create a 'factual situation' in which marketing a product without complying with the standard is 'impossible'. But even if it is conceded that such a situation is possible, it is clear that there are situations possible where a private standards body, *in reality*, holds the power to regulate entry to a market without this being caused by governmental action. Otherwise, the second prong of the test would be useless. But if that much is true, it is obvious that the dissent has to take for granted that there is no necessary relation at all between the existence of a trade restriction and the outcome of the test. That hardly seems sensible trade law policy. The standards body, by the same token, is held responsible for its standards only if they become de facto mandatory in a way that is attributable to the government – in other words, in a way unconnected with the activities of the standards body itself. What exactly a private standards body then has to answer for is unclear.

In the unlikely case the second prong can be satisfactorily applied and leads to a positive answer, the dissent's approach casts its net *wider* than the Panel majority and would hold a de facto mandatory private standard such as the one in *Fra.bo* to be a 'technical regulation'. This result, then, is this: where the Majority would hold an exclusive reference to the requirements laid down in a particular private standard even in a non-binding regulatory scheme to constitute a 'technical regulation', the dissent would hold even an indicative reference to the requirements laid down a private standard in a binding regulatory scheme to be a 'technical regulation'. In the dissent's approach, it is the power of the standard on the market that creates the problem. The panel majority, on the other hand, at least puts responsibility where it belongs: with the government deciding to grant a monopoly to a private standards body.

D. Conclusion

If the regulatory technique of referring to standards is accepted as legitimate in principle and encouraged as useful for the elimination of unnecessarily discriminatory trade barriers, the legal imperative for public authorities must be to encourage and monitor the procedural integrity of private standards bodies at every level. What this requires is attention to the space *between* regulation and standards, on the rules and procedures of recognition and acceptance of private standards as suitable for use in public law. The legitimacy of private rule-making requires standards bodies to be kept at arm's length from the legal system, not pushed asunder in the realm of the purely private or drawn kicking and screaming into the sphere of public law. What the concept of 'de facto mandatory standards' in the two leading cases discussed here achieves is, of course, the latter: it seeks to establish criteria for deciding whether standards belong in one or the other. This makes for messy and dysfunctional jurisprudence.

10

Competition and cooperation in the market of voluntary sustainability standards

AXEL MARX AND JAN WOUTERS

A. Introduction

In the spring of 2013 the United Nations Forum on Sustainability Standards (UNFSS), a joint initiative by five UN agencies (Food and Agriculture Organization, UN Industrial Development Organization, International Trade Centre, UN Environment Programme and UN Conference on Trade and Development), was launched.[1] The UNFSS is a platform created to generate knowledge and information on voluntary sustainability standards (VSS) with a particular focus on their potential contribution to development. The initiative is a recognition of the increasing importance of VSS in international trade. As the website of the UNFSS notes, VSS, although legally non-binding, can or in some areas already have become a de facto market-entry hurdle.[2] Indeed, they potentially affect

The authors thank Colleen Carroll for her help in generating Figures 10.6 and 10.7 and Wout Van Doren and Anna-Luise Chané, Petros Mavroidis, Panos Delimatsis and Ulrich Hoffman for comments on an earlier draft. Part of the chapter originates from research conducted in the context of the PULSE and KLIMOS Research Platforms funded by VLIR-UOS/Acropolis.

[1] A note on the use of the terminology. Throughout the literature and policy community, one can find reference to many terms including 'private standards', 'voluntary standards', 'eco-certification', 'eco-labels', etc. In this chapter, we refer to 'voluntary sustainability standards' in line with UNFSS. We prefer not to use the term 'private standards' because some VSS are initiated by public bodies. In addition, many so-called private standards are based on international intergovernmental 'public' standards laid down in conventions and international agreements. For example, many private standards rely for the social part extensively on International Labor Organization (ILO) conventions. This mix of public and private blurs the distinction of the two. We do also no precisely define sustainability because it varies according to sector. For example, in agriculture very specific and refined definitions of sustainability have been proposed that do not necessarily apply to other sectors.

[2] See also J. Wouters and D. Geraets, 'Private Food Standards and the World Trade Organisation: Some Legal Considerations', (2012) 3 *World Trade Review* 11, 479–489.

international trade in different ways. First, they can act as new non-tariff barriers diminishing export opportunities for those who face constraints in complying with stringent standards. Second, they can generate relocation effects. In several manufacturing industries, factories, or capital sustaining them, are highly mobile and are constantly searching for locations with the lowest input costs. As Levi et al. note: 'When challenged by workers forming unions or pressured by MNCs [multi-national corporations] trying to induce compliance with private regulatory schemes, many factories will simply shut their doors without paying severance to workers and re-locate'.[3] Third, as Maertens and Swinnen argue, they might also create opportunities for trade, opening up high-standards markets for those who comply with the standards.[4]

The rise of transnational VSS has also been well documented in the academic literature. The literature includes references to the rise of civil regulation[5] or to the 'certification revolution'.[6] This chapter first maps this revolution. It shows that over the past few decades, the number of VSS proliferated, resulting in more than four hundred systems, of which many are operating transnationally. The first part of the chapter analyses the emergence and proliferation of VSS and then assesses the transnational diffusion of VSS. It ends with a discussion of the main drivers of VVS development. The second part focuses on the diversity of systems. The term 'sustainability standards' captures a variety of initiatives. The main distinguishing features are introduced and discussed and an empirical analysis of the standards' institutional diversity is presented. What emerges out of the first two parts is a market of many different VSS. This proliferation and diversification of VSS is creating problems such as a credibility gap, increased costs to gain certification by multiple VSS systems and a potential race to the bottom of VSS systems. These problems point to the need for increased cooperation between VSS systems. The third part of the chapter discusses these problems and looks at initiatives that aim to

[3] M. Levi, C. Adolph, D. Berliner, A. Erlich, A. Greenleaf, M. Lake and J. Noveck, *Aligning Rights and Interests: Why, When and How to Uphold Labor Standards. Background Paper for the World Development Report 2013* (2012), 22.

[4] M. Maertens and J. Swinnen, 'Private Standards, Global Food Supply Chains and the Implications for Developing Countries', in A. Marx, M. Maertens, J. Swinnen and J. Wouters (eds), *Global Governance and Private Standards. Interdisciplinary Perspectives* (2012), 153–171.

[5] D. Vogel, 'Private Global Business Regulation', (2008) *Annual Review of Political Science* 11, 261–282.

[6] M.E. Conroy, *Branded! How the Certification Revolution Is Transforming Global Corporations* (2007).

foster cooperation in the market of sustainability standards. In this context, the chapter focuses on two mechanisms specifically suited to address these problems: mutual recognition and meta-regulation. Mutual recognition is especially suited to address cost issues, while meta-regulation might contribute to leveling the playing field and preventing a race to the bottom. The chapter analyses these mechanisms and argues that cooperation in the market of VSS is currently more developed with regard to addressing the credibility gap (meta-regulation) than with regard to addressing the cost issue (mutual recognition). Throughout the chapter, we provide some examples and an empirical analysis on the basis of the Ecolabel Index Database.[7]

B. The emergence, proliferation and global diffusion of VSS

I. The emergence and proliferation of VSS

In 1924, Rudolf Steiner gave eight lectures at Koberwitz (Silesia) titled 'Spiritual Foundations for the Renewal of Agriculture'. He outlined the basic principles of biodynamic farming. An experimental circle of anthroposophical farmers implemented Steiner's principles in their daily farming practices.[8] In 1927, a co-operative was formed to market biodynamic produce. One year later, in 1928, the first standards for quality control were formulated and the first sustainability label, the Demeter symbol, was introduced. When a farm conformed to the standards, the products could use the Demeter label. Three years later almost a thousand biodynamic firms were certified in Germany. Voluntary sustainability certification was born. Seventy-five years later, in 1999, a new Demeter logo was developed, and Demeter was operating worldwide, certifying

[7] The Ecolabel Index database contains data on more than four hundred eco-labels operating worldwide. The database is currently the most exhaustive database on eco-labels that is available for research purposes. This database was chosen because it contains data on many eco-labels, which allows for a more comprehensive analysis of trends and variation. The realm of voluntary standards is larger than the population of eco-labels. However, eco-labels are a significant subset of the phenomenon of sustainability standards and hence are an interesting research population to empirically explore sustainability standards. The authors gratefully acknowledge access to the database for research purposes to Big Room Inc.

[8] J. Paull, 'Attending the First Organic Agriculture Course: Rudolf Steiner's Agriculture Course at Koberwitz, 1924', (2011) 2 *European Journal of Social Sciences* 21, 64–70; J. Paull, 'The Secrets of Koberwitz: The Diffusion of Rudolf Steiner's Agriculture Course and the Founding of Biodynamic Agriculture', (2011) 1 *Journal of Social Research & Policy* 2, 19–29.

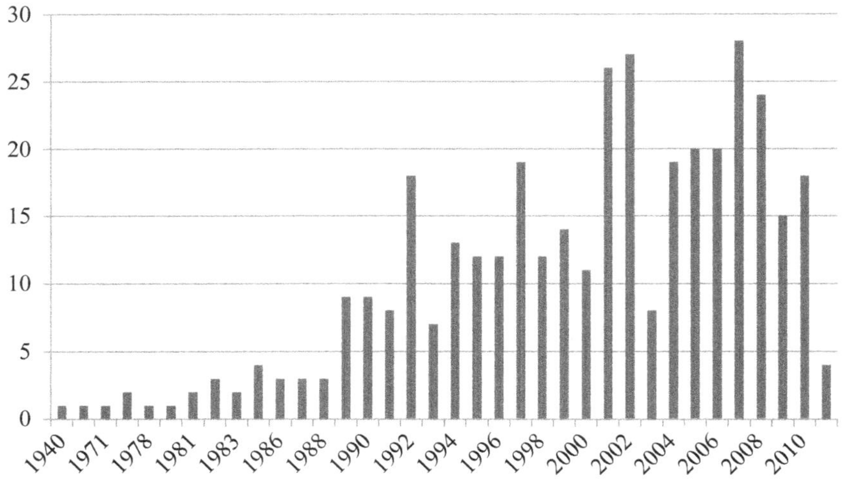

Figure 10.1. Number of VSS established per year
Source: Ecolabel Index Database (N = 380)

more than 3500 products including bread, wines, clothes and toilet articles. It took many years before new labels emerged. In 1954, Spiel Gut was established to promote sustainable toys. In the 1970s, some additional biological agricultural labels emerged, such as Bioland in 1971, California Certified Organic Farmers (1973) and the Soil Association organic standard (1973). In 1978, the first government-led voluntary sustainability standard emerged in Germany, Blue Angel. In the early 1980s, one could observe the emergence of three to four new initiatives on an annual basis (see Figure 10.1).

From 1989 onward, VSS were established more frequently. From the mid-1990s onward, one can observe a strong increase in the emergence of the number of VSS (Figure 10.1) and the survival/persistence of VSS (Figure 10.2). Indeed, as Figure 10.2 shows, one can observe not only an annual increase in the emergence of VSS in the past two decades but also that many VSS survive, persist and were still active in 2011.

Moreover, VSS are becoming active in many economic sectors, and several are active in more than one economic sector, which explains the fact that Figure 10.3 has 1218 observations. Most VSS are active in the food sector (145), followed by textiles (103), building products (101) and forest and paper products (80). In other words, for most products, several

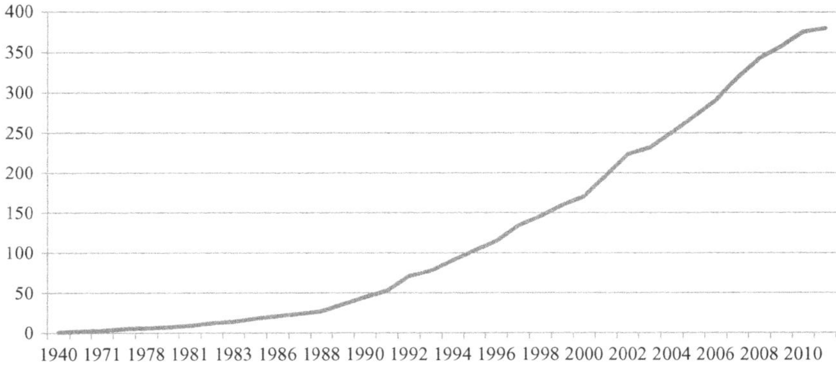

Figure 10.2. Number of cumulative VSS: 1940–2011
Source: Ecolabel Index Database (N = 380)

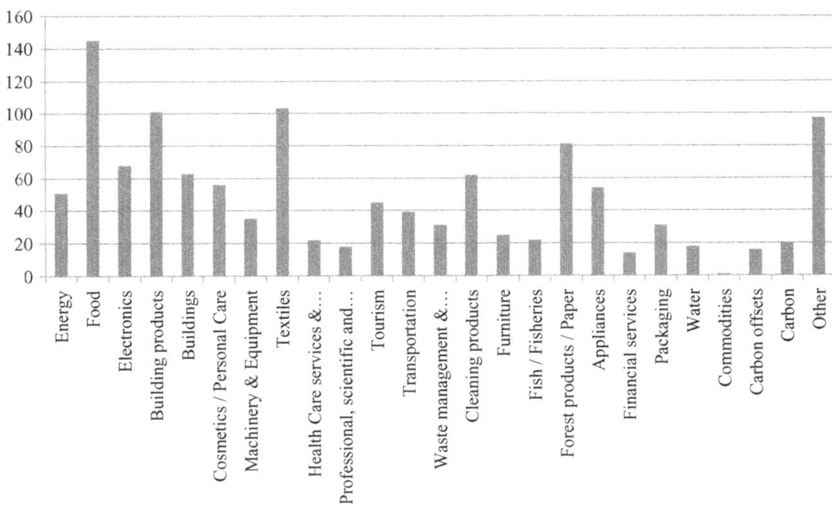

Figure 10.3. VSS distribution across economic sectors
Source: Ecolabel Index Database (N = 1218)

VSS are available from which to choose if one intends to apply for a certificate.

The various initiatives differ from one another (as discussed subsequently). Most initiatives, however, are characterised by the fact that an organisation defines social and ecological standards and that there is a procedure to check that products or production processes conform to these standards (i.e. conformity assessment). When products or production processes comply with the defined standards, a certificate is awarded that may or may not be used for external communication ('label').

This collection of voluntary standards comprises many initiatives. A few examples illustrate the diversity. Some initiatives are government-driven, such as the Organisation for Economic Co-operation and Development (OECD) Guidelines for Multinational Enterprises, which were most recently revised in 2011. These are recommendations, accepted by governments, made to multinational companies regarding socially responsible business. The Guidelines define voluntary principles and standards for responsible behavior of companies, relating to matters such as the environment, fighting corruption, labour relations and competition.

Other initiatives are driven by industry (associations) or non-governmental organisations (NGOs), such as Responsible Care in the chemical sector, the charter of employment rights of the clothing giant GAP, and the Clean Clothes Campaign, which strives to achieve better terms of employment in textile production plants in developing countries. Other initiatives result from a collaboration between various types of actors such as the International Labor Organization (ILO)'s Tripartite Declaration of Principles Concerning Multinational Enterprises and Social Policy.

Some initiatives that have generated significant academic interest are non-state multi-stakeholder sustainability certification initiatives. Some authors consider them 'one of the most innovative and startling institutional designs of the past 50 years'.[9] Among the most prominent and representative examples of these private regulatory initiatives are the Fairtrade Labelling Organisation (FLO), the Forest Stewardship Council (FSC), the Fair Labour Association (FLA), Social Accountability International (SAI) and the Marine Stewardship Council (MSC). The FLO, established in 1997, was founded to enable producers and workers in developing countries to evolve from a position of vulnerability to a position of economic security and self-sufficiency. The core of the system focuses on the concept of a fair price. The FSC, set up in 1993, is an international, multi-stakeholder, consensus-based sustainable forestry initiative. It guarantees that a wood or paper product has been made using material from a sustainably managed forest. SAI is an non-governmental, international, multi-stakeholder and nonprofit organisation with the mission of promoting

[9] B. Cashore, G. Auld and D. Newsom, *Governing through Markets. Forest Certification and the Emergence of Non-state Authority* (2004).

the rights of workers worldwide and improving working conditions by applying socially responsible standards. The SAI standards are based on internationally recognised guidelines, including various ILO conventions, the UN Convention on the Rights of the Child and the Universal Declaration of Human Rights. The FLO grew out of the Apparel Industry Partnership (AIP) initiative of the Clinton administration to protect workers worldwide and provide firms and consumers with the information they need to make informed purchasing decisions. The partnership was composed of apparel and footwear firms, human rights groups, labour and religious organisations, and consumer advocates. The FLA now represents a multi-stakeholder coalition of business enterprises, colleges and universities, and NGOs (human rights, labour, religious and consumer groups). Its mission is to combine the efforts of these stakeholders to promote adherence to international labour standards and improve labour rights worldwide. A final example is the MSC, which grew out of a partnership between the World Wildlife Federation and Unilever and aims to sustainably manage oceans and fisheries.

II. The transnational diffusion of VSS

Besides their exponential growth, VVS are also strongly diffused transnationally because they mainly aim to govern supply chains that operate across borders. Figure 10.4 assesses the global adoption of four leading VSS systems. These are FSC; PEFC (Program for the Endorsement of Forest Certification); SAI and GLOBALG.A.P. This selection is not representative but presents the use throughout the world of four of the most established and academically analysed schemes. These schemes operate on a global scale[10] and make information on their adoption publicly available. To identify trends, Figure 10.4 reports the results of two measurement waves with a gap of three years. The first mapping was conducted in October 2010 and a second mapping in October 2013. The countries in dark grey are based on the mapping of 2010 and indicate that the VSS in question is active in that particular country in 2010 and 2013; the countries in white are based on the mapping of 2013 and presents the countries where no VSS were active in 2010 but became active in 2013.

[10] See also K. W. Abbott and D. Snidal, 'Strengthening International Regulation Through Transnational New Governance: Overcoming the Orchestration Deficit', (2009) 2 *Vanderbilt Journal of Transnational Law* 42, 501–578.

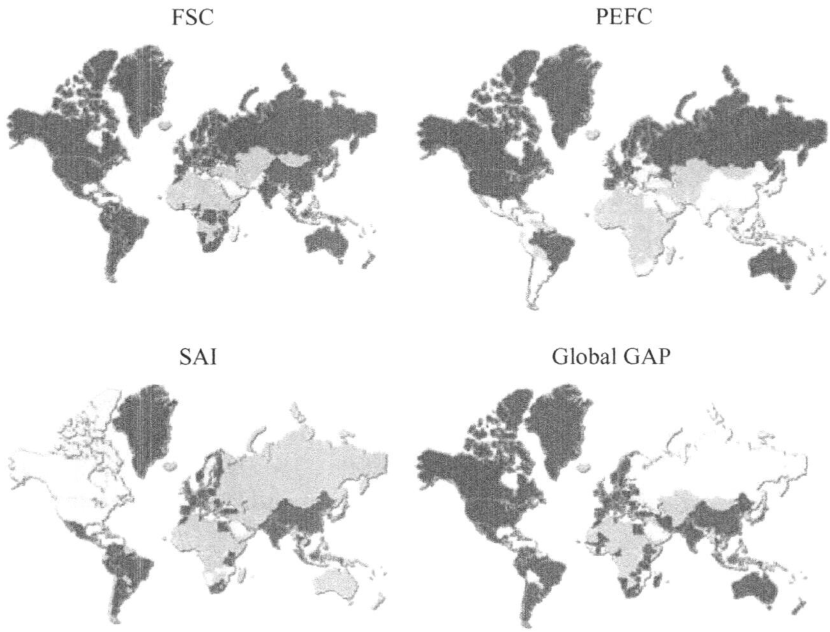

Figure 10.4. Transnational diffusion of selected schemes
Note: PEFC: annual reports accessed 13 October 2010–16 October 2013; SAI online database: accessed October 2010–16 October 2013; FSC: Facts and Figures: accessed October 2010–16 October 2013; GLOBALG.A.P.: countries with certified farms – newsletter 2008 – annual report 2012 16 October 2013. First wave of measurement in 2010 are shown in dark; second wave of measurement in 2013 are shown in white.

Figure 10.4 reveals several interesting facts. First, VSS continue expanding across the globe, with the only exception of Cuba, which lost its GLOBALG.A.P.-certified entities. Second, the truly global diffusion of VSS schemes is limited. In a truly global scenario each of the systems would operate in each country. Figure 10.4 shows the transnational diffusion of VSS schemes. In 2013, VSS schemes are present in 352 of 904 possible cases (38.9 percent). It should be noted that this constitutes a strong increase compared with 2010 when 254 of possible 904 countries (28 percent) had a VSS present. The figure also shows that there are several countries in which no VSS scheme is active, while in other countries, all four are active. A closer look reveals that VSS schemes are mainly adopted in developed countries. The twenty-eight EU Member States account for 103 of the 352 (29 percent) cases in which the selected VSS are active. This is a small decrease compared with 2010, when they took up 33 percent. In many least developed countries, no VSS systems are active. In addition, in countries where VSS can be observed, there is a significant variation in the number of entities certified. For SAI this varies from one facility being certified in several countries such as Tanzania, Nepal, Kenya or Jordan, to a few hundred in Italy. Similar figures hold for GLOBALG.A.P., which certifies only one or a few farms in Zimbabwe and Zambia and a few thousand farms in Germany and Greece. In general, the increase of countries in which VSS are active is mostly explained by the fact that the four VSS are getting a hold on the ground in more countries, albeit often with only one or two facilities certified. But once active within a country, the adoption of VSS by other facilities can go quickly. Indeed, in some cases the number of certified entities increased enormously over the three-year period. Moreover, overall figures of adoption of VSS on a global scale indicate a strong demand for VSS and a strong increase in VSS adoption. For example, in 2012, GLOBALG.A.P. had 123,115 facilities certified worldwide (Annual Report, 2012), compared with 112,576 in 2011 (Annual Report, 2011). The FSC has two types of certificates, those that certify forests (FM) and those that certify all operators in the supply chain (Chain-of-Custody, CoC). The number of forests (FM) certified grew from 1000 in January 2011 to 1225 in September 2013. The number of CoC-certified facilities grew from 18,000 in July 2010 to 26,886 in September 2013, an increase of almost 50 percent (FSC Facts and Figures, 2013). This rapid growth is driven by a strong demand for products that are certified according to sustainability standards. This demand is especially driven by large retailers. Marks & Spencer, for example, committed 'to become the world's most sustainable major retailer by 2015'.[11] This strategy, by 2020, will include building in environmental and social principles in all 2.7 billion individual products. In most cases, this will be done by using recognised external environmental or social certificates, i.e. using a VSS system.[12] Overall, the figures show a wide diffusion of VSS and a significant demand for VSS.

This adoption overview does not provide information on the nature of organisations (small versus large) that are participating in these schemes. From previous research,[13] we know that, in the case of FSC, for example,

[11] Marks & Spencer, *Our Plan A Commitments 2010–2015*, (2010), 3. [12] Ibid, 14.

[13] A. Marx and D. Cuypers (2010) 'Forest Certification as a Global Environmental Governance Tool. What Is the Macro-impact of the Forest Stewardship Council?', (2010) 4 *Regulation and Governance* 4, 408–434; A. Marx, E. Becault and J. Wouters, 'Private

mainly large-scale producers were participating in FSC. As a result, recent focus has turned to small-scale forest owners and wood producers to join FSC. These efforts to engage small-scale forest owners are mainly driven by FSC itself or technical cooperation units in the context of development cooperation. FSC is providing for group CoC certification that enables small-scale producers to join forces and financial and technical assistance to small forest owners for forest management certification. Large donors, like GiZ, are providing support to engage small-scale forest owners in VVS dynamics. These efforts might result in significant increased adoption by small-scale producers. Similar trends and developments might occur in other VSS. It should be noted that this scaling up should focus not only on providing assistance to small-scale owners but also on removing perverse incentives that generate market distortions in some markets (i.e. subsidies favouring large-scale industrialised production).

III. The Drivers of VSS Emergence

The rise and proliferation of VSS initiatives is driven by several interrelated factors. Some of these standards were developed to address certain policy gaps and failures, such as a lack of social and environmental protection or food safety,[14] fulfil consumer demand (ethical consumerism) or to protect key business interests such as reputation or liability. In the social and environmental field, NGO boycotts and reputation protection provided the main incentives to establish VVS.[15] Brand protection is a key issue for many leading companies. The proliferation, diversification and internationalisation of NGOs establishing themselves as genuine counter-powers through the use of media strategies, have forced companies to take civil society concerns into account.[16] Gereffi et al. noted, for example, that

Standards in Forestry: Assessing the Legitimacy and Effectiveness of the Forest Stewardship Council', in A. Marx et al. (eds), *Private Standards and Global Governance* (2012).

14 N. Hachez and J. Wouters, 'A Glimpse at the Democratic Legitimacy of Private Standards. Assessing the Public Accountability of GlobalG.A.P.', (2011) *Journal of International Economic Law* 14, 677–710; Levi et al., supra note 2.

15 G. Gereffi, R. Garcia-Johnson and E. Sasser, 'The NGO–Industrial Complex', (2001) *Foreign Policy* 125, 56–65; T. Bartley, 'Institutional Emergence in an Era of Globalisation: The Rise of Transnational Private Regulation of Labor and Environmental Conditions', (2007) 2 *American Journal of Sociology* 113, 297–351; A. Marx, 'Limits to Non-state Market Regulation: A Qualitative Comparative Analysis of the International Sport Footwear Industry and the Fair Labor Association', (2008) 2 *Regulation and Governance* 2, 253–273.

16 T. Bartley, 'Certifying Forests and Factories: States, Social Movements, and the Rise of Private Regulation in the Apparel and Forest Products Fields', (2003) 3 *Politics and Society*

protests and direct actions against brand-name retailers were only fifteen years old at the time their research was published but were regarded as extremely powerful tools to force retailers to take environmental, social and safety issues into consideration.[17] Joining a voluntary sustainability initiative can be a strategy to protect a firm's reputation, especially when independent external parties certify products and production processes and hence provide assurance on the efforts made by firms to manage for example social issues.[18]

In the food sector, a sector with many VSS, the globalisation of food markets and the increasing need to market fully safe food has also contributed to the proliferation of VSS.[19] It has become increasingly difficult for single governments to keep track of the range of products present on their domestic markets and to keep up with the assessment of all the risks associated therewith. In the recent past, public food safety regulations have not been able to prevent some major food crises.[20] Consequently, states have adopted new legislation providing for a system of liability for food chain actors in case of food safety problems. An early example of such regulatory design is the due diligence requirement contained in the UK Food Safety Act (UKFSA) of 1990. The act puts the liability mainly with retailers and provides that food retailers are not liable for non-compliance with food safety laws if they can demonstrate that they have taken all precautions in this regard. The possibility of being held liable for food safety issues prompted a response from the food industry, resulting in the development of many voluntary standards that acted as proof of due diligence.[21] Mirroring the UKFSA, EU Food Law Regulation (EC) No 178/2002 of 2002 also provides for a quite stringent responsibility

31, 433–464; D. O'Rourke, 'Outsourcing Regulation: Analysing Nongovernmental Systems of Labor Standards and Monitoring', (2003) 1 *Policy Studies Journal*, 31, 1–29.

17 Gereffi et al., supra note 15; see also Bartley, supra note 16.

18 Marx, supra note 15.

19 J. Wouters, A. Marx and N. Hachez, 'Private Standards, Global Governance and International Trade – The Case of Global Food Safety Governance' in A. Marx, M. Maertens, J. Swinnen and J. Wouters (eds), *Global Governance and Private Standards. Interdisciplinary Perspectives* (2012), 255–292.

20 C. Ansell and D. Vogel, *What's the Beef? The Contested Governance of European Food Safety* (2006).

21 Henson, Spencer (2008), 'The Role of Public and Private Standards in Regulating International Food Markets', *Journal of International Agricultural Trade and Development*, 4, 63–81; Food and Agriculture Organisation (2007), 'Private Standards in the United States and European Union Markets for Fruit and Vegetables – Implications for Developing Countries', FAO Commodity Studies No. 3. Available at ftp://ftp.fao.org/docrep/fao/010/a1245e/a1245e00.pdf

threshold for commercial actors involved with food products.[22] Given the important role and position of retailers in current-day supply chains,[23] this liability, which rests on them, results in pushing through 'voluntary' standards down the supply chain to all actors involved in the production process. Major retailer groups have particular influence as they are the 'gatekeepers' of the food markets and concentrate significant bargaining power. The World Bank refers in this context to the 'supermarket revolution'[24] and Hamilton et al. to the 'Market Makers'.[25]

From a more theoretical perspective, three major drivers can be identified that focus on pre-empting government regulation, addressing information asymmetries and direct targeting of firms by NGOs. Bartley has elaborated two possible explanations for the demand of VSS systems and identified two main, partially complementary, theoretical approaches.[26] A first theoretical approach, mainly rooted in political institutionalism, approaches the emergence of VSS from the perspective of social actors (NGOs) and sees VSS as an NGO-driven institutional capacity-building project to achieve their aims. Rather than being confrontational toward firms or trying to influence firm behaviour via lobbying governments to implement regulatory standards, NGOs are increasingly using a co-operative strategy toward firms, of which VSS are the most prominent example. As Bartley notes, 'private efforts have also been perceived by many NGOs as a way to bypass political roadblocks'.[27] A second theoretical approach, rooted in institutional economics, starts from the perspective of firms and highlights the importance of information asymmetries that arise via NGO actions such as 'naming and shaming' and consumer demand for ethical products. In this perspective, firms have an incentive to solve information problems by creating or joining VSS initiatives.[28] In addition, firms may have incentives to engage in standard-setting and adoption because it increases consumer confidence in product safety and quality and hence increases consumer demand; or because VSS can be

[22] J. Wouters, 'Private Law, Global Governance and the European Union' in A.L.M. Keirse and M.B.M. Loos (eds), *Alternative Ways to a New Ius Commune* (2012), 21–32.

[23] G. Hamilton, M. Petrovic and B. Senauer, *The Market Makers. How Retailers Are Reshaping the Global Economy* (2012).

[24] World Bank (2008), *Agriculture for Development. World Development Report 2008*. Washington, DC: World Bank.

[25] Hamilton et al., supra note 23.

[26] T. Bartley,'Certification as a Mode of Social Regulation' in D. Levi-Faur (ed), *Handbook on the Politics of Regulation* (2011) pp. 444–446.

[27] Ibid, 445. [28] Ibid.

used as a strategic tool to differentiate products, thereby creating market segmentation and softening competition.[29] A third theoretical perspective comes from economics which, following the work of Nobel laureate George Stigler, consider VSS as mechanisms that pre-empt government regulation.[30] Private firms may choose a level of standards that minimise their own costs before governments or international organisations set standards. McCluskey and Winfree argue, in addition, that an important advantage of voluntary over public mandatory standards is that the former are more flexible in response to changes in consumer tastes and preferences and to changes in technology.[31] The discussion of these drivers for the emergence and development of VSS shows that several actors have strong incentives to develop and work with VSS, which structurally embeds them in a bigger constellation of transnational regulatory governance.[32]

C. The institutional diversity of VSS

Notwithstanding the similarities in the process of granting sustainability certificates there is significant variation in how this process is organised and which actors and stakeholders are involved. Several authors have analysed this variation.[33] For the purpose of this chapter, we focus on four dimensions of variation: who sets standards and how, who assesses the standards *ex ante* and how is *ex post* verification facilitated.

The first dimension focuses on who sets the standards and how they are set. Here attention goes to the process of rule development and standard-setting, including the actors involved and decision-making processes. The International Social and Environmental Accreditation and Labelling (ISEAL) Alliance Code of Good Practice for Setting Social and Environmental Standards (2010) identifies at least five components that are important in the setting of standards: the identification of stakeholders

[29] J. Swinnen and T. Vandemoortele, 'Trade and the Political Economy of Food Standards', (2011) 2 *Journal of Agricultural Economics* 62, 259–280.

[30] Ibid; J. McCluskey, 'Public and Private Food Quality Standards: Recent Trends and Strategic Incentives' in J.F.M. Swinnen (ed), *Global Supply Chains, Standards and the Poor* (2007).

[31] J. McCluskey and J.A. Winfree, 'Pre-empting Public Regulation with Private Food Quality Standards', (2009) 4 *European Review of Agricultural Economics* 36, 525–539.

[32] Abbott and Sindal, supra note 10.

[33] Ibid; A. Marx, 'Varieties of Legitimacy: A Configurational Institutional Design Analysis of Eco-labels', (2013) 3 *Innovation: European Journal for Social Science Research* 26, 268–287.

via stakeholder mapping; the development of a strategy to proactively approach and involve the identified stakeholders; the bringing together of several major stakeholders on a more or less equal representative basis in a process of decision-making; the opening up of the decision-making process to all interested parties not initially identified in the first round of the stakeholder mapping; and the deployment of consensus-based decision-making to ensure that all interests are included. The Ecolabel Index database provides data on how many eco-labels have an open and consensus-based standard-setting and decision-making procedure. In this context, open and consensus-based decision-making is defined and operationalised following the American National Standards Institute's (ANSI; 2012) due process requirements for American National Standards. The ANSI code (2012) includes specific requirements for due process, meaning that any person (organisation, company, government agency, individual, etc.) with an interest has a right to participate in the process. These requirements stipulate inter alia that the process should be open, not be dominated by any single interest category, balance several interests and operate on the basis of a consensus vote.

Concerning *ex ante* verification, several authors have argued that a standard-setting process can be open and transparent but that the procedures to assess conformity with the rules can differ significantly, affecting the legitimacy of the systems.[34] Different verification mechanisms exist. In this context, a first classification was proposed by Gereffi et al.[35] It focussed on the number of independent actors involved in the certification process. First-party conformity assessment is characterised by the fact that a firm enforces compliance with standards defined by the respective firm. In second-party conformity assessment one party develops the standards, and a second party demands the standards. Conformity assessment with standards is conducted by either one of the two parties. In the case of third-party conformity assessment, the standards are developed by an independent organisation that also accredits organisations to carry out conformity assessments. Accreditation guarantees that the organisation that carries out the conformity assessment is competent to do so and operates independently. In this case, three independent parties are involved:

[34] Gereffi et al., supra note 15; M. Blair, C. Williams and L. Lin, 'The New Role for Assurance Services in Global Commerce', (2008) 2 *Journal of Corporation Law* 33, 325–360.

[35] Gereffi et al., supra note 15, 58ff; see also A. Marx, 'Global Governance and the Certification Revolution: Types, Trends and Challenges' in D. Levi-Faur, *Handbook on the Politics of Regulation* (2011); F. Van Waarden, 'Varieties of Private Market Regulation: Problems and Prospects', in D. Levi-Faur, *Handbook on the Politics of Regulation* (2011).

an organisation applying for a standard, a body setting the standards and an accredited party verifying conformity with the standards.

Concerning *ex post* verification two components have been highlighted as being relevant – namely, the provision of information to the public on the implementation of standards and the accessibility of systems to raise complaints or disputes. The most extensive treatment of transparency and accountability in this context can be found in Auld and Gulbrandsen.[36] They distinguish between two forms of transparency: procedural transparency and outcome transparency. Procedural transparency refers to openness of the decision-making process. This form of transparency is closely linked to the standard-setting process outlined earlier. Outcome transparency refers to openness about the outcome of the standard-setting process and is important in the context of accountability mechanisms where stakeholders use disclosed information to hold actors accountable to their commitments. In organisational terms, this is operationalised in two ways. First, VSS need to be transparent by providing accurate and objective information. Information disclosure procedures can inform different stakeholders on the implementation of standards. Publicly available information should include specific information about certification procedures, auditing reports, reports on violations and reports on corrective action plans. This allows stakeholders to assess whether the reported information mirrors real conditions. Second, a well-developed complaint procedure should be put in place to challenge decisions.[37]

The Ecolabel database allows us to explore empirically the degree to which these dimensions are present in VSS. Each dimension is either present (P) or absent (A) in a VSS. For a population of 426 VSS Figure 10.5 provides the results. The figure brings the four dimensions together in a property space of sixteen cells. Because a dimension is either present (1) or absent (0), four dimensions lead to sixteen possible combinations (2 to the power 4) of dimensions. Figure 10.5 maps the 426 VSS over these sixteen cells and shows that many cells contain existing VSS indicating a significant variation in their institutional design.

Figure 10.5 highlights five interesting characteristics of the population of VSS. First, there is significant variation in how VSS are designed because they spread out over several cells of the matrix. If they were uniform, they

[36] G. Auld and L. H. Gulbrandsen, 'Transparency in Nonstate Certification: Consequences for Accountability and Legitimacy', (2010) 3 *Global Environmental Politics* 10, 97–119.

[37] N. Ascoly and I. Zeldenrust, *Considering Complaint Mechanisms. An Important Tool for Code Monitoring and Verification* (2003).

				Open and Consensus-Based Standard-Setting			
				Present (P)		Absent (A)	
				Third Party		Third Party	
				P	A	P	A
Public Information Transparency	A	Dispute	A	155	173	7	3
			P	33	8	0	0
	P	Dispute	A	19	12	1	0
			P	12	3	0	0

Figure 10.5. The institutional diversity of VSS

would be concentrated in one or two cells of the property space. Second, most VSS have an open and consensus-based standard-setting procedure. Only 11 of 426 do not have this (two columns on the right). The latter can probably be explained by the fact that there are several institutional pressures to adopt open and consensus-based decision-making procedures. For example, several regulatory approaches toward sustainable public procurement, a key market for VSS, provide specific guidelines on this. Article 23 of EU Directive 2004/18/EC on public procurement procedures[38] and article 41 of the proposed new EU directive[39] lay out the criteria with which VSS have to comply to be recognised in the context of public procurement. Article 23 of Directive 2004/18/EC stipulates that the standards in VSS systems are 'adopted using a procedure in which all stakeholders, such as government bodies, consumers, manufacturers, distributors and environmental organisations can participate, and they are accessible to

[38] Directive (EC) 2004/18 of the European Parliament and of the Council of 31 March 2004 on the coordination of procedures for the award of public works contracts, public supply contracts and public service contracts [2004] OJ L 134/114.

[39] Proposal for a Directive of the European Parliament and of the Council on public procurement (December 2011). Available at http://eur-lex.europa.eu/LexUriServ/LexUriServ.do?uri=COM:2011:0896:FIN:EN:PDF

all interested parties'.[40] However, third, many VSS only have an open and consensus-based standard-setting procedure and no credible ex ante and ex post enforcement mechanisms. This is highlighted by the 173 VSS in the dark-grey cell. This cell contains all VSS for which open and consensus-based standard-setting procedures are present but the three other components (third-party conformity assessment, the provision of public reports and dispute settlement) are absent. Fourth, many systems (155) have open and consensus-based standard-setting procedures and third-party conformity assessment, but lack ex post verification tools, as is indicated by the black cell. Finally, only 12 VSS (light-grey cell) have all four components and hence an elaborated institutional design for developing standards, assessing conformity and providing mechanisms for ex post verification. In sum, one-third of the systems have no enforcement mechanisms, and only 3 percent have a fully elaborated standard-setting and enforcement mechanism in place. In-between, a significant number of systems have an open and consensus-based standard-setting procedure and well-developed third-party conformity assessment procedures. This diversity generates a clear credibility gap and several references can be found that VSS are pure 'green-washing'. One commentator even refers to the 'wild west of labelling'.[41]

D. Cooperation in the market of VSS

What emerges is a diverse and growing (both on the supply and demand side) market of VSS. This proliferation of initiatives leads to several (unintended) consequences. First, users (both consumers and businesses) of VSS are getting increasingly confused about the number of VSS and for what they stand. A survey with opinion makers in the field of sustainability standards concluded by the ISEAL Alliance in 2010 showed that 'the large number of standards has a series of implications: Overlaps between standards systems, confusion and difficulties in differentiating between them and a lack of consumer awareness and demand for specific certification schemes'.[42] Similar results were recently published in a

[40] See also Article 41 of the newly proposed directive.

[41] J. Entine, *Ecolabels – The Wild West of Labelling*. Ethical Corporation (2013). Available at http://www.ethicalcorp.com/environment/ecolabels-wild-west-labelling

[42] Iseal Alliance, *The ISEAL 100: A Survey of Thought Leader Views on Sustainability Standards* (2010), 15. Available at http://www.isealalliance.org/sites/default/files/ISEAL100.pdf

report by Seifert and Comas who surveyed several producers demanding VSS.[43] This results in a credibility and legitimacy gap on the side of the VSS systems. The well-developed systems increasingly need to distinguish themselves from the non-credible systems.

Second, as indicated, several retailers are demanding compliance with VSS as a prerequisite to buy certain products. However, retailers and other intermediaries differ on the preferences they hold with regard to VSS. Some, for example, will ask for wood products based on the FSC label, while others might prefer the PEFC label. Because they do not mutually recognise each other's systems (infra), producers or forest owners often need to apply for the two labels, thereby increasing the costs of getting certified. Many systems also do not allow for multiple labeling with regard to labels that in essence certify the same standards. As a result, sometimes different packaging needs to be developed, which adds to the increased costs.

Finally, competition between systems can lead to a race to the bottom of standards and enforcement mechanisms to increase market share. Little information is available on how competitive dynamics influence the stringency of systems, but the rise of many VSS in certain sectors indicates that some of these VSS aim to capture quick gains. Van Waarden and van Dalen show that in the sector of Halal certification several VSS were recently introduced that were not very stringent and basically granted certificates with hardly any conformity assessment.[44]

These concerns raise the issue of how (more) cooperation in the world of VSS can be achieved. Cooperation in this context refers to cooperative efforts to harmonise practices and reduce confusion and information gaps. Two possible cooperation mechanisms are discussed next: equivalence or mutual recognition, and meta-regulation. Equivalence and mutual recognition mechanisms are well-suited to address the issue of increased costs following multiple certification requirements. Meta-regulation is best suited to address the credibility/legitimacy gap and prevent a race to the bottom of VSS systems. Given the competitive and dynamic nature of the VSS market, it is hypothesised that equivalence or

[43] R. Seifert and J. Comas, *Have Ecolabels Had Their Day? The Truth behind Sustainability Labels from the People Who Integrate Them* (2012). Available at http://www.imd.org/research/challenges/sustainability-ecolabels-effectiveness-ralf-seifert-joana-comas.cfm

[44] F. Van Waarden and R. van Dalen,'Halal and the Moral Construction of Quality. How Religious Norms Turn a Mass Product into a Singularity', pp. 197–222 in Jens Beckert and Christine Musselin (eds), *Constructing Quality. The Classification of Goods in the Economy*, (2013). Oxford: Oxford University Press.

mutual recognition currently holds little potential to act as a cooperative mechanism to reduce costs in the market, while meta-regulation might generate more interest to address the credibility gap.

I. Equivalence or mutual recognition

One way to cooperate within the market of VSS is via a system of equivalence recognition whereby one voluntary sustainability standard-setting organisation formally recognises the criteria and requirements of another. If such an agreement exists, entities that have been awarded the label in one program may therefore be able to register in another without undertaking certification and verification procedures again. This might significantly reduce costs for those applying for certificates. This system of recognition is directional and not necessarily symmetrical. One standard might recognise another without the other one recognising the former. In other words, a VSS system can recognise another system or be recognised by another system. When there is symmetry or reciprocity, one can refer to mutual recognition. For the field of eco-labels, we can analyse the degree and nature of equivalence or mutual recognition because data are available.

In a highly coordinated and cooperative field, many VSS would consider other VSS as equivalent. The field of eco-labels, however, is highly fragmented and isolated. Figure 10.6 maps which VSS recognise other VSS as equivalent. The direction of the arrow indicates which VSS is considering another VSS as equivalent. The figure shows that the network, on the basis of the data available in the Eco-label Index database, is largely unconnected and that it has a very low density, a measure of the degree to which members of a network are connected. The circle shows the more than three hundred VSS that are not recognised by another VSS and also do not recognise other VSS. The overall density, which is the proportion of equivalence recognitions on the potential total of recognitions, is 0.005, which is extremely low. This means that, even though there are a large number of VSS, very few of them utilise the system of mutual recognition. Although 426 VSS initiatives are included in the dataset, only 90 instances are registered where one system recognises the system employed by another VSS. Moreover, only 64 of the 426 VSS are recognised by others; this indicates that certain VSS are recognised by multiple VSS. Furthermore, only very few VSS apply mutual recognition. Mutual recognition occurs only seven times among the total of ninety unique relationships (arrow on both sides of the line).

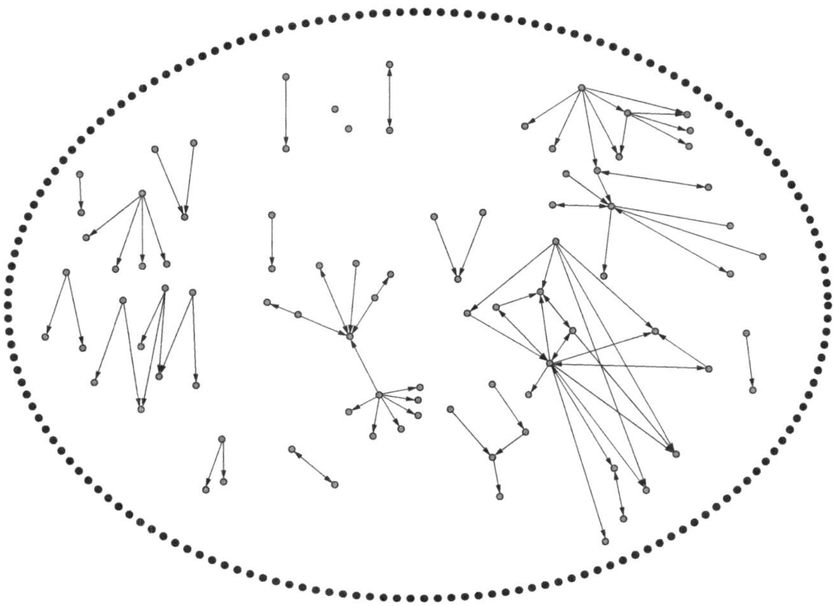

Figure 10.6. Network graph of equivalence recognition of 426 VSS

When we zoom in on the ones that are connected (see Figure 10.7), we can identify some small clusters of VSS that play a role in equivalence recognition. The magnitude of the dots (a VSS system) indicates to how many other VSS systems they are connected. The bigger the dot, the more VSS systems recognise that particular VSS or are recognised by that VSS. Figure 10.7 presents the VSS initiatives that recognise or are recognised by other VSS initiatives. An arrowed line connects two VSS initiatives in the diagram. The system at the start of the line is the labeling system that 'recognises' and the VSS at the end of the line – indicated by an arrow pointing to its name – is the initiative 'recognised'.

The figure shows that five VSS play a central role in equivalence recognition of which Environmental Choice New Zealand is the most central. It is recognised by several other VSS and recognises several other VSS. Environmental Choice New Zealand is the official environmental label of New Zealand owned and endorsed by the government of New Zealand. Other central VVS are PEFC, Greener Product Certification Seal, Energy Star USA and Bio Hotels. Energy Star USA is a government-backed program that is the national symbol for energy efficiency. Bio Hotels is an organic

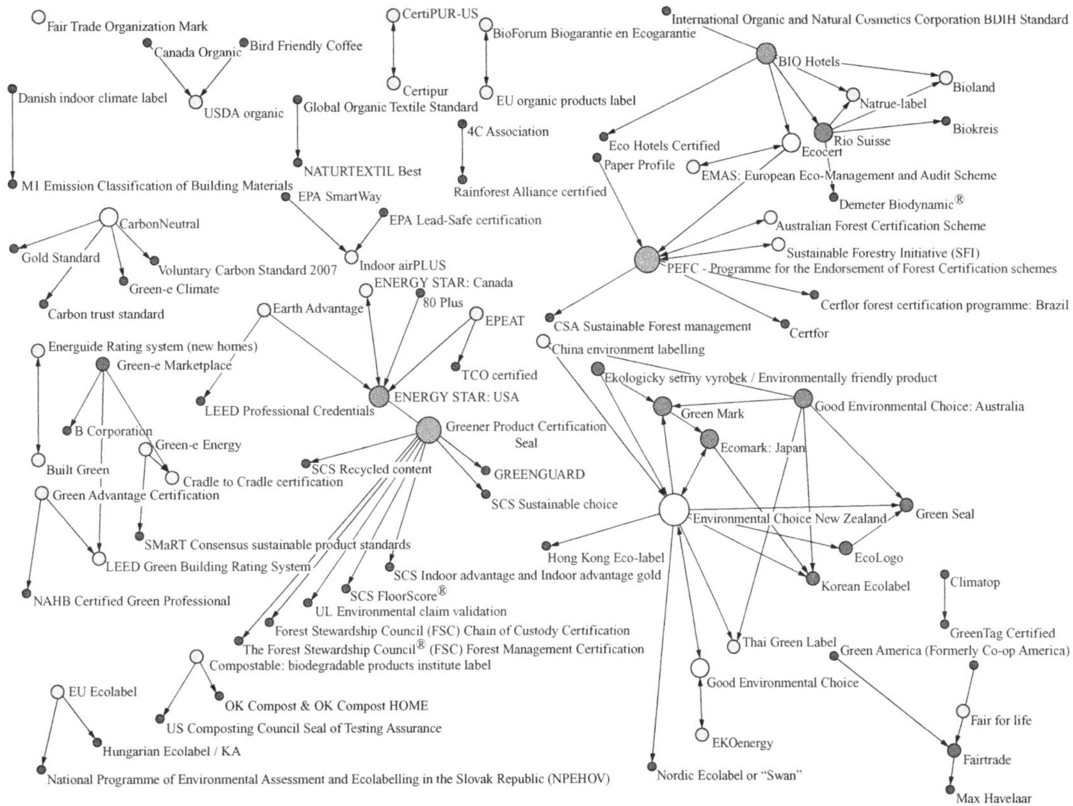

Figure 10.7. Network graph of the VSS with equivalence recognition

certification system and identifies hotels that use only organic-certified products that meet the standard of some other VSS such as Bioland or similar VSS. In case of PECF equivalence, recognition mostly refers to other forest certification systems, which previously operated independently but now operate under the global umbrella of PEFC and are in the process of merging into PEFC. In total eleven VSS are recognised by two other VSS, and forty-eight VSS are recognised by one other VSS. The limited equivalence recognition that takes place in the market of VSS shows that this is mainly driven by government-led initiatives or incentives to merge and does not apply to the many private VVS systems operating. The latter remain highly competitive and operate in an uncoordinated fashion.

This highly fragmented picture corresponds to results from case studies that show limited cooperation and coordination between several similar VSS. For example, in 2005, an initiative was launched in relation to labour rights and workers' rights with a view to organising cooperation between the Clean Clothes Campaign, the Ethical Trading Initiative, the Fair Labor Association, the Fair Wear Foundation, Social Accountability International and the Worker Rights Consortium in a pilot project in Turkey. The 'Jo-In' pilot project (Joint Initiative) aimed to promote closer cooperation between the various initiatives so as to increase the impact of the systems. This pilot project did not lead to closer cooperation, and the various systems continue to operate separately.[45] The obstacles for closer cooperation related to several issues including the stringency of standards, monitoring procedures and sanctioning. As Fransen notes, when analysing private labour governance systems, no real steps are taken to address competition between systems and have convergence on standards and enforcement.[46] In his comparative study, Fransen shows that the prominent systems for private labour governance differ significantly in how they organise auditing in terms of stringency, who does the audits, and so forth.[47] Similar efforts for cooperation, and their subsequent failure, have been reported in other sectors such as the cut-flowers sector, which is highly competitive.[48]

[45] L. Fransen, 'Why Do Private Governance Organisations Not Converge? A Political-Institutional Analysis of Transnational Labor Standards Regulation', (2011) 2 *Governance*, 359–387; A. Marx, 'Global Governance and the Certification Revolution: Types, Trends and Challenges in D. Levi-Faur, *Handbook on the Politics of Regulation* (2011).

[46] Ibid.

[47] Ibid; see also L. Fransen, *Corporate Social Responsibility and Global Labor Standards: Firms and Activists in the Making of Private Regulation* (2012).

[48] L. Riisgaard, 'Global Value Chains, Labor Organisation and Private Social Standards: Lessons from East African Cut Flower Industries', (2009) 2 *World Development*, 37,

Even in areas of VSS that have been long established and have significant convergence in standards over time no full mutual recognition materialises. A case in point is the organic agriculture and organic certification. David Winickoff and Kendra Klein compare transatlantic organic food labelling and make the case that organic cultures converged transatlantically to recognise similar process-based and precaution-based standards.[49] However, despite 'relative convergence', relatively small but intransigent peculiarities remain, which result in some important differences between the US and EU standards with three in particular creating obstacles to equivalence: the approach to the farm unit, the lists of prohibited and allowed inputs and criteria for livestock operations. Hence, notwithstanding a great potential for standards harmonisation in organic, equivalence negotiations have foundered on these issues. The complexity of international accreditation and certification that has emerged in lieu of equivalence adversely affects the development of organic markets and trade according to Winickoff and Klein. As a result, cooperative measures that aim to reduce costs via mutual recognition are very little developed, and there are only few signs of significant incentives to do so in the near future, especially with regard to competing private systems. The latter is probably explained by the fierce competition between the schemes and the reluctance of public authorities to intervene and structure the market of VSS.

II. *Meta-regulation of VSS*

A second mechanism to foster cooperation in the VSS market that is recently receiving increasing attention is meta-regulation[50] or the development of meta-standards.[51] There is no consensus on what meta-regulation is,[52] but, as Verbruggen and Havinga point out, the concept

326–40; L. Riisgaard, 'How the Market for Standards Shapes Competition in the Market for Goods: Sustainability Standards in the Cut Flower Industry', DIIS Working Paper no. 2009/07 (2009).

[49] D. Winickoff and K. Klein, 'Food Labels and the Environment: Towards Harmonisation of EU and US Organic Standards' in D. Vogel and J. Swinnen, *Transatlantic Regulatory Cooperation. The Shifting Roles of the EU, the US and California* (2011), 229–248.

[50] P. Verbruggen and T. Havinga, 'The Meta-regulatory Turn in Transnational Private Regulation', paper presented at the international conference 'The Ways and Means of Transnational Private Regulation', 11–12 October 2013, Dublin, Ireland.

[51] A. Loconto and E. Fouilleux, 'Politics of Private Regulation: ISEAL and the Shaping of Transnational Sustainability Governance', (2013) *Regulation and Governance*, 8, 2, pp. 166–185.

[52] C. Coglianese, C and E. Mendelson, 'Meta-regulation and Self-regulation' in M. Cave, R. Baldwin and M. Lodge (eds), *The Oxford Handbook of Regulation* (2010), 146–168, 147.

involves the use of control and steering mechanisms between different regulatory actors.[53] Several authors have argued that due to the proliferation and diversification of systems, there is an increasing need for the independent certification of the 'certifiers' by a multilateral organisation or by a private organisation. The idea is that meta-regulators determine the rules of the game and provide for minimum requirements to which VSS should adhere. This form of meta-regulation can address the legitimacy gap and the potential problem of a race to the bottom.[54]

A number of authors argue in favour of multilateral organisations taking on a pro-active role in this field.[55] Abbott and Snidal, in this context, refer to the orchestrating role international organisations can play in transnational regulatory governance.[56] To date, however, no initiatives have been launched at the multilateral level with a view to achieving this objective. The UNFSS is the first multilateral initiative addressing sustainability standards. The UNFSS could potentially play this role, but it is currently not in its mandate and its institutional structure. The UNFSS mainly addresses the potential trade or development obstacles VSS may create, with a particular emphasis on their impact on small-scale producers and less developed countries. It will mainly focus on analysing trade effects.

Within the realm of private VSS, some meta-regulatory systems are emerging. Within the food sector meta-regulatory initiatives have been taken such as the Global Food Safety Initiative (GFSI). The GFSI aims to harmonise food safety rules between different initiatives to reduce costs for suppliers. GFSI is a standard for other VVS in the food sector and is mainly focusing on the harmonisation of the different retail specific standards.[57] One of the most developed meta-regulatory systems in the realm of VSS is the International Social and Environmental Accreditation and Labelling (ISEAL) Alliance, which is discussed in the next section.[58]

[53] Verbruggen and Havinga, supra note 50.

[54] S. Bernstein, 'Legitimacy in Intergovernmental and Non-state Global Governance', (2011) 1 *Review of International Political Economy* 18, 17–51.

[55] C. Sabel, A. Fung and D. O'Rourke, *Ratcheting Labor Standards. Regulation for Continuous Improvement in the Global Workplace* (2000); van Waarden, F. (2009), 'Governing global commons: public–private-protection of fish and forests', in J. Swinnen, D. Vogel, A. Marx, H. Riss and J. Wouters (eds), *Handling Global Challenges*, Leuven: Leuven Centre for Global Governance Studies, pp. 140–175.

[56] Abbott and Sindal, supra note 10.

[57] P. Verbruggen and T. Havinga, 'The Meta-regulatory Turn in Transnational Private Regulation', paper presented at the international conference 'The Ways and Means of Transnational Private Regulation', 11–12 October 2013, Dublin, Ireland.

[58] Bernstein, supra note 54; Loconto and Fouilleux, supra note 51.

III. *Meta-regulation: the case of the ISEAL Alliance*

The ISEAL Alliance was established in 2000 by eight existing VSS including FLO, FSC and SAI. It is an umbrella members' organisation that codifies best practices in the design and implementation of certification initiatives. ISEAL has drawn up a Code of Good Practice for Setting Social and Environmental Standards – that is, a standard for organisations that set standards, against which existing initiatives can be tested. In addition, members must embrace the credibility principles of the ISEAL Alliance, which include sustainability, continuous improvement, relevance, rigour, engagement, impartiality, transparency, accessibility, truthfulness and efficiency. This set of credibility principles illustrates the importance that ISEAL, as a meta-regulator, attaches to addressing the legitimacy/credibility gap.

Full members of the ISEAL Alliance are VSS that have demonstrated a high level of compliance with different codes of good practice that have been developed by ISEAL. Granting of membership is based on independent evaluation and peer-review processes. Currently the ISEAL Alliance has fourteen full members, coming from eight members.

In addition, ISEAL Alliance also has a system of associate members, which are VSS that are on a pathway to becoming full members and have met some baseline requirements; and subscribers, which are mainly VSS that align with ISEAL's vision. In total, ISEAL now has seven associate members and approximately forty VSS subscribers such as GLOBALG.A.P. The interest of several VSS in ISEAL Alliance is becoming evident and its role as a meta-regulator is increasing.

According to Loconto and Fouilleux, ISEAL Alliance attempts to develop an organisational field that legitimises specific actors, tools and practices in the area of VSS.[59] They aim to create distinctions in the VSS market with the aim of distinguishing between credible and legitimate VSS on the one hand and illegitimate ones on the other hand. As Reinecke et al. point out, ISEAL's coordinating role makes it 'a market watchdog protecting the legitimacy of the sustainability standards movement.'[60] As a result, several VSS systems are interested in becoming a member to distinguish them from others. On the one hand, this creates more diversity in the market of VSS between members and non-members. On the other

[59] Loconto and Foullieuw, supra note 51.

[60] J. Reinecke, S. Manning and O. von Hagen O, 'The Emergence of a Standards Market: Multiplicity of Sustainability Standards in the Global Coffee Industry', (2012) *Organisation Studies* 33, 791–814, 804.

hand, it creates more harmonisation and coordination for the sub-group that is a member.

One possible explanation why this form of meta-regulation generates interest is that it creates status hierarchies between VSS systems in the VSS market. Being an ISEAL member generates a status that VSS systems can use in the market. Following the work of Joel Podolny on status-based models of market competition, one could hypothesise that VSS systems seek status, via ISEAL Alliance, in a market characterised with significant confusion on the quality of the services they are delivering.[61] High status, as Joel Podolny and many others applying his status-based model show, will lead to competitive advantages, which is of key importance in a quickly expanding market of VSS and related CSR initiatives. Hence, the interest in meta-regulation, which establishes high minimum standards and procedures, might be driven by self-interest to gain market share. This status-seeking dynamic is also supported by studies that focus on competition between systems. Contrary to some expectations, they do not necessarily lead to a race to the bottom of standards and procedures. Christine Overdevest, for example, showed how FSC is playing a pivotal role in ratcheting up forest standards of rival, industry-sponsored certification initiatives in the forest sector.[62]

E. Conclusion

Voluntary sustainability standards have become a prominent instrument to govern supply chains according to a set of social, environmental and safety standards. Their rise is remarkable, and they have generated significant attention in terms of their potential both to 'regulate' markets and to have an impact on trade.[63] This chapter engaged in the debate on the development of these systems from the perspective of a whole group of VSS, contrary to focusing on a single VSS. It first mapped the emergence, proliferation and global diffusion of VSS. This part showed an exponential growth of these systems as well as a significant, and fast-growing, global

[61] J. Podolny, 'A Status-Based Model of Market Competition', (1993) 4 *American Journal of Sociology* 98, 829–872; J. Poldony, J., *Status Signals: A Sociological Study of Market Competition* (2005).

[62] C. Overdevest, 'Comparing Forest Certification Schemes: The Case of Ratcheting Standards in the Forest Sector', (2010) *Socio-Economic Review* 8, 47–76.

[63] A. Marx, M. Maertens, J. Swinnen and J. Wouters, 'Private Standards: The Emergence of a Global Governance Tool?' in A. Marx, M. Maertens, J. Swinnen and J. Wouters (eds), *Global Governance and Private Standards. Interdisciplinary Perspectives* (2012).

diffusion. In the second part, we argued that VSS significantly differ in their design in how they set and enforce standards. This institutional diversity analysis is important in the current debates on the credibility of these systems.

What emerges is a diverse and growing market (both on the supply and demand sides) of voluntary sustainability standards. This proliferation and diversification of VSS is creating market problems that we identified in the third part of the chapter. These problems concern a credibility gap, increased costs to gain certification by multiple VSS-systems and a potential race to the bottom of VSS systems. Next, the chapter assessed two mechanisms that can facilitate more cooperation to address these market problems: mutual recognition and meta-regulation. It was argued that mutual recognition is especially suited to address cost issues, whereas meta-regulation might contribute to leveling the playing field within the market of VSS. The chapter analysed these mechanisms and showed that cooperation in the market of VSS is currently more developed with regard to addressing the credibility gap (meta-regulation) than the cost issue (mutual recognition).

11

Harmonizing non-tariff measures

Always a blessing?

OLIVIER CADOT, ANNE-CÉLIA DISDIER AND
LIONEL FONTAGNÉ

A. Introduction

Non-tariff measures (NTMs) are increasingly prevalent in world trade. While traditional command-and-control NTMs such as quotas, prohibitions and the like have been receded, being effectively disciplined by World Trade Organization (WTO) agreements, technical and sanitary regulations have proliferated.

These regulations are often driven by non-trade concerns such as the preservation of the environment or human health. However, they can also be hijacked by special interest to serve as surrogate protectionist instruments, and when poorly designed, they can damage trade and economic efficiency. For instance, in 2009, faced with a flood of steel beams from China and pressed to protect its ailing steel giant before privatization, Indonesia chose to adopt an idiosyncratic steel standard covering both "long" products for the construction sector and "flat" products for the automobile sector. Poorly designed, the regulation infuriated Japanese automakers, threatening a sector that provided hundreds

We thank Scott Baier, Céline Carrère, Aaditya Mattoo and participants at the TILEC Conference, CEPII-PSE Workshop on Trade Policy Analysis, Paris Trade Seminar, World Bank Seminar, ETSG 2010 Conference, Canadian Economics Association 2011 Conference, and EEA-ESEM 2011 Conference for helpful suggestions. Cadot gratefully acknowledges support from France's Agence Nationale de la Recherche under program ANR-10-LABX-14-0.1, and from Switzerland's NCCR under Work Package 6. Disdier gratefully acknowledges support received from the ANR-12-JSH1-0002-01 project and from the European Union's Seventh Framework Programme under Grant Agreement no. 290693 FOODSECURE. The authors only are responsible for any omissions or deficiencies. Neither the FOODSECURE project and any of its partner organizations nor any organization of the European Union or European Commission are accountable for the content of the chapter.

of thousands of manufacturing jobs, while less damaging instruments, such as trade remedies, were available and could have been used instead at lower costs. Because NTMs are only weakly covered by WTO disciplines, tit-for-tat regulatory wars to export unemployment may easily become reality among emerging countries where manufacturing jobs are key political considerations.

In industrial countries, manufacturing jobs have been dwindling as a proportion of the workforce and trade wars to retain them may not be so likely. The danger for the multilateral trading system is rather that risk-averse regulatory system, over-reacting to consumer scares, impose made-to-measure regulations that seriously harm trade in order to assuage real or imaginary fears among consumers. For instance, the United States (US) runs a severely restrictive system of "positive certification" for exporting countries in the fruit and vegetable sector (Jouanjean 2011), and recent research has shown that food alerts at both the US and the European Union (EU) borders have persistent effects and negative externalities on the sales of similar products (Jaud, Cadot and Suwa-Eisenmann 2013). The political economy here is essentially regulatory agencies (and politicians) pushing over the risk of legal liability to third parties (foreign producers) by imposing super-safe regulations.

Whether it be for political-economy reasons or risk aversion, NTMs have the potential to disrupt the multilateral trading system, and this potential is compounded by disciplines that, although already viewed by public opinion in some countries as overly stringent, are weaker than in the case of pure trade instruments such as tariffs.

As a result, NTM streamlining and harmonization has been taken up as a key part of the "deep-integration" agenda of many regional trade agreements. Whereas South-South agreements have typically attempted to reduce non-tariff barriers (NTBs) through an identify-negotiate-eliminate approach, North-South agreements have sometimes involved deep integration through harmonization clauses for either technical regulations or conformity-assessment procedures, or both. Harmonization might be viewed as a shortcut way of preventing regulatory fragmentation and the proliferation of idiosyncratic standards such as Indonesia's steel one, as harmonization of regulations with those of a large bloc, for a Southern partner, ensures access to a large and presumably competitive market and may, at least in some cases, facilitate the adoption of "best-practice" regulations while at the same time obtaining technical assistance.

However, for Southern countries, harmonization on stiff Northern standards can be a double-edged sword. For instance, when Mauritius aligned its Dangerous Chemicals Act on EU standards, it made it impossible for its paint manufacturers to use pigments that were outlawed in the EU. As a result, they had to turn to more expensive ones, raising the price of their paints, which did not actually improve their market access in the EU (they were too small to gain access to supermarket shelves) while pricing them out of markets in continental Africa where those standards would confer neither regulatory acceptance nor consumer recognition. We show in this chapter that the "Mauritius paint syndrome" is actually an empirical regularity, so strong as to show up in an aggregate gravity equation. That is, North-South harmonization tends to lock Southern partners in hub-and-spoke trade patterns with their Northern partners at the expense of South-South trade.

The policy implications of our analysis are not straightforward. In the face of the danger of regulatory fragmentation and use of NTBs as surrogate protectionist instruments in bad times – at least in emerging countries where manufacturing jobs are a central policy concern – harmonization may have a strong advantage as a commitment device. However, the dangers of "premature harmonization" on stiff Northern standards should be taken into account. The only clear way out of this trade-off would be the generalization of worldwide standards, but then it is not clear that the societal choices embodied in technical regulations should necessarily be the same everywhere.

The rest of this chapter is organized as follows. Section B discussed the policy issues surrounding NTMs, including transparency, capture, and existing WTO disciplines. Section C presents our results on the trade-diversion effects generated by North-South trade agreements. Section D concludes the chapter.

B. Non-tariff measures: the broader issues

I. Definitions and sources

NTMs are all measures other than tariffs that affect imported or exported products. So-called technical measures include sanitary and phytosanitary (SPS) measures as well as technical regulations. Nontechnical measures include, inter alia, quantitative restrictions (prohibitions, quotas and voluntary export restraints), price measures, forced channels and anticompetitive measures (such as forced used of national operators for transit operations), and a host of other regulatory measures. The 2012 Multi-Agency Support Team (MAST)/WTO classification is shown in Figure 11.1.

Import measures	Technical measures	A	Sanitary and Phytosanitary (SPS) measures
		B	Non-sanitary technical regulations (TBT)
		C	Customs formalities, including procedures for import
	Non-technical measures	D	Contingent protection (Antidumping, CD, safeguards)
		E	QR and non-automatic licensing
		F	Price-control measures, including additional taxes & surcharges
		G	Finance measures
		H	Measures affecting competition
		I	Trade-related investment measures (TRIMs)
		J	Distribution restrictions
		K	Restrictions on post-sales services
		L	Subsidies (excluding export subsidies)
		M	Government procurement restrictions
		N	Intellectual property
		O	Rules of origin
		P	Export-related measures

Figure 11.1. The 2012 MAST/WTO classification of NTMs
Source: UNCTAD (2013)

According to the conceptual definition implicit in the MAST classification, all NTMs are trade-relevant, but regulations can affect trade without being NTMs. Production standards, in particular, are not NTMs, even though they may affect trade. For instance, regulations prohibiting the discharge of polluting effluents in rivers (production standards) may affect the competitiveness of the companies concerned compared with foreign competitors not subject to the same regulations and hence affect trade. However, if one were to count such regulations as NTMs, every domestic regulation could become an NTM.

Transparency is limited worldwide as far as NTMs are concerned. Because of their diverse nature, authority over NTMs is fragmented over a large number of government agencies and ministries, including, for example, health, agriculture, trade, industry, and standards and metrology bureaus. These agencies typically do not coordinate, or do so only weakly. One key source of information is the WTO's a database of notifications, available through the iTIP,[1] the WTO's trade intelligence portal. WTO members must notify all new trade-relevant regulations if they have the potential to restrict trade, unless they are based on international standards. The WTO also maintains a database of "specific trade concerns" raised on the regulations of one member by others. The database

[1] Available at http://www.wto.org/english/res_e/statis_e/itip_e.htm.

was recently used to estimate the effect of NTMs at the firm level by Fontagné et al. (2013). However, several issues constrain the use of the WTO's data for research. One is that although notification is mandatory, countries seldom notify and may be expected to notify primarily the most innocuous measures for fear of attracting attention to more problematic ones, thus creating a selection bias. Second, the data sent by countries is of variable quality, with regulations sometimes described in imprecise terms and without always providing a full set of Harmonized System (HS) codes for all covered products.

The primary alternative source of data on NTMs is UNCTAD's Trade Analysis and Information System (TRAINS) database,[2] but TRAINS data were collected in most cases for only one year (2001) and according to a classification of measures that is now obsolete. In 2009, the new MAST classification of measures was agreed on, and a pilot project was launched to collect new data worldwide. The project's rate of data collection accelerated in 2009 when the World Bank invested substantial resources to improve country coverage, and in 2011 several regional banks (the African Development Bank and the Inter-American Development Bank) joined the project, together with the WTO, which adopted an amended version of the MAST classification for its NTM notification system. By June 2012, more than sixty-seven countries had been covered (counting EU members individually) and new data were collected in several additional ones, including West Africa.

II. Measuring regulatory fragmentation

In this section, we propose two measures of cross-country "regulatory distance," one based on differences in the application of measures over products, which we the "extensive margin" of regulatory distance, and one based on differences in the stringency of a particular measure (say, maximum residual levels of chemicals), which we call the "intensive margin" of regulatory distance.

1. Extensive margin

Let c index countries, p HS6 products, and n NTM types, and let

$$I_{cnp} = \begin{cases} 1 & \text{if country } c \text{ applies NTM type } n \text{ to product } p \\ 0 & \text{otherwise} \end{cases} \tag{1}$$

[2] Available at http://www.unctad.info/en/Trade-Analysis-Branch/Key-Areas/TRAINSWITS/.

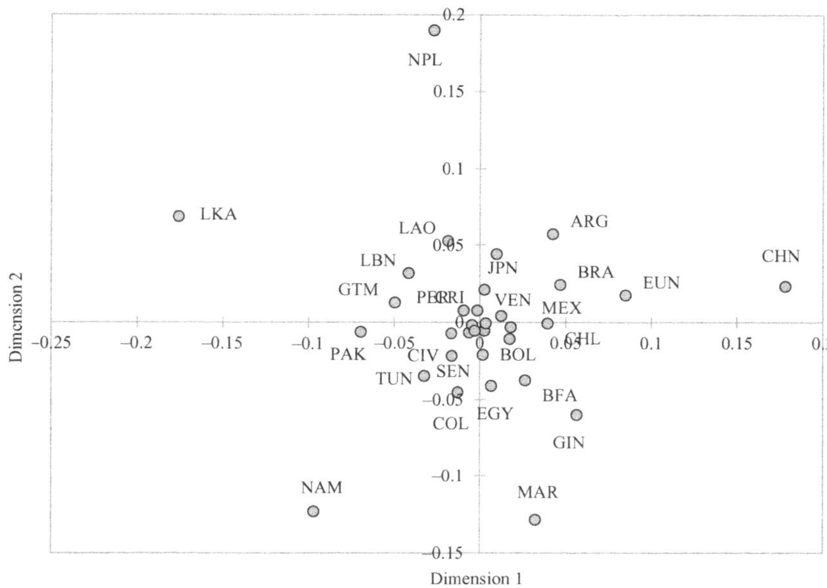

Figure 11.2. Two-dimensional representation of regulatory distance
Note: Kruskal stress value 0.208
Source: Authors' calculations based on UNCTAD multilateral NTM database

Regulatory distance at the measure-product level is

$$D_{cc'np} = \left| I_{cnp} - I_{c'np} \right| .$$

Letting $N = \max\{N_c; N_{c'}\}$ be the total number of NTMs used by any of the two countries and $P = \max\{P_c; P_{c'}\}$ the total number of products covered in any of the two countries, aggregate regulatory distance between c and c' is

$$D_{cc'} = \frac{1}{NP} \sum_n \sum_p D_{cc'np} \qquad (2)$$

that is, the sum of the absolute values of the differences in NTM application status. Because regulatory distance is normalized by the grand total of product-NTM combinations, it lies between zero and one and is typically a small number. The complete matrix of bilateral regulatory distances between countries in the NTM database is shown in Figure 11.2 and in Appendix Table 11.A1. Table 11.A1 shows overall regulatory distances (all NTMs together). Figure 11.2 shows a projection of bilateral regulatory distances on a plane with two arbitrary dimensions (see the Appendix for details). A small number of countries (Nepal, Sri Lanka, China, Morocco and Namibia) stand out for unusual patterns of NTM imposition,[3] whereas a "core" of countries share similar patterns of NTM imposition at the product level, providing a sort of "common pattern."[4]

Do regional trade agreements (RTAs) foster regulatory convergence? As a first pass, we regress our regulatory-distance measure on a dummy variable equal to one when both countries belong to one of the RTAs identified by Piermartini and Budetta (2009). Results are shown in Table 11.1.

The coefficient in the first column of Table 11.1 is negative and highly statistically significant, suggesting that, on average, the RTAs overlapping with our NTM database do reduce regulatory distance between their members. The effect is quantitatively large. As average regulatory distance between country pairs in our sample is 0.079, the average RTA cuts it by 41 percent (0.033/0.079). The second column breaks down this effect by individual agreement. The estimated effect for ALADI is also negative and highly significant, but other agreements have too few observations to be significant.

2. Intensive margin

While the "extensive margin" of regulatory differences evaluates how patterns of imposition of particular types of measure, as defined by the MAST nomenclature, are applied to different products, the "intensive margin" evaluates how the stringency of a given type of measure, applied to a given product, differs between countries. Consider the case of oranges (HS080510) in four ALADI countries: Argentina, Brazil, Chile and Colombia. According to the NTM database, the pattern of NTMs imposed by those four countries on product HS080510 is as shown in Table 11.2.

The matrix of bilateral extensive-margin regulatory distances for HS080510 is as shown in Table 11.3. Argentina and Colombia have the smallest distance (0.2), as both report tolerance limits (also called maximum residual levels and coded as A200) and neither impose conformity assessment (A800) and labeling requirements (B310). Brazil and Chile

3 We recoded Chinese data to transform all NTMs erroneously coded as B for products other than agri-food products (chapters HS01 to HS24) into A, keeping the last three digits the same.

4 Morocco's position as an outlier is surprising in view of the regulatory harmonization envisaged by the Morocco-EU Association agreement and stated in its Article 51.

Table 11.1. *Regression results, regulatory distance and RTAs*

	(1)	(2)
Both in the same RTA (any)	– 0.033[a]	
	(8.07)	
Both in ALADI		– 0.029[a]
		(2.83)
Both in Andean Comm.		– 0.023
		(0.77)
Both in CACM		– 0.049
		(0.72)
Both in COMESA		– 0.033
		(0.85)
Both in SADC		– 0.045
		(1.14)
Both in SAFTA		0.018
		(0.46)
Constant	0.086[a]	0.083[a]
	(24.33)	(26.15)
Observations	992	992
R^2	0.01	0.01

Note: Estimator: ordinary least squares; dependent variable: bilateral regulatory distance. Robust t-statistics in parentheses. ALADI: Latin American Integration Association; CACM: Central American Common Market; COMESA: Common Market for Eastern and Southern Africa; SADC: Southern African Development Community; SAFTA: South Asian Free Trade Area.
[a] Significant at 1%.

are also close, but with the opposite pattern (no MRL but conformity assessment and labeling requirements). The largest distances are between Brazil and Argentina (0.9) and between Brazil and Colombia (0.9).

However, other sources indicate that all four countries impose maximum residual levels in oranges for a large number of pesticides. For instance, the US Department of Agriculture's (USDA's) Foreign Agricultural Service maintains a restricted-access International Maximum Residue Level (IMRL) Database with information on MRL levels by country as well as according to the Codex Alimentarius, an international standard for food regulations.[5] For oranges, 213 chemical substances

[5] See http://login.mrldatabase.com.

Table 11.2. *NTMs on oranges (HS080510) in four ALADI countries*

MAST code	Argentina	Brazil	Chile	Colombia	Description of measure
A110	0	1	0	0	Temporary prohibitions for SPS reasons
A200	1	0	0	1	Tolerance limits for residues and restricted use of substances
A800	0	1	1	0	Conformity assessment related to SPS
A820	1	0	0	1	Testing requirements
B310	0	1	1	0	Labeling requirements
C300	0	0	1	1	Requirement to pass through specified port of customs
E140	0	1	0	0	*Not in 2012 nomenclature*
E190	1	0	0	1	*Not in 2012 nomenclature*
E390	1	0	0	1	*Not in 2012 nomenclature*
G400	1	0	0	0	Regulations concerning terms of payments for imports

Source: UNCTAD multilateral NTM database.

Table 11.3. *Bilateral regulatory distances for oranges, extensive margin*

	Argentina	Brazil	Chile	Colombia
Argentina	0	0.9	0.8	0.2
Brazil		0	0.3	0.9
Chile			0	0.6
Colombia				0

Source: Authors' calculations using UNCTAD multilateral NTM database.

are concerned. Out of those, Argentina imposes 79 MRLs, Brazil 101, Chile 110 and Colombia 72. The Codex Alimentarius has 83. Interestingly, the Codex's own home page lists only 10 MRLs for pesticides on foods and feeds under category FC0004, "oranges, sweet, sour (including

Table 11.4. *Bilateral regulatory distances for oranges, intensive margin: MRL level*

	Argentina	Brazil	Chile	Columbia	Codex
Argentina	0	0.44	2.66	1.18	1.94
Brazil		0	1.76	0.78	1.27
Chile			0	0.41	0.08
Columbia				0	0.27
Codex					0

Note: Distances do not lie between zero and one because they are averages of the absolute values of differences between MRLs expressed in milligrams per kilogram.
Source: Authors' calculations using USDA IMRL database.

orange-like hybrids): Several cultivars."[6] Using the USDA's IMRL database, one can calculate a first intensive-margin regulatory distance measure defined by formula (2) but applied to individual pesticides.

Taking advantage of the fact that MRLs are all expressed in a homogeneous metric (mg per kg), one can pushing the logic of intensive-margin distance one step further and assess regulatory distance in terms of stringency. That is, let x_{ck} be the MRL of pesticide k imposed on oranges by country c, in mg per kg. Then

$$\tilde{D}_{cc',\text{oranges}} = \frac{1}{N_k} \sum\nolimits_p |x_{ck} - x_{c'k}| \,, \tag{3}$$

where N_k is the number of pesticides. When a country imposes no MRL for a given pesticide, the IMRL database codes it as missing value; then x_{ck} is undefined and pesticide k drops out of the sample when taking the differences in (3), which only include cases where both c and c' impose MRLs on the same pesticide. The matrix of intensive-margin bilateral distances is shown in Table 11.4.

The smallest distance is bar far between Chile and the Codex Alimentarius, reflecting the adoption by Chile of many standards based on the Codex. Bilaterally, the smallest distance is between Chile and Colombia, followed closely by Argentina-Brazil.

[6] Those are on spinetoram, carbosulfan, cyhexatin, azocycltin, carbofuran, kresoxim-methyl, amitraz, carbendazim, dithiocarbamates and methidation. See http://www.codexalimentarius.net/pestres/data/commodities/details.html?id=127.

III. Multilateral disciplines as "minimal harmonization"

For nontechnical measures, the most basic discipline is set out in Articles XI and XIII of the GATT. Essentially, Article XI prohibits the use of quantitative restrictions (QRs), and Article XIII requires that licensing schemes (whether they be used to enforce QRs or other measures) should be nondiscriminatory; that is, some categories of users should not be preferred to others in the granting of licenses (except based on performance criteria such as compliance with technical regulations).

For technical measures, the general guiding principles of WTO rules are based on the necessity and proportionality tests. The term "necessity test" means that government intervention should be justified by a reasonably compelling case of market failure; "proportionality test" indicates that government intervention should not inflict more damage on trade than needed to achieve the measure's objective.

Specific rules for the design of Technical Regulations (TRs) other than sanitary ones are laid out in the WTO's TBT Agreement.[7] Beyond the necessity and proportionality tests, the disciplines contained in the TBT Agreement are based on the principles of nondiscrimination and national treatment. That is, regulations should apply to all products irrespective of their origin (they could not, for instance, be relaxed on a preferential basis, which would make no sense anyway), and they should not penalize imports more than domestic products. In particular, a regulation cannot impose restrictions on imported products based on their method of production: only the product's characteristics that are relevant for public health or the environment count.[8]

Rules for the design of Sanitary and Phytosanitary (SPS) measures are laid out in the WTO's SPS Agreement.[9] The agreement's most essential rule is that the protection of the population or the environment against sanitary risks should be motivated by scientific evidence. When scientific evidence is ambiguous, Article 5.7 allows the application of the precautionary principle (imposing restrictions in anticipation of possible harm rather than after the harm has been established) provided that the measures are maintained only temporarily while further scientific evidence is actively sought.

Whether WTO disciplines are too weak or too strong is a matter of judgment. On one hand, the phase-out of EU restrictions on the

[7] The TBT acronym stands for "Technical Barriers to Trade."

[8] There are subtleties in complex cases like activities that affect endangered species. See, inter alia, Cadot, Maliszewska and Saez (2010).

[9] See http://www.wto.org/english/tratop_e/sps_e/spsagr_e.htm.

commercialization of hormone-treated beef or genetically modified organisms (GMOs), following decisions of the WTO Appellate Body, was widely perceived in European public opinion as an intrusion into issues that had more to do with societal choices than with free trade. On the other hand, WTO transparency requirements on NTMs, as reflected in the notifications process, are undoubtedly weak (few NTMs are effectively notified); as for constraints on the discriminatory use of technical regulations, precisely because of the technical nature of the measures, the "gray zone" is wide. Moreover, even without malicious intentions, technical regulations are typically adopted at the national level through concertation with manufacturing interests and naturally lead to fragmented standards, a problem that the WTO is not equipped to deal with.

As a result, many regional blocs have embarked in their own NTM-streamlining processes, involving harmonization, mutual recognition or the negotiated elimination of NTBs based on private-sector complaints. We now turn to a brief overview of those efforts.

C. Harmonization, trade and market structure

I. Coordination in NTM streamlining: the regional experience

Because countries typically have large numbers of regulations on the books, many with potential trade effects, streamlining them is a long and difficult undertaking. Regional trade agreements (RTAs) can help in several regards. First, as in the EU, supranational institutions – like the European Commission and the European Court of Justice – can play a leading role in fostering the elimination of NTBs, at least on intra-bloc trade, and a general streamlining and harmonization of regulations. Second, as in North American Free Trade Agreement (NAFTA), the agreement itself can serve as a political anchor to reform-minded politicians with liberalizing agendas. That is, if the RTA enjoys strong multi-partisan backing, it can be used as a justification for domestic reforms even when they hurt particular interests (e.g. the elimination of NTBs benefitting domestic producers). Third, as in the Association of South-East Asian Nations (ASEAN), regional institutions can provide a useful forum to exchange experiences, suggest directions for further moves, and motivate reformers by overcoming their isolation at home.

The EU experience suggests an especially powerful route to the streamlining of product standards. Harmonizing them across the European Union to achieve a single market was an almost impossible task (it would have been impossible to require Member States consensus for all new regulations). Instead, the EU cut the Gordian knot with a legal approach.

When a German spirit importer complained in 1979 to the European Court of Justice (ECJ) that German authorities were preventing him from importing a French spirit called cassis de Dijon, the ECJ responded with a landmark decision. It ruled that a product – in this case, a spirit – that had been cleared for sale in one Member State had no reason not to be cleared, automatically, in any other Member State. In its ruling, the ECJ rejected the German government's defense that "mutual recognition" (as the principle has come to be known) would lead to a race to the bottom. In fact, it did not, and the European Commission has since enforced this principle in a wide range of areas.

Although less ambitious, NAFTA contributed to give political traction to an NTM-streamlining agenda in Mexico through a different mechanism – by providing a political anchor. NAFTA was important for Mexico in terms of securing stable access to the US market and had high political visibility. Mexico's partner was a large, Northern country, which, in addition, had recently pursued a program of privatization and regulatory rollback under the Reagan administration. Thus, the political power in Mexico reckoned that business with the United States was inconsistent with old-fashioned policies. Moreover, because of its high stakes, NAFTA provided a guarantee against reform reversal, which was important to gather support around them.

ASEAN did not provide such a strong political anchor. However, it has consistently pursued an agenda of NTB elimination. In 1987, the Memorandum of Understanding on Standstill and Rollback of NTBs sought the elimination of General Agreement on Tariff and Trade (GATT)-inconsistent measures and the preferential reduction of others. The 1992 AFTA-CEPT[10] Agreement called for the immediate elimination of all quantitative restrictions, and in 1997 the AFTA Council designated priority areas for product standards harmonization. Since then, the AFTA Council and the ASEAN Council of Ministers have repeatedly emphasized the need to streamline NTMs and to avoid a substitution of NTBs for tariffs as the latter are phased out (World Bank 2008). Although the elimination of NTBs and the harmonization of standards are still distant objectives, ASEAN is relatively more advanced in the process than many other free-trade areas. More recently, the "ASEAN Scorecard" listed three broad objectives: (i) the virtual elimination of tariffs on goods imports from ASEAN member countries; (ii) moves underway to address NTBs

[10] AFTA stands for ASEAN Free Trade Area and CEPT stands for Common Effective Preferential Tariff.

(e.g. through harmonization of standards, streamlining of customs procedures, and improvements in logistics); and (iii) greater liberalization in services and investment provisions. It has reported substantial progress, with an estimated 73.6 percent implementation rate of the measures, activities, and sectoral agreements scheduled to be adopted during the first two years.

Preferential agreements involving the European Union and the United States often contain measures aimed at the reduction or elimination of NTMs, reviewed systematically in Horn, Mavroidis and Sapir (2009). They divide them into two broad types of commitments:

Article I. Those going beyond the WTO, but building on WTO commitments ("WTO+"), including, for example, SPS and TBT commitments
Article II. Those covering areas not covered by the WTO ("WTO-X"), including, for example, labor or environment commitments.

Many US and EU agreements have WTO+ clauses. For instance, all fourteen EU agreements reviewed include TBT provisions, but those are enforceable in only five of them (Cariforum,[11] Mexico, Chile and European Economic Area [EEA] and Turkey). These commitments are typically deeper than in the case of US agreements, which only restate the WTO obligations of preferential partners. Eight EU agreements include SPS provisions, and only three are legally enforceable (EEA, Chile and Cariforum). As for the United States, twelve agreements include SPS provisions, but only two are legally enforceable through dispute settlement (Israel and NAFTA).

In another broad-ranging review, Lesser (2007) covers twenty-eight North-South and South-South PTAs signed by Chile, Mexico and Singapore. Most rely on mutual recognition of conformity assessment results, an approach considered less costly than harmonization, and on transparency requirements. Members are, for instance, asked to notify each other about the introduction of new measure or the modification of existing ones. Most of the reviewed agreements also call for the establishment of joint bodies to monitor the implementation of TBT provisions and facilitate cooperation. Most agreements also include dispute-settlement mechanisms for TBT-related disputes.

In terms of depth of TBT commitments, few PTAs go beyond the TBT Agreement. However, they often display "WTO+" characteristics,

[11] Cariforum is the Forum of the Caribbean Group of African, Caribbean and Pacific (ACP) States.

using the term coined by Horn, Mavroidis and Sapir (2009). The most far-reaching involve mutual recognition of conformity assessment procedures and bodies, meaning that parties are required to justify non-equivalence and non-recognition. A few blocs and sub-blocs (e.g. the EU and the more developed ASEAN and Asia-Pacific Economic Cooperation [APEC] members) have adopted mutual recognition arrangements for conformity assessment in particular sectors such as telecoms, electrical, electronic and medical equipment.

Three key factors influence the depth of regional TBT commitments. The first is the level of development of the parties. Standards harmonization and even more mutual recognition of conformity assessment results are much easier among countries with similar levels of development. The second factor is the agreement's degree of integration. Deeper agreements such as customs unions and common markets can go more easily beyond WTO commitments. The third factor is the presence of the European Union or United States as one of the parties to the agreement. Agreements involving the United States often include acceptance of partner technical regulations as equivalent, alignment on international standards and mutual recognition of conformity assessment. Agreements involving the European Union often rely on alignment on the EU's own regulations, standards and conformity assessment procedures, especially with close partners such as Mediterranean countries. For instance, Article 51 of the European Community (EC)-Morocco RTA states:

> The Parties shall cooperate in developing: (a) the use of Community rules in standardisation, metrology, quality control and conformity assessment; (b) the updating of Moroccan laboratories, leading eventually to the conclusion of mutual recognition agreements for conformity assessment; (c) the bodies responsible for intellectual, industrial and commercial property and for standardization and quality in Morocco.

Article 51 of the European Community (EC)-Tunisia RTA is identical. Article 40 of the EC-Palestinian Authority RTA contains a harmonization clause worded in similar language: "The objective of cooperation will be to narrow the gap in standards and certification. In practical terms cooperation will take the form of the promotion of the use of Community technical regulations and European standards and conformity assessment procedures." Likewise, Article 68 of the EC-Jordan agreement states that "cooperation in this field will be aimed in particular at: (a) increasing the application of Community rules in the field of standardization, metrology, quality standards, and recognition of conformity."

EU trade agreements with more distant countries like Chile, with which it does not have deep integration agendas, contain less stringent clauses on TBTs. For instance, Article 18 of the EC-Chile agreement states that "cooperation between the Parties will seek to promote efforts in (a) regulatory cooperation; (b) compatibility of technical regulations on the basis of international and European standards."

In the EC-Mexico nor the EC-Egypt agreements, harmonization is expected to take place on the basis of international standards. For instance, Article 19 of the EC-Mexico merely states that the Parties "shall work towards: . . . (c) promoting the use of international standards, technical regulations and conformity assessment procedures on the basis of international agreements; (d) facilitating the adoption of their respective standards, technical regulations and conformity assessment procedures on the basis of international requirements." Similar clauses can be found in other North-South agreements. For instance, Article 705 of the Australia-Thailand RTA states that "the Parties shall, where appropriate, endeavor to work towards harmonization of their respective technical regulations, taking into account relevant international standards, recommendations and guidelines, in accordance with their international rights and obligations." However, there is a nuance in the scope of harmonization. In the latter case (Australia-Thailand), Chapter 7, to which Article 705 belongs, applies to "all goods traded between the parties," implying that goods not traded bilaterally could potentially remain uncovered, whereas no such scope limitation can be found in the EC-Mexico clause on harmonization. Therefore, if one accepts the idea that even when the letter of the agreement does not prescribe convergence on the Northern standard, de facto this is what is likely to happen, the EC-Mexico harmonization clause can be taken as more encompassing than the Australia-Thailand one, which leaves regulations that are irrelevant to bilateral trade outside of the agreement's scope. Similar scope limitations can be found, for example, in Article 7.2 of the US-CAFTA (Dominican Republic – Central America) agreement and in Article 7.1 of the US-Bahrain agreement.

In a review more than seventy PTAs covering several regions, levels of development and depth of integration, fifty-eight of them with TBT provisions, Piermartini and Budetta (2009) also find that harmonization is more frequent than mutual recognition for technical regulations (twenty-nine agreements against fifteen), but mutual recognition of conformity assessment is the most frequent approach (thirty-nine agreements) followed by harmonization of conformity assessment procedures (twenty-five agreements). Harmonization of technical regulations is a characteristic of EU agreements, sometimes, as noted, implying adoption of the EU acquis communautaire by RTA partners. Transparency provisions are also found in thirty agreements. Most of the agreements with TBT provisions also include the establishment of a monitoring committee or body for standard-related matters, and twenty-four have provisions for the resolution of disputes among members.

Harmonization of technical regulations in the context of North-South agreements is not free of risks regarding their compatibility with the broader aim of multilateral liberalization, as they can lead to specifications that are overly complex or burdensome from the point of view of many developing countries. As argued by Maur and Shepherd (2011), different economic and social conditions may call for different levels of "strictness" of technical regulations. Indeed, in the next section, we discuss how North-South RTAs with harmonization provisions can contribute to hub-and-spoke trade patterns, based on the findings of Disdier, Fontagné and Cadot (2015). The findings suggest that harmonization issues in North-South RTAs should be viewed by Southern governments strategically rather than as a technical issue.

South-South agreements have only recently turned their attention to NTMs, at least on paper. For instance, Article 6 of the SADC Trade Protocol calls for the elimination of all existing forms of NTBs and for Member States to refrain from imposing new ones. While implementing this article remains a major challenge, SADC ministers of trade have identified ten categories of NTBs for "immediate" action: (i) cumbersome customs documentation and procedures, (ii) cumbersome import and export licensing/permits, (iii) import and export quotas, (iv) unnecessary import bans and prohibitions, (v) import charges not falling within the definition of import duties, (vi) restrictive single channel marketing, (vii) prohibitive transit charges, (viii) complicated visa requirements, (ix) preshipment inspections, and (x) national food security restrictions. In some of these areas there has been progress, but in most barriers still remain.

Most of the tangible efforts have tended to focus on improving the monitoring and reporting of NTBs rather than elimination. Monitoring has taken two main forms:

> Audits on the implementation of the SADC Protocol on Trade have been undertaken every year since 2007. Their main focus has been on progress in removing tariffs facing regional trade, as per countries' commitments, but they also review some NTBs, in particular those relating to rules of origin.

An SADC Trade Monitoring and Compliance Mechanism (TMCM) was established in mid-2008. This has two distinct elements: an online NTB Monitoring Mechanism (NTBMM) that records reported NTBs by firms and the elimination and reduction of barriers (both tariffs and NTBs) following bilateral negotiation or outcomes from the various Dispute Settlement Mechanisms.

The publication of NTBs under the auspices of the NTBMM is a major step forward. However, while the NTBMM is now well established, there are problems with it including misidentification of some of the barriers reported and, most important, slow progress in resolving the barriers once they have been notified. Just half of the complaints received by SADC, and 20 percent received by COMESA have been resolved under the Tripartite Monitoring Mechanism. The main reason is the absence of obligation for countries to remove their barriers once notified by others. It relies purely on moral suasion.

Similarly, as discussed by Imani Development (2009), Article 49 of the COMESA Treaty obliges Member States to remove all existing NTBs to imports of goods originating from the other Member States and thereafter refrain from imposing any further restrictions or prohibitions, stating that

> except as may be provided for or permitted by this Treaty, each of the member States undertakes to remove immediately upon entry into force of this Treaty, all the then existing non-tariff barriers to the import into that member State of goods originating in the other member States and thereafter refrain from imposing any further restrictions or prohibitions.

However, like the WTO, the COMESA Treaty recognizes the right of Member States to impose trade-restricting regulations for health, safety or the environment.

In the East African Community (EAC), as discussed in Kirk (2010), practically all the NTMs listed under Category A and targeted for "immediate removal" are still in place. Implementing the EAC agenda on NTMs has proved difficult for a variety of reasons, including lack of political traction but also lack of administrative capacity because the issues can quickly become technical. Recognizing this, the EAC Secretariat has tried to set up information and monitoring mechanisms supported by capacity building and training and to favor the creation of country-level structures such as NTB monitoring committees. This is a promising area of cooperation between Member State governments and Regional Economic Communities, in the EAC and elsewhere, and also clearly an area where donor support would have a high payoff.

There is no doubt that beyond capability issues, many political economy issues lurk in the background, and it would probably be naïve to assume that training officials in implementing agencies would be enough to move forward ambitious NTB-elimination agendas at the regional level. As rightly pointed out by Kirk (2010), "Experience from other regional groupings, such as the EU and ASEAN, points to the importance of establishing effective enforcement and compliance mechanisms." However, establishing such mechanisms presupposes the existence of a political drive for deep integration. Thus, in a way it presupposes that the problem of political motivation is solved. What can be hoped for from capacity-building efforts is to tackle at least those problems that can have technical solutions and to gather momentum for reform from observed successes.

As for ASEAN, the approach adopted by the Secretariat consisted in classifying NTMs in broad classes using several criteria. NTMs were first classified on the basis of WTO principles: transparency, non-discrimination, science basis (for SPS measures), proportionality/necessity. The Secretariat tried to strike a balance between non-trade regulatory objectives such as revenue generation and protection of health and safety of consumers and trade costs measured through a Trade Impact Criterion. The Criterion used a number of indicators, including number of private sector complaints, difference between domestic and world prices, sectoral importance and trade value, and was used to group NTMs into three categories:

Article I. Red Box: NTMs that impede trade in ASEAN and which require immediate elimination;
Article II. Amber Box: NTMs that could not be clearly identified or classified as barriers; and
Article III. Green Box: NTMs that could be justified, including measures that have a scientific basis and are applied to both domestic and imported goods.

The categorization into boxes naturally led to a prioritization in terms of streamlining. In addition, the Secretariat called for eliminating first the non-transparent and discriminatory measures, then turn to those that were transparent but discriminatory. NTMs deemed unnecessary would be removed without being replaced with alternative measures (e.g. automatic licensing). For NTMs with protective objectives, a reexamination was suggested in view of the commitment to promote intraregional trade. In such a case, the replacement with tariffs should also be set, initially at rates with equivalent impact to the NTM, and gradually reduced to be less discriminatory against imports. Moreover, any less-trade distorting measure replacing an existing NTM would need to take into account the regulatory objectives of the original measures.

The ASEAN Secretariat identified that the greatest positive impact on trade is likely to come from removing the following NTMs: administrative pricing, non-automatic licensing, quotas, enterprise-specific restrictions and pre-shipment inspection. These should be replaced with tariffs, fiscal incentives or risk management with post-entry audit systems at customs. NTMs that are transparent but discriminate between imports may be considered next although their immediate removal would also yield trade benefits (e.g. prohibitions on "nonsensitive" goods and single channel for imports). The ASEAN Secretariat also sought to remove NTMs in nine priority sectors including electrical equipment, organic chemicals, motor vehicles, pharmaceuticals, cosmetics, beverages, edible fruit and nuts, cocoa and dairy products. Tariff quota duties, antidumping measures, restrictive foreign exchange allocation are not included in the ASEAN scheme, the last two affect products outside of the nine priority sectors, whereas antidumping is covered by WTO rules. Prohibitions are usually imposed on sensitive goods for national security, religious or moral, health and safety, or environmental reasons, hence those covering "nonsensitive" goods should be tackled first.

On the compliance side, the ASEAN Secretariat clearly took into consideration that enforcement of the rules may take place at different levels. Self-compliance is highly likely where the net benefits of the proposed arrangement are unequivocal for the member, which would serve as the impetus to implementation. Second- or third-party enforcement will require bodies with clear mandates, rules that are flexible yet stable, and quality information, but more important, members' political will to deliver on their commitments. Nevertheless, formal mechanisms and arrangements within ASEAN were considered essential to institution building as they improve on informal practices as well as instill a sense of obligation to the agreement by bringing countries under the same jurisdiction. Nevertheless, the elimination of NTMs has been moving slowly, due mainly to difficult measurement and methodological issues. This reflects a challenge that many regional grouping are facing when it comes to tackling NTMs. Pursuing the "elimination" of NTMs in the same way than tariffs may actually be not feasible. Governments should rather aim at harmonizing or streamlining NTMs to reduce bureaucratic procedure and compliance costs bearing on trade between member countries.

II. *From coordination to harmonization*

Although the coordination of NTM policies in regional groupings has proved challenging, a number of trade agreements, either regional or bilateral, have adopted a more binding route, putting harmonization at the heart of their agendas. This may accelerate the convergence of measures, but it may also raise challenging and complex policy issues, to which we now turn.

1. Harmonization and trade patterns: What does the gravity tell us?

In this section, we explore how TBT provisions in North-South RTAs affect the trade patterns of their Southern partners, both North-South (i.e. within the bloc) and South-South (i.e. with non-bloc Southern partners). This second question is particularly important in view of the rising importance of South-South trade and the relative decline, since decolonization, of the trade dependence of Southern countries with their former colonial powers. Our starting point is a gravity equation. Let Y_{it} and Y_{jt} be respectively the GDP's of two trading countries i and j; Y_{Wt} be the world's GDP, V_{ijt} the dollar value of exports from i to j, and τ_{ij} the bilateral trade cost between the two. Subscript t stands for year. Let also P_{jt} be the aggregate price index in destination country j and Ω_{it} a weighted average of country i's bilateral trade costs with all of its partners, variables sometimes referred to respectively as inward and outward "multilateral resistance" terms; the gravity equation states that[12]

$$V_{ijt} = {\tau_{ijt}}^{1-\sigma} \frac{1}{\Omega_{it} {P_{jt}}^{1-\sigma}} Y_{it} Y_{jt}, \tag{4}$$

where $\sigma > 1$ is the elasticity of substitution between varieties imported from different countries, and

$$\Omega_{it} = \sum_{j=1}^{m} {\tau_{ijt}}^{1-\sigma} \left(\frac{E_{jt}}{{P_{jt}}^{1-\sigma}} \right). \tag{5}$$

An empirical specification consistent with theory involves the use of fixed effects by importer-year and exporter-year to control for variation in Ω_{it} and P_{jt} (Baldwin and Taglioni, 2006) as well as GDPs Y_{it} and Y_{jt}. The fixed effects incorporate size effects as in gravity but also the other origin and destination determinants seen above, the price and the number of

12 Head and Mayer (2014) refer to this formulation with the appropriate handling of the multilateral resistance terms as 'structural gravity'.

varieties of the exporting country and the demand size and price index (often referred to as a remoteness term) of the importing country.

The simplest approach is to apply ordinary least squares (OLS) on a log-linearized version of (4). However, as shown by Santos Silva and Tenreyro (2006), the resulting estimates are biased in the presence of heteroskedasticity; accordingly, we use the Poisson estimator suggested by Santos Silva and Tenreyro (2006). This estimator provides estimates that are comparable to elasticity estimates from the standard linear-in-logs specification.

We split our sample into two subsamples corresponding respectively to North-South and South-South trade relations,[13] dropping North-North relations. We cover forty-three RTAs and use the template of Piermartini and Budetta (2009), updating it with recent North-South RTAs. For each RTA, we focus on harmonization of technical regulations, disentangling the impact of the RTA as such from that of harmonization provisions.

For North-South trade relations, our basic treatment is a "North-South RTA" dummy taking the value of one when *i* and *j* are members of a common regional North-South agreement (zero otherwise). We interact the basic treatment with several dummies picking up harmonization clauses. We first control whether the common North-South RTA includes a TBT provision involving standards harmonization. One dummy is defined for the harmonization of technical regulations.

We then explore whether harmonization explicitly promotes regional and/or international standards, by including a dummy equal to one if the RTA promotes the use of regional standards only (0 otherwise); a second dummy is set to one if the RTA promotes international standards (alone or in addition to regional standards) and zero otherwise.

For South-South trade relations, our basic treatment is whether the exporter or the importer has signed an RTA with a Northern country. We interact the basic treatment for the exporting country with a dummy picking up, again, standards harmonization clauses in the North-South agreement in question. We control whether the RTA signed by the Southern exporting partner with the North involves standards harmonization.

Finally, for South-South trade, we also control for the existence of a South-South RTA between trading partners by including a "South-South RTA" dummy equal to one if *i* and *j* are members of a common regional South-South agreement (0 otherwise).

[13] In addition, a Chow test suggests that estimated coefficients on both subsamples differ significantly and confirms this divide.

Trade data come from the CEPII's BACI database, which provides reconciled free on board trade values.[14] Our dependent variable is dollar value of bilateral imports of country *j* from country *i* in year *t*. Our sample period runs from 1990 to 2006. GDPs are from the World Bank's World Development Indicators. Bilateral distances and other bilateral trade-costs determinants are all from the CEPII's database.[15]

As emphasized earlier, we expect different impacts of standards harmonization within North-South RTAs on South-North trade and South-South trade. Accordingly, we will first focus on North-South trade and then discuss the results for South-South trade.

Table 11.5 presents results for North-South trade, meaning Northern imports from Southern trade partners. Columns (1) and (2) show the mean impact of an RTA between a Northern and a Southern country on their bilateral trade. The first uses separate year, importer and exporter fixed effects, the second importer-year and exporter-year and country-pair fixed effects. The trade impact of deeper integration is analyzed in column (3) where harmonization is included. Column (4) distinguishes whether, in addition to harmonization, the RTA promotes the use of regional or/and international standards.

Column (1) suggests that the existence of a RTA between the Northern importing country and the Southern exporting country increases bilateral exchanges. An RTA raises trade by a factor of 1.43 (exp[0.36]), *ceteris paribus*. Regarding traditional covariates, distance negatively influences bilateral imports, whereas countries' gross domestic product and common border have a positive effect on trade. The cultural proximity variables suggest that imports are higher if both countries share the same language. The existence of a past colonial relationship has no significant influence.

Column (2) replaces time-invariant importer and exporter effects by interacted importer-year, exporter-year, and country-pair effects in order to control for the potential endogeneity of the RTA variable. The positive estimate on RTA in column (1) becomes insignificant in column (2), suggesting both that there is a strong omitted-variable bias in the former and that the latter goes some way toward correcting it.

14 http://www.cepii.fr/anglaisgraph/bdd/baci.htm. This database uses original procedures to harmonize the United Nations COMTRADE data (evaluation of the quality of country declarations to average mirror flows, evaluation of cost, insurance and freight rates to reconcile import and export declarations).

15 http://www.cepii.fr/anglaisgraph/bdd/distances.htm. These distances are calculated as the sum of the distances between the biggest cities of both countries, weighted by the share of the population living in each city.

Table 11.5. *Regression results, North-South trade*

Specification	Poisson maximum likelihood			
Dependent variable	Bilateral imports of the Northern country from the Southern partner			
Years	1990–2006			
Model	(1)	(2)	(3)	(4)
Ln GDP exporter	0.71[a]			
	(0.08)			
Ln GDP importer	0.51[a]			
	(0.09)			
Ln distance	−0.61[a]			
	(0.10)			
Common border	1.37[a]			
	(0.33)			
Common language	0.42[a]			
	(0.12)			
Past colonial links	0.10			
	(0.19)			
North-South RTA	0.36[a]	−0.06		
	(0.12)	(0.07)		
N-S RTA with standards harmonization			0.06	
			(0.08)	
N-S RTA with standards harmonization and promotion of regional standards				−0.20[b]
				(0.08)
N-S RTA with standards harmonization and promotion of international standards				0.52[a]
				(0.15)
Observations	49,522	52,448	1,731	1,731
Log pseudolikelihood	−13,618.21	−10,396.75	−889.30	−889.23
FE exporter	Yes			
FE importer	Yes			
FE year	Yes			
FE exporter × year		Yes	Yes	Yes
FE importer × year		Yes	Yes	Yes
FE dyad		Yes	Yes	Yes

Note: Constant and fixed effects not reported. Robust standard errors (importer-exporter clustered) in parentheses. In columns (3) and (4), the sample is restricted to the subsample of Southern exporter countries having an RTA with a Northern partner. GDP: gross domestic product. RTA: Regional Trade Agreement. FE: fixed effects.

[a,b] Significance at the level of 1% and 5%, respectively.

Column (3) highlights the trade effect of the harmonization of technical regulations between the two partners (conditional on the existence of an RTA) by adding an interaction term between the RTA variable and the harmonization of technical regulations. The estimated coefficient on the interaction term is positive but not significant.

Column (4) suggests that harmonization to regional standards is trade-impeding, whereas harmonization to international standards is trade-enhancing. These results show that the detrimental effect on North-South trade of harmonization contained in North-South RTAs is falling on harmonization of regional standards only. When harmonization authorizes the use of international standards, the negative impact on trade vanishes.

Our results suggest that the cost linked to regional standards harmonization – that is, the adoption by developing exporters of regional standards imposed on developed markets – may be too high for some of these exporters, which are therefore excluded from the market. According to our results, the worst situation for a Southern country in terms of trade with the North is the signature of a North-South RTA involving the harmonization of technical regulations on regional standards. In such case, the positive effects of trade preferences are more than offset.

Results for South-South trade are reported in Table 11.6. Columns (1) and (2) estimate the determinants of bilateral flows between Southern countries without controlling for the potential existence of RTAs between Southern countries and Northern partners. We then investigate the trade impact of North-South RTAs and standards harmonization on South-South trade (columns (3) and (4)). Column (3) tests for potential diversion effects by investigating the impact on South-South trade of the signature by either one or both Southern partners of a RTA with the North. Column (4) examines the additional trade impact linked to the harmonization of technical regulations.

In column (1), GDPs have a positive and significant impact on trade. Distance has a negative and significant impact on bilateral flows, while common border, common language increase trade. Furthermore, the dummy variable controlling for the existence of a South-South RTA is not significant.

Once we control for the endogeneity issue by introducing country-pair fixed effects (column (2)), the dummy variable controlling for the existence of a South-South RTA becomes significant, suggesting that South-South integration has significant trade effects in our sample period.

Table 11.6. *Regression results, South-South trade*

Specification	Poisson maximum likelihood			
Dependent variable	Bilateral imports of the Northern country from the Southern partner			
Years	1990–2006			
Model	(1)	(2)	(3)	(4)
Ln GDP exporter	0.42[a]	0.41[a]	0.37[a]	0.11
	(0.07)	(0.06)	(0.07)	(0.11)
Ln GDP importer	0.61[a]	0.58[a]	0.56[a]	0.60[a]
	(0.09)	(0.09)	(0.09)	(0.9)
Ln distance	−0.79[a]			
	(0.05)			
Common border	0.71[a]			
	(0.11)			
Common language	0.47[a]			
	(0.10)			
Past common colonizer	0.23[c]			
	(0.12)			
South-South RTA	0.06	0.41[a]	0.42[a]	0.30[a]
	(0.12)	(0.07)	(0.07)	(0.06)
N-S RTA for the exporter			−0.20[a]	
			(0.04)	
N-S RTA for the importer			−0.12[b]	−0.11[a]
			(0.06)	(0.04)
N-S RTA and standards harmonization with the North for the exporter				−0.22[b]
				(0.10)
Observations	296,376	245,069	245,069	24,803
Log pseudo-likelihood	−16,296.97	−9606.60	−9592.48	-1061.93
FE exporter	Yes			
FE importer	Yes			
FE year	Yes	Yes	Yes	Yes
FE dyad		Yes	Yes	Yes

Note: Constant and fixed effects not reported. Country fixed effects could not be interacted with year dummies in this table because standards harmonization variables are country-time variant. Robust standard errors (exporter-year clustered) in parentheses. In column (4), the sample is restricted to the subsample of Southern exporter countries having an RTA with a Northern partner. RTA: Regional Trade Agreement. FE: fixed effects.

[a,b,c] Significance at the level of 1%, 5% and 10%, respectively.

Column (3) shows that the signature of an RTA with the North by one – or both – Southern partners reduces their bilateral trade, suggesting trade-diversion effects. But whether such effects are of the traditional kind or conditional to the presence of standards harmonization is not controlled at that stage.

Column (4) controls for the harmonization of technical regulations in the North-South agreement signed by the exporter. Results suggest that the signature by the exporting country of an RTA involving the harmonization of its standards with the North has a negative and significant impact on its trade with other Southern countries. This result shows the presence of an "unconventional" trade-diversion effect. Harmonization raises production costs, reducing competitiveness on third markets in the South, possibly pricing exporters out of other Southern countries.

D. Concluding remarks

Harmonization is a strong step toward deeper integration, reducing regulatory fragmentation and potentially increasing the competitiveness of regional markets. For South-South blocs, this is a key argument given the small size of many of them. For North-South blocs, it is a way of speeding up the adoption of best practices and technologies by firms in the Southern partner. Although Southern firms exporting to the Northern partner must have already adopted the North's standards, harmonization typically extends the requirement to the South's domestic market, thus forcing all firms to comply. If accompanied by technical assistance, this may help overcome information barriers. This is indeed the EU's approach in the Euromed agreements.

Does this approach work? Our empirical results are worldwide averages derived from a gravity equation and thus cannot properly account for any particular case study. With this caveat in mind, they do suggest that North-South harmonization is a dangerous tool because it may price the exports of the Southern partner out of third markets, in particular in the South, where the Northern partner's stiff standards confer no competitive advantage. This danger is particularly important when the standards in question are regional (idiosyncratic) rather than international.

In view of the opportunities that lie in rapidly expanding South-South trade, one may wonder why Southern governments would accept agreements such as some of the EU's agreements with their Mediterranean partners that force them in advance to accept standards that may be well beyond what is appropriate given the development of their domestic markets. One possible explanation, at least in some cases, may lie in the recent results of Augier, Cadot and Dovis (2014) who show that stiff standards in Morocco in sectors where they were harmonized effectively locked out imports from low-income countries, raising the profitability of Moroccan firms. Whereas these effects do not come out in our gravity equation and therefore do not seem to be prevalent, they may apply in some cases.

Appendix

This appendix details the method used to generate the two-dimensional projection of regulatory distances in Section B. Consider a set of m countries indexed by c and c' and let $D_{cc'}$ stand for the distance between c and c'. The dissimilarity matrix

$$\Delta = \begin{bmatrix} 0 & \dots & D_{1m} \\ \dots & \dots & \\ D_{m1} & \dots & 0 \end{bmatrix} \tag{6}$$

is a square, symmetric matrix with zeroes on the diagonal and bilateral distances off the diagonal. The matrix is shown in Appendix Table 11.A1. Multidimensional scaling (MDS) consists of finding m coordinate vectors $\mathbf{x}_i$ (one for each country) such that, using an appropriate distance metric (noted $\| \|$),

$$D_{cc'} \simeq \|\mathbf{x}_c - \mathbf{x}_{c'}\| , \tag{7}$$

that is, the projection of the individuals onto a space of less than m dimensions represents reasonably well their true dissimilarity. If the space had m dimension, the representation would be perfect; as the number of dimensions shrinks (e.g. to two in a plane projection), the distortion potentially grows. The most usual way of formulating the problem is to minimize a quadratic loss function:

$$\min_{\mathbf{x}_1,\dots,\mathbf{x}_m} \sum_{i<j} \left(D_{ij} - \|\mathbf{x}_i - \mathbf{x}_j\|\right)^2 \tag{8}$$

Figure 11.A1 shows the distortion imposed by MDS onto a two-dimensional space for our regulatory-distance measure by plotting true

Table 11.A1. *Overall regulatory distance matrix*

	ARG	BFA	BOL	BRA	CHL	CHN	CIV	COL	CRI	ECU	EGY	EUN	GIN	GTM	JPN	KAZ	LAO	LBN	LKA	MAR	MDG	MEX	MUS	NAM	NPL	PAK	PER	PRY	SEN	TUN	TZA	URY	VEN
ARG	–	0.10	0.07	0.07	0.08	0.19	0.08	0.09	0.08	0.08	0.10	0.12	0.13	0.10	0.10	0.08	0.11	0.11	0.21	0.17	0.07	0.07	0.07	0.19	0.21	0.12	0.08	0.06	0.09	0.10	0.07	0.07	0.08
BFA	0.10	–	0.05	0.09	0.08	0.20	0.05	0.08	0.05	0.06	0.07	0.11	0.07	0.09	0.08	0.06	0.08	0.08	0.21	0.14	0.04	0.06	0.04	0.17	0.21	0.09	0.06	0.05	0.06	0.07	0.04	0.05	0.06
BOL	0.07	0.05	–	0.06	0.05	0.17	0.03	0.05	0.03	0.03	0.05	0.08	0.08	0.06	0.06	0.03	0.06	0.06	0.18	0.12	0.02	0.03	0.02	0.14	0.18	0.07	0.03	0.02	0.04	0.05	0.02	0.03	0.04
BRA	0.07	0.09	0.06	–	0.06	0.19	0.07	0.08	0.06	0.07	0.08	0.10	0.11	0.09	0.08	0.07	0.10	0.10	0.21	0.16	0.06	0.06	0.06	0.18	0.22	0.11	0.06	0.05	0.07	0.09	0.06	0.06	0.07
CHL	0.08	0.08	0.05	0.06	–	0.18	0.06	0.07	0.05	0.05	0.07	0.09	0.10	0.08	0.07	0.06	0.09	0.09	0.20	0.14	0.05	0.05	0.05	0.16	0.21	0.09	0.05	0.04	0.06	0.07	0.04	0.05	0.06
CHN	0.19	0.20	0.17	0.19	0.18	–	0.18	0.19	0.17	0.17	0.19	0.17	0.21	0.18	0.17	0.17	0.19	0.21	0.29	0.25	0.17	0.17	0.17	0.27	0.30	0.21	0.17	0.17	0.18	0.18	0.17	0.17	0.18
CIV	0.08	0.05	0.03	0.07	0.06	0.18	–	0.06	0.04	0.04	0.05	0.09	0.08	0.07	0.06	0.04	0.07	0.06	0.19	0.12	0.02	0.04	0.03	0.15	0.19	0.07	0.04	0.03	0.04	0.06	0.02	0.04	0.05
COL	0.09	0.08	0.05	0.08	0.07	0.19	0.06	–	0.05	0.05	0.07	0.11	0.10	0.08	0.08	0.06	0.09	0.09	0.19	0.15	0.05	0.06	0.05	0.16	0.21	0.10	0.05	0.05	0.05	0.08	0.05	0.05	0.06
CRI	0.08	0.05	0.03	0.06	0.05	0.17	0.04	0.05	–	0.04	0.05	0.09	0.08	0.06	0.06	0.04	0.07	0.06	0.18	0.12	0.03	0.04	0.03	0.15	0.19	0.07	0.04	0.03	0.04	0.06	0.02	0.04	0.04
ECU	0.08	0.06	0.03	0.07	0.05	0.17	0.04	0.05	0.04	–	0.05	0.09	0.08	0.07	0.06	0.04	0.07	0.07	0.19	0.13	0.03	0.04	0.03	0.15	0.19	0.08	0.03	0.03	0.04	0.06	0.03	0.04	0.04
EGY	0.10	0.07	0.05	0.08	0.07	0.19	0.05	0.07	0.05	0.05	–	0.10	0.09	0.08	0.08	0.05	0.08	0.08	0.20	0.14	0.04	0.05	0.04	0.16	0.20	0.09	0.05	0.04	0.06	0.07	0.04	0.06	0.06
EUN	0.12	0.11	0.08	0.10	0.09	0.17	0.09	0.11	0.09	0.09	0.10	–	0.14	0.11	0.10	0.09	0.12	0.12	0.22	0.18	0.08	0.08	0.08	0.20	0.23	0.13	0.09	0.08	0.10	0.11	0.08	0.09	0.10
GIN	0.13	0.07	0.08	0.11	0.10	0.21	0.08	0.10	0.08	0.08	0.09	0.14	–	0.11	0.11	0.08	0.12	0.11	0.23	0.17	0.07	0.09	0.08	0.19	0.24	0.12	0.08	0.08	0.09	0.10	0.07	0.08	0.09
GTM	0.10	0.09	0.06	0.09	0.08	0.18	0.07	0.08	0.06	0.07	0.08	0.11	0.11	–	0.08	0.07	0.09	0.10	0.18	0.15	0.06	0.07	0.06	0.18	0.19	0.08	0.06	0.05	0.07	0.09	0.05	0.07	0.07
JPN	0.10	0.08	0.06	0.08	0.07	0.17	0.06	0.08	0.06	0.06	0.08	0.10	0.11	0.08	–	0.07	0.08	0.09	0.20	0.15	0.05	0.06	0.05	0.17	0.21	0.10	0.06	0.05	0.07	0.08	0.05	0.06	0.07
KAZ	0.08	0.06	0.04	0.07	0.06	0.17	0.04	0.06	0.04	0.04	0.05	0.09	0.08	0.07	0.07	–	0.07	0.07	0.19	0.13	0.03	0.04	0.03	0.15	0.19	0.08	0.04	0.03	0.04	0.06	0.03	0.04	0.05
LAO	0.11	0.08	0.06	0.10	0.09	0.19	0.07	0.09	0.07	0.07	0.08	0.12	0.12	0.09	0.08	0.07	–	0.10	0.21	0.15	0.06	0.07	0.06	0.18	0.22	0.11	0.07	0.06	0.07	0.08	0.05	0.07	0.07
LBN	0.11	0.08	0.06	0.10	0.09	0.21	0.06	0.09	0.06	0.07	0.08	0.12	0.11	0.10	0.09	0.07	0.10	–	0.21	0.15	0.05	0.07	0.05	0.18	0.21	0.10	0.07	0.06	0.07	0.09	0.05	0.07	0.07
LKA	0.21	0.21	0.18	0.21	0.20	0.29	0.19	0.19	0.18	0.19	0.20	0.22	0.23	0.18	0.20	0.19	0.21	0.21	–	0.27	0.18	0.19	0.18	0.27	0.26	0.17	0.19	0.18	0.19	0.21	0.18	0.19	0.20
MAR	0.17	0.14	0.12	0.16	0.14	0.25	0.12	0.15	0.12	0.13	0.14	0.18	0.17	0.15	0.15	0.13	0.15	0.15	0.27	–	0.11	0.13	0.11	0.23	0.25	0.16	0.13	0.12	0.13	0.14	0.11	0.13	0.13
MDG	0.07	0.04	0.02	0.06	0.05	0.17	0.02	0.05	0.03	0.03	0.04	0.08	0.07	0.06	0.05	0.03	0.06	0.05	0.18	0.11	–	0.03	0.03	0.14	0.18	0.06	0.03	0.02	0.03	0.05	0.01	0.03	0.04
MEX	0.07	0.06	0.03	0.06	0.05	0.17	0.04	0.06	0.04	0.04	0.05	0.08	0.09	0.07	0.06	0.04	0.07	0.07	0.19	0.13	0.03	–	0.03	0.15	0.19	0.08	0.04	0.03	0.05	0.06	0.03	0.04	0.04
MUS	0.07	0.04	0.02	0.06	0.05	0.17	0.03	0.05	0.03	0.03	0.04	0.08	0.08	0.06	0.05	0.03	0.06	0.05	0.18	0.11	0.02	0.03	–	0.14	0.17	0.06	0.03	0.02	0.03	0.05	0.01	0.03	0.03
NAM	0.19	0.17	0.14	0.18	0.16	0.27	0.15	0.16	0.15	0.15	0.16	0.20	0.19	0.18	0.17	0.15	0.18	0.18	0.27	0.23	0.14	0.15	0.14	–	0.30	0.18	0.15	0.14	0.13	0.16	0.13	0.15	0.16
NPL	0.21	0.21	0.18	0.22	0.21	0.30	0.19	0.21	0.19	0.19	0.20	0.23	0.24	0.19	0.21	0.19	0.22	0.21	0.26	0.25	0.18	0.19	0.17	0.30	–	0.20	0.19	0.18	0.19	0.21	0.17	0.19	0.17
PAK	0.12	0.09	0.07	0.11	0.09	0.21	0.07	0.10	0.07	0.08	0.09	0.13	0.12	0.08	0.10	0.08	0.11	0.10	0.17	0.16	0.06	0.08	0.06	0.18	0.20	–	0.08	0.07	0.08	0.09	0.06	0.08	0.08
PER	0.08	0.06	0.03	0.06	0.05	0.17	0.04	0.05	0.04	0.03	0.05	0.09	0.08	0.06	0.06	0.04	0.07	0.07	0.19	0.13	0.03	0.04	0.03	0.15	0.19	0.08	–	0.03	0.04	0.06	0.03	0.03	0.04
PRY	0.06	0.05	0.02	0.05	0.04	0.17	0.03	0.05	0.03	0.03	0.04	0.08	0.08	0.05	0.05	0.03	0.06	0.06	0.18	0.12	0.02	0.03	0.02	0.14	0.18	0.07	0.03	–	0.04	0.05	0.02	0.03	0.03
SEN	0.09	0.06	0.04	0.07	0.06	0.18	0.04	0.05	0.04	0.04	0.06	0.10	0.09	0.07	0.07	0.04	0.07	0.07	0.19	0.13	0.03	0.05	0.03	0.13	0.19	0.08	0.04	0.04	–	0.06	0.03	0.05	0.05
TUN	0.10	0.07	0.05	0.09	0.07	0.18	0.06	0.08	0.06	0.06	0.07	0.11	0.10	0.09	0.08	0.06	0.08	0.09	0.21	0.14	0.05	0.06	0.05	0.16	0.21	0.09	0.06	0.05	0.06	–	0.04	0.06	0.07
TZA	0.07	0.04	0.02	0.06	0.04	0.17	0.02	0.05	0.02	0.03	0.04	0.08	0.07	0.05	0.05	0.03	0.05	0.05	0.18	0.11	0.01	0.03	0.01	0.13	0.17	0.06	0.03	0.02	0.03	0.04	–	0.03	0.03
URY	0.07	0.05	0.03	0.06	0.05	0.17	0.04	0.05	0.04	0.04	0.06	0.09	0.08	0.07	0.06	0.04	0.07	0.07	0.19	0.13	0.03	0.04	0.03	0.15	0.19	0.08	0.03	0.03	0.05	0.06	0.03	–	0.05
VEN	0.08	0.06	0.04	0.07	0.06	0.18	0.05	0.06	0.04	0.04	0.06	0.10	0.09	0.07	0.07	0.05	0.07	0.07	0.20	0.13	0.04	0.04	0.03	0.16	0.17	0.08	0.04	0.03	0.05	0.07	0.03	0.05	–

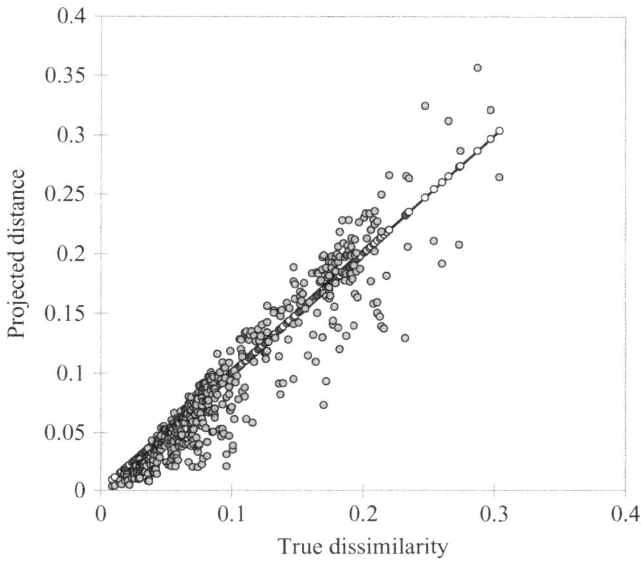

Figure 11.A1. Distortion due to two-dimensional projection of regulatory distances

dissimilarities (true values of the regulatory distance) on the horizontal axis and represented ones on the vertical axis.

References

Augier, P., O. Cadot, and M. Dovis (2014), "NTM Harmonization, Profits, and Productivity: Firm-level Evidence from Morocco," Unpublished Manuscript, University of Lausanne.

Bairoch, P. and R. Kozul-Wright (1996), "Globalization Myths: Some Historical Reflections on Integration, Industrialization and Growth in the World Economy," WIDER Discussion Paper No. 113, Geneva: United Nations.

Baldwin, R. and D. Taglioni (2006), "Gravity for Dummies and Dummies for Gravity Equations," NBER Working Paper 12516, Cambridge, MA: National Bureau for Economic Research.

Cadot, O., M. Maliszewska, and S. Saez (2010), "Non-tariff Measures: Impact, Regulation, and Trade Facilitation," in G. McLinden, E. Fanta, D. Widdowson, and T. Doyle (eds), *Border Management Modernization*, Washington, DC: World Bank.

Chen, M. X. and A. Mattoo (2008), "Regionalism in Standards: Good or Bad for Trade?," *Canadian Journal of Economics* 41(3), 838–863.

Disdier, A-C., L. Fontagné, and O. Cadot (2015), "North-South Standards Harmonization and International Trade," *The World Bank Economic Review* 29(2), 327–352.

Fontagné, L., G. Orefice, R. Piermartini, and N. Rocha (2013), "Product Standards and Margins of Trade: Firm Level Evidence," CEPII Working Paper No. 2013-06, Paris: CEPII.

Head, K. and T. Mayer (2014), "Gravity Equations: Workhorse, Toolkit, and Cookbook," in G. Gopinath, E. Helpman and K. Rogoff (eds), *Handbook of International Economics* (volume 4), Elsevier, pp. 131–195.

Horn, H., P. Mavroidis, and A. Sapir (2009), "Beyond the WTO? An Anatomy of EU and US Preferential Trade Agreements," Bruegel Blueprint Series, vol VII, Brussels: Bruegel Foundation.

Imani Development (2009), *Corporate Social Responsibility in Sub-Saharan Africa – Built In or Bolted On? A Survey on Promoting and Hindering Factors*, Eschborn, Germany: Deutsche Gesellschaft für Technische Zusammenarbeit [GTZ] GmbH/BMZ/British High Commission South Africa.

Jaud, M., O. Cadot, and A. Suwa-Eisenmann (2013), "Do Food Scares Explain Suppliers Concentration: An Analysis of EU Agri-food Imports," *European Review of Agricultural Economics* 40(5), 873–890.

Jouanjean, M-A. (2011), "Standard, Reputation and Trade: Evidence from US horticultural imports refusals," LICOS Discussion Paper 281/2011, Leuven, Belgium: Katholieke Universiteit Leuven.

Kirk, R. (2010), "Addressing Trade Restrictive Non-Tariff Measures on Goods Trade in the East African Community," World Bank Africa Trade Policy Note #7, Washington, DC: The World Bank.

Lesser, C. (2007), "Do Bilateral and Regional Approaches for Reducing Technical Barriers to Trade Converge towards the Multilateral Trading System?," OECD Trade Policy Working Paper No. 58, Paris: Organisation for Economic Co-operation and Development.

Maur, J-C. and B. Shepherd (2011), *Preferential Trade Agreements and Development: Issues and Implications*, Washington, DC: The World Bank.

Piermartini, R. and M. Budetta (2009), "A Mapping of Regional rules on technical barriers to trade," in A. Estevadeordal, K. Suominen and R. Tech (eds), *Regional Rules in the Global Trading System*, Cambridge, UK: Cambridge University Press, pp. 250–315.

Santos Silva, J.M.C. and S. Tenreyro (2006), "The Log of Gravity," *Review of Economics and Statistics* 88, 641–658.

UN Conference on Trade and Development (UNCTAD) (2013), *Classification of Non-Tariff Measures – February 2012 Version*, Geneva: UNCTAD

World Bank (2008), *A Survey of Non-Tariff Measures in the East Asia and Pacific Region*, Washington, DC: The World Bank.

PART IV

International standardisation and development

12

You'd be surprised how much it costs to look this cheap! A case study of ISO 26000 on social responsibility

STEPHANIE BIJLMAKERS AND GEERT VAN CALSTER

A. Introduction

The International Organisation for Standardisation (ISO) is commonly viewed as an organization by and for business. Hence, heads turned when ISO signaled its intention to embark on a project that many considered to lie outside its purview. ISO's core business is the developing of technical standards. In this area, it has built a global reputation and credibility over the course of recent decades. The out-of-the-box project was the development of ISO 26000, an international guidance document in the area of social responsibility (SR), a public policy issue that has been hotly debated in the recent decade.

ISO 26000, officially titled the 'International Guidance Standard on Social Responsibility – ISO 26000' was published in November 2010.[1] It joined a large number and diversity of existing SR initiatives providing voluntary guidance on SR. These initiatives have come to govern the SR of organizations, and in particular companies, in the absence of international law effectively disciplining their activities and operations in the interest of human rights and the environment. ISO 26000's added value within a complex, diverse and dense field of SR governance initiatives was, and remains, anything but clear-cut.

The outcome document is in many ways different from the management systems standardization (MSS) approach to SR. This approach was

[1] ISO defines a standard as 'a document, established by consensus and approved by a recognized body that provides, for common and repeated use, rules, guidelines or characteristics for activities or their results, aimed at the achievement of the optimum degree of order in a given context'. ISO/IEC Guide 2:2004 Standardization and related activities – General Vocabulary [2004], definition 3.2.

initially intended by ISO in this domain. This chapter argues that the MSS approach would have given ISO edge in the SR domain, which it currently lacks. ISO 26000 was developed and adopted as a 'guidance standard' only. The standard focuses on the substantive elements of SR. Clause 2.18 of the standard defines SR as:

> responsibility of an organization for the impacts of its decisions and activities on society and the environment, through transparent and ethical behaviour that contributes to sustainable development, including health and the welfare of society; takes into account the expectations of stakeholders; is in compliance with applicable law and consistent with international norms of behaviour; and is integrated throughout the organization and practised in its relationships.

ISO 26000 is 'not intended or appropriate for certification purposes or regulatory or contractual use' (Clause 1). Hence, organizations cannot claim or seek compliance with ISO 26000 through, for instance, assurance or certification. This move was welcomed by non-governmental organizations (NGOs). They were concerned ISO 26000 might otherwise facilitate misrepresentations of lukewarm SR performances. National standards based on ISO 26000 have, however, surfaced to which organizations can get certified. At least thirteen cases of rogue ISO 26000 certificates have also been identified.[2] Such misuse adds to the confusion about what ISO 26000 is and what it is set out to be.

ISO 26000 proclaims that it aims to assist companies in contributing to the overall objective of sustainable development (SD), which it phrases as follows (Clause 3.3.5):

> Social responsibility has the organization as its focus and concerns an organization's responsibilities to society and the environment. Social responsibility is closely linked to sustainable development. Because sustainable development is about the economic, social and environmental goals common to all people, it can be used as a way of summing up the broader expectations of society that need to be taken into account by organizations seeking to act responsibly. Therefore, an overarching objective of an organization's social responsibility should be to contribute to sustainable development.

To push forward the SD agenda, in a manner considered legitimate and credible, ISO recognized that its work needed to engage nontraditional

[2] A Henriques, 'Standards for Change? ISO26000 and Sustainable Development', International Institute for Environment and Development, Shaping Sustainable Market Series (2012), 18.

stakeholders. It also had to be attentive to the needs of the developing countries, for which carefully carved procedural arrangements were put in place. The joint leadership of the Working Group (WG) is a case in point. It fell to the Swedish Standards Institute and Brazil's Association of Technical Standards: a neat division between an 'established' economy and one in transition.

ISO 26000 was developed over the course of about nine years, through an intensive process of negotiations, involving more than seven hundred experts from different stakeholder groups. It shaped ISO 26000 into an 'odd one out' among ISO standards – and an expensive one as well. Millions of dollars were spent on expert participation. ISO 26000 itself is sold on ISO's website for CHF 198 (more or less Euro 190). Most influential SR standards, including the UN Global Compact, the OECD Guidelines for Multinational Enterprises and the Equator Principles, are publicly available to all, free of charge. Charging for accessing ISO 26000 aligns with ISO's business model. It raises eyebrows, however, where the public policy nature of SR and its overall purpose of SD are concerned.

This chapter examines the cost-benefit for ISO 26000. The first section of the chapter goes back to the origins of the standard to determine what made ISO embark on the project and why ISO 26000 was developed as a non-certifiable guidance standard. The second section examines whether ISO 26000, and standardization of SR in the broader sense, is appropriate to further the SD agenda, and what role ISO can and cannot/should not play in it. The argument holds that the money spent on developing and procuring the standard may have been more efficiently spent elsewhere – that is, on alternative processes of SR governance that promise to, and actually make, companies' clock tick: the pursuit of civil tort liability cases against subsidiaries and holding companies liable before jurisdictions in the 'developed' world.

B. The origin of ISO 26000

I. The recommendations of COPOLCO

In 2001, ISO adopted a resolution requesting ISO Committee on Consumer Policy (COPOLCO) to consider the viability of an International Standard in the area of social accountability. It was the first step toward the establishment of ISO 26000 and, arguably, one of strategic relevance. The resolution was adopted at a time when consumers had shown an increasing concern about the social integrity of corporations in the global market place. ISO was keen on getting involved in the domain of SR. COPOLCO, the body responsible for promoting consumer interests, seemed the right body to identify a market demand that could provide justification for a future ISO deliverable and, more generally, a role for ISO in the SR domain. COPOLCO delegated the responsibility to prepare the report to the Consumer Protection in the Global Market Working Group, which delivered its findings at the COPOLCO Plenary Meeting in Trinidad and Tobago, June 2002.

ISO mandated COPOLCO to consider the desirability and feasibility of a 'standard'. This language exempted COPOLCO from having to consider the suitability of alternative approaches than a standard. It gave COPOLCO considerable discretion to present a case in favor of the MSS approach, which COPOLCO did. The COPOLCO report identified a market need for an international standard providing globally relevant criteria on the basis of which consumers and other interested parties (including investors and shareholders), could make meaningful assessments and better informed decisions with respect to a firm's corporate responsibility practices. An MSS could be valuable for companies seeking to attain and measure improvements to their internal management in function of obtaining corporate responsibility objectives, the report noted. It would provide a systematic approach to internal management that could translate into verifiable and measurable outputs to which firms could self-declare compliance and get certified by authorized third parties.[3]

COPOLCO also reasoned that an internationally agreed-on standard could introduce clarity and certainty amid an expanding and increasingly complex landscape of corporate SR (CSR) initiatives. At the time of drafting the report, many actors had taken concrete action toward establishing CSR-related initiatives.[4] COPOLCO identified 'a tremendous range in quality, content, comprehensiveness and operability among them', which could cause confusion and disengagement by actors from trying to 'reward good behaviour'.[5] A future standard could facilitate trade and, through the harmonization of divergent management approaches between countries, remove barriers and promote equal opportunities in cross-border trade, COPOLCO noted.

[3] ISO Consumer Policy Committee, 'The Desirability and Feasibility of ISO Corporate Social Responsibility Standards', report prepared by the Consumer Protection in the Global Market Working Group of the ISO Consumer Policy Committee (2002), V.

[4] For an overview of relevant initiatives, see ISO Consumer Policy Committee, ibid.

[5] ISO Consumer Policy Committee, supra note 3, IV.

COPOLCO recommended the development of an MSS on corporate responsibility (CR). This presented justification not only for ISO to get involved in the domain of CSR but also to develop the *type* of standard in which ISO had a unique competence. ISO had obtained such competence through the development of ISO 9000 quality MSS and ISO 14000 environmental MSS. The report noted how ISO could build on 'the expertise, infrastructure, and widespread usage' of these standards'. The integration of a future MSS on CR with existing MSS could make for a holistic and global approach to CR. COPOLCO envisaged 'ISO CR MSS approach as being a fully compatible addition to ISO 9000 and ISO 14000 MSS approach, with the same status, profile, and operational objectives as ISO 9000 and ISO 14000 MSSs'.[6]

A new MSS on CR could draw from the great appreciation for the value of ISO 14001 and ISO 9001 standards that existed at that time, as illustrated by their global uptake.[7] The occasion of a half million firms having been ISO 9001 certified presented new momentum to building on this success. It also created 'an immediate base of potential clientele that would be in a position to adopt ISO CR MSS to their existing ISO 9000 and ISO 14000 operations', the report noted.[8] The implementation of a future ISO CR MSS would also require few changes from companies because those that are compliant with ISO 9001 and ISO 14001 would already have the required architecture and operational culture in place.[9]

The COPOLCO report provided justification for the *type* of standard in which ISO not only had a unique competence but also in which it could distinguish itself from other initiatives, hence establish its own relevance and authority in the CSR domain. The opportunity to develop a new MSS on CR came at an opportune time, especially after ISO had missed the boat in relation to the development of MSS for occupational health and safety. Previously, ISO had considered the initiation of an ISO standard on Occupational Safety and Health Management Systems. This initiative did not materialize within the ambit of ISO then, however. The former Occupational Safety and Health Branch of the International Labour Organisation (ILO), was deemed a more appropriate body for the project. In co-operation with the International Occupational Hygiene Association,

6 Ibid, VI.

7 International Institute for Sustainable Development (IISD), 'A Background Paper to the International Organization for Standardization's (ISO) Strategic Advisory Group on Corporate Social Responsibility', ISO CSR Briefing #1(2003). Available at: http://www.iisd.org/pdf/2003/standards_csr_background.pdf. Last retrieved 11 January 2015.

8 ISO Consumer Policy Committee, supra note 3, 56.

9 Ibid, VII.

it had published the ILO Guidelines on Occupational Safety and Health Management Systems in December 2001.[10]

The relevance of a future MSS was reaffirmed by COPOLCO in relation to the SD agenda. The Plan of Implementation of the World Summit on SD (WSSD), held in September 2002, explicitly mentioned ISO standards in relation to actions that governments should take 'to improve social and environmental performance through voluntary initiatives, including environmental management systems, codes of conduct, certification and public reporting on environmental and social issues'.[11] COPOLCO recommended the establishment of a Strategic Advisory Group on CR, with similar terms of reference as the Strategic Advisory Group on the Environment (SAGE), established after the 1992 Rio Conference.[12]

II. The recommendations of the Advisory Group on Social Responsibility

An ISO Strategic Advisory Group on Social Responsibility (SAG) was established in September 2002, by ISO TMB resolution 78/2002,[13] with the terms of reference 'to determine whether ISO should proceed with the development of ISO deliverables in the field of corporate social responsibility' and 'if so, to determine the scope of the work and the type of deliverable'. On 30 April 2004, the SAG presented its recommendations to ISO TMB,[14] as well as a working report (WR).[15] The SAG noted that the SR deliverable should be 'A guidance document, and therefore not a specification document against which conformity can be assessed'. The SAG thus did not recommend an MSS, as COPOLCO had suggested.[16]

This shift can be explained in relation to the composition of the SAG, which reflected a greater diversity of stakeholder views than was the

[10] The British Standards Institution (BSI) sought to give new vigor to ISO's potential role in this area by launching an official proposal to establish a new field of ISO technical activity on occupational health and safety management. This counter-initiative was abandoned, however, amid strong protests and campaigning. ISO Consumer Policy Committee, supra note 3. It was later re-introduced however. In 2013, ISO began the development of ISO 45001, a standard to set requirements for occupational health and safety management systems.

[11] IISD, supra note 7. [12] ISO Consumer Policy Committee, supra note 3, VII.

[13] Technical Management Board resolution 78/2002, September 2002 (2002).

[14] ISO Strategic Advisory Group on Social Responsibility, Recommendations to ISO Technical Management Board, ISO/TMB AG CSR N32 [2004].

[15] ISO Strategic Advisory Group on Social Responsibility, Working Report on Social Responsibility (2004).

[16] ISO Strategic Advisory Group on Social Responsibility, supra note 14.

case for COPOLCO. The SAG represented a balance and compromise among diverse interests from business, labour, environmental groups, consumer groups, civil society, UN bodies and developing countries.[17] COPOLCO was composed of a homogenous group of member standardization bodies, thus reflecting mainly the interests of industry, ISO's traditional constituency.[18] According to Castka and Balzarova (2008), the decision to recommend a guidance standard was largely politically driven and reflected a move toward an approach more prone to find consensus among the different and competing interests.[19]

Reaching a consensus on a certifiable MSS setting fixed requirements would have proved problematic at the time. The concept of SR and its substantive issues had not sufficiently matured yet. Its complexity would complicate efforts to attain a commonly held vision on SR. For standard-setting to be feasible, communications on provisions reflecting such a common vision had to satisfy a certain measure of clarity, comprehensiveness, consistency and credibility, the report notes.[20] That public interest issues fell outside ISO's core business of developing technical standards seemed to provide additional support for the development of a guidance standard clarifying the controversies surrounding SR. The SAG set the condition that 'ISO recognizes that social responsibility involves a number of subjects and issues that are qualitatively different from the subjects and issues that have been already dealt with by ISO'.[21]

The desirability of certification as a market-based practice in the area of SR was questioned. The WR of the SAG identified various views in relation hereto.[22] Certification was said often to be 'prohibitively costly for small and medium organizations'. Also, it was noted how 'the quality of oversight provided by certifiers is not always consistent' and that 'there may be other more constructive ways to provide assurance'.[23] At the International Conference on Social Responsibility in Stockholm, 21–22 June

[17] Technical Management Board resolution 35/2004, 24–25 June 2004, Stockholm, Sweden (2004).

[18] IISD, 'CSR Briefings and Documents' (2013). Available at: http://www.iisd.org/standards/csr_documents.asp. Last retrieved 11 January 2015.

[19] P Castka and M Balzarova, 'The Impact of ISO 9000 and ISO 14000 on Standardization of Social Responsibility – An Insider Perspective' (2008), *International Journal of Production Economics* 113, at 84–85.

[20] ISO Strategic Advisory Group on Social Responsibility, supra note 15, 67.

[21] ISO Strategic Advisory Group on Social Responsibility, supra note 14.

[22] W Baue, 'Stakeholder Engagement, Verification, and Transparency in ISO Social Responsibility Standards: An Interview with ISO Secretary-General' (2004). Available at http://www.socialfunds.com/news/save.cgi?sfArticleId=1484. Last viewed 11 January 2015.

[23] ISO Strategic Advisory Group on Social Responsibility, supra note 15, 70.

2004, NGOs expressed concern about the potential misuse of certification by companies to create false representations of their performance and the malpractices that have come to taint the certification industry.[24] The SAG recommended that a future SR guidance document should exclude third-party certification.

If ISO were to embark on the development of a new SR guidance standard, it would have to legitimize its decision-making in relation to substantive as well as procedural issues in the development of the standard.

First, ISO would have to act with a certain restraint in terms of defining the substantive elements of SR. The authority and legitimacy of ISO as a private standard-setting body engaging in public policy standard-setting was not self-evident. The WR notes how ISO is not an international organization and may lack the legitimacy and competence to define its own substantive expectations in the area of SR. Some argued that SR standards should incorporate already existing universally applicable standards for SR, yet respect a delicate balance between recognizing and promoting these standards, without reducing or obscuring their intent, scope and applicability. It suggests that ISO should show deference to the relevant standards defined by international law, which are of a more authoritative nature, due to their democratic foundations. The SAG set the condition that 'ISO recognizes that it does not have the authority or legitimacy to set social obligations or expectations which are properly defined by governments and intergovernmental organizations'.[25]

The SAG also implicitly called on ISO to be cautious, that is, to take into account the potential effects of its activities on similar standardizing activities by other nongovernmental SR initiatives. The SAG WG held that the activities of other bodies that had developed or were developing SR standards, norms, guidelines and tools, including nongovernmental international SR initiatives (e.g. the Global Reporting Initiative [GRI], the Forrest Stewardship Council [FSC] and the Fair Labor Association [FLA]) had to be considered. Because the private standard-setting bodies were not subject to democratic decision-making processes, unlike the UN and its agencies, some argued that they should not be treated as equivalent or be given the same recognition as public institutions.[26]

In other words, these private initiatives lacked the authority deriving from public legal mandates. Their decisions therefore could be considered

[24] Castka and Balzarova, supra note 19, 84.

[25] ISO Strategic Advisory Group on Social Responsibility, supra note 14.

[26] ISO Strategic Advisory Group on Social Responsibility, supra note 15, 68.

to carry an equivalent weight as similar decisions by ISO or, where not reflecting the universal principles set out in authoritative international legal instruments, to be of lesser weight. Deference to the standards set by these SR initiatives seemed uncalled for. Engaging and drawing from the expertise of these initiatives in the development of a new ISO standard could, however, give legitimacy to the standard. The SAG set the condition that 'ISO recognizes the difference between on the one hand, instruments adopted by authoritative global inter-governmental organizations . . . and on the other hand, private voluntary initiatives that may or may not reflect the universal principles contained in the above instruments'.[27]

Certain deference to the special authority of the ILO was also deemed warranted. The SAG included one representative of the ILO, the Director of ILO's Multinational Enterprises Program.[28] It appeared that the ILO was averse to ISO entering its domain of labor standard-setting.[29] The ILO allegedly did not believe ISO would respect and uphold the benchmarks that ILO standards had set.[30] The SAG noted that the 'ISO recognizes through a formal communication the ILO's unique mandate as the organization that defines, on a tripartite basis, international norms with respect to a broad range of social issues'.[31] The ILO was one of the first bodies to negotiate a MoU on SR with ISO, on 4 March 2005, acknowledging the ILO as 'the internationally-recognized, representative body competent to establish and supervise international labour standards, relating notably to rights at work, employment, social protection, and social dialogue'.[32]

Second, ISO had to be proactive in ensuring that appropriate structures would be in place to give its standard-developing processes legitimacy.

[27] ISO Strategic Advisory Group on Social Responsibility, supra note 14.

[28] IISD, supra note 18.

[29] R Chhabara, 'ISO 26000 – Responsibility's New Guiding Light', (2010) Ethical Corporation. Available at: http://www.ethicalcorp.com/business-strategy/iso-26000-responsibilitys-new-guiding-light. Last retrieved 11 January 2014.

[30] S Wood, 'Will ISO 26000 Corner the Market for International Social Responsibility Standards? Competition for Transnational Regulatory Authority', Law and Society Annual Meeting, Denver, CO (2009). Available at: http://citation.allacademic.com/meta/p_mla_apa_research_citation/3/0/4/2/1/pages304218/p304218–1.php. Last retrieved 11 January 2015.

[31] ISO Strategic Advisory Group on Social Responsibility, supra note 14.

[32] Memorandum of understanding between the International Labour Organisation and the International Organisation for Standardization in the field of social responsibility (2005). Available at: http://training.itcilo.org/actrav/courses/2005/A3–50909_web/resources/technical_files/3.4%20ILO-ISO_agreement.pdf. Retrieved last 11 January 2015.

This called for engagement with stakeholders other than those involved in previous ISO processes, especially from developing countries, and the organization of stakeholder engagement around dimensions of legitimacy, expertise and representation. The SAG notes, 'ISO reviews its processes and where necessary makes adjustments to ensure meaningful participation by a fuller range of interested parties'.[33]

ISO would have to take account of the transboundary effects of its standardization activities, especially on developing countries. The WR recognized the relevance of providing a platform for marginalized voices, especially in countries with less economic development. SR initiatives existing then did not reflect their priorities and concerns, due to their low involvement in their development, risking unwarranted costs being inflicted on these countries, it noted. An international SR standard could level the playing field, 'not just for business organizations, but also for developing and developed countries, and for stakeholders of all types'. The SAG recommended that 'ISO should make every effort to ensure that developing countries can meaningfully participate in this work'.[34]

The recommendations by the SAG were significant. Through their adoption in Resolution 35/2004[35] of ISO's Technical Management Board (TMB), which kick-started the development of ISO 26000, they contributed to shaping the type, substance and scope of ISO 26000. They were sanctioned by the TMB as a main reference in the development of the guidance document and were subsequently integrated in the New Work Item Proposal adopted in January 2005. Moreover, as elaborated later in the chapter, some controversial issues that the SAG recognized in its WR, involving the legitimacy of ISO' involvement in the SR domain, later resurfaced during the WG negotiations. In this chapter, they serve as a point of reference in the argument put forth that ISO 26000 was an expensive endeavor for the value it added to the SR domain.

C. The value of ISO 26000

I. *Standards and costs*

The development of ISO 26000 was a cost-intensive enterprise. Running over the course of five years, the negotiation process was intense and

[33] ISO Strategic Advisory Group on Social Responsibility, supra note 14. [34] Ibid.

[35] Technical Management Board resolution 35/2004, supra note 17.

complex. It aimed to build consensus around a future SR standard among the widest range of stakeholders.[36] Eight plenary meetings were held across different continents, in addition to numerous committee and subcommittee meetings at ISO and national level. The process was designed carefully to ensure an appropriate balance of stakeholder representation and North-South representation. The WG SR was organized around six stakeholder categories reflecting the main interests affected: consumers, government, industry, labor, NGO and 'service, support, research and others'. The members of the WG SR (1 May 2010)[37] counted a total of 450 nominated experts, 210 nominated observers, 99 countries represented and 42 international organizations and other interested organizations. Special arrangements were put in place to secure funding for the participation of smaller organizations, notably NGOs from the South. To ensure the transparency of the process, translations of successive drafts were made available in multiple languages – English, French, Spanish, Arabic and Russian. By the time consensus on ISO 26000 was reached on 13 September 2010,[38] millions of dollars had been spent.[39]

An important source of revenue for ISO for financing such projects is the sale of access to ISO 26000 standards. This practice, as noted, has been controversial. As a nongovernmental organization, ISO has a legal mandate to sell access to standards in support of its objectives.[40] The revenues obtained thereby, supplemented by the income from royalties

[36] ISO defines a consensus as 'general agreement, characterized by the absence of sustained opposition to substantial issues by any important part of the concerned interests and by a process that involves seeking to take into account the views of all parties concerned and to reconcile any conflicting arguments'. Consensus need not imply unanimity. ISO/IEC Directives, Part 1. Procedures for the technical work (2012).

[37] ISO, 'Report of the Secretariat, Agenda Item 6, Dokument N 183' (2010).

[38] Sixty-six P-members (participating members) voted in favor and five voted against. Abstaining countries were Algeria, Australia, Austria, Bangladesh, Germany, Hungary, Iceland, Islamic Republic of Iran, New Zealand, the Former Yugoslav Republic of Macedonia and Vietnam. The countries that issued a negative vote were the United States, Cuba, India, Turkey and Luxembourg.

[39] There is no official financial overview of the ISO 26000 project from ISO. Some estimate that the costs for the participation of the experts in the WG and the national meetings combined, which are born by national member bodies and stakeholders, rose up to at least $72 million US. 'Cost Estimation' (2013). Available at: http://www.26kestimation.com/html/cost_estimation.html. Last retrieved 11 January 2015.

[40] ISO statute defines the organization's objective as 'to promote the development of standardization and related activities in the world with a view to facilitating international exchange of goods and services and to developing cooperation in the spheres of intellectual, scientific, technological and economic activity.' ISO, ISO Statutes (2003). Available at: http://www.iso.org/iso/statutes.pdf. Last retrieved 11 January 2015.

on copyrights and services, are used to finance the operation of ISO Central Secretariat and make up 30 percent of its total budget.[41] ISO's total income consists of revenues obtained by member body subscriptions; of sales, copyrights and services; funding for developing countries; and gains on financial investments. These used 'to meet the direct costs of the organization, investments in specific projects and developing country programmes'.[42] The standardization work itself is financed by the Member Bodies, which provide the experts and resources for the technical committees to function. Stakeholders, and in particular industry and business sectors, provide financial support for the participation of professionals.[43]

The business model of ISO seems to put forth two justifications for why ISO standards cost money.

First, there is the public relevance of the standardization work. The Danish National Standards Organizations indicates that the public relevance of standards is one of the main reasons why the authorities and interests groups support ISO generously with millions of euros annually. Reasoning from the traditional standardization policy perspective, it notes how standardization can serve as a tool to 'eliminate technical barriers to trade globally and in Europe to create a common platform'. In the absence of the revenues obtained from the sale of standards, ISO could still operate but would not be able to reach its maximum activity.[44]

Second, the costs are born by those who use the standards. The price is justified on the basis of the value it provides to these users. The value is linked to the help that it offers organizations 'with development and completion of technical requirements on health, safety and environment'. The value is not in 'the words or the number of pages it contains', but it is said to be in 'the safety the standards give you, regarding compliance with contract and Directive requirements, the export market it opens for you, the many cost savings because experts have given you the best solutions, compliance and improvements to the environment and much, much more'.[45]

41 E Shamir-Borer, 'Legitimacy without Authority in Global Standardization Governance: The Case of the International Organization for Standardization (ISO)' in S Cassese, B Carotti, L. Casini, E. Cavalieri and E MacDonald (eds), *Global Administrative Law: The Case Book* (2012), 166.

42 ISO, 'Annual Report 2012: Our Finances', 2013. Available at: http://www.iso.org/iso/annual_report_finances-2012.pdf. Last retrieved 11 January 2015.

43 Shamir-Borer, supra note 41, 166.

44 Danish Standards, 'The Reason Standards Cost Money' (2013). Available at: http://www.ds.dk/en/publications/the-reason-standards-cost-money. Last retrieved 11 January 2015.

45 Ibid.

These justifications for why ISO standards cost money do not seem to uphold as far as ISO 26000 is concerned.

First, the justification put forth is ill suited because ISO 26000 was developed not from a traditional standardization policy perspective but from a public policy perspective. If one were to assess whether the money spent on the negotiation processes of ISO 26000 was justified, one would have to assess the public relevance of ISO 26000 in terms of its contribution to more responsible companies and the overall objective of SD.

Second, by its nature, scope, structure and content, ISO 26000 does not necessarily provide the aforementioned material benefits to organizations. These benefits are linked to standards that aim to bolster company's internal capacity, and in particular MSS, which ISO 26000 is not. This seems to undermine the justification for why accessing ISO 26000 costs as much as it does.

As a starting point to analyzing why ISO 26000's contribution to the SR falls below expectations and hence was an overly costly exercise for that what it offers today, the following section focuses on the second argument. It outlines the benefits of MSS to companies and the opportunities missed by ISO to develop an MSS that could have made corporate incentives work in the interest of promoting SR.

II. *Driving business toward SR through MSS*

1. The value of MSS to companies

First, MSS is designed for companies to achieve, measure and communicate externally their internal capacity to meet substantive objective and goals.[46] ISO 9000 family on effective quality management systems and ISO 14000 family on environmental management systems are successful examples. They contain fundamentals, guidelines and guidance with respect to creating, improving and enhancing management systems that can assist companies in operationalizing quality and environmental requirements. They are generic – that is, they can be applied to all organizations, irrespective of sector or industry – and process based. The focus is on the management of the application, identification and interaction of processes, and systems of processes in an organization. Companies have the discretion to determine their own substantive objective and goals. MSS assists companies in the establishment and identification of objectives, the achievement of these objectives and the demonstration of the ability

[46] ISO Consumer Policy Committee, supra note 3.

to achieve these objectives. Had ISO 26000 been established as a management system standard, it would have assisted companies in enhancing their ability 'to attain and measure improvements in CR performance'.[47]

Second, companies can save costs by implementing MSS. MSS guides companies on how to systemize their approach to management – to introduce a form of order into how they go about business. This can result in a more efficient and effective use of resources. MSS can also remove the costs involved in finding the most valuable approach. MSS is said to offer a model that provides the best solutions, reflecting that what experts consider the state of the art. Being anchored on the notion of 'continual improvement', management systems are subject to continuant revision to identify areas of improvement.

Third, MSS can give companies a powerful means to create representations of their practices, which can be subjected to verification, assurance and certification. MSS promotes the operationalization of CR requirements in a manner that is systemic, coherent and credible. Companies are recommended to use the 'plan-do-check-act' (PDCA) systematization methodology, which seems particularly suited for this purpose. To verify their progress and results, companies can submit their approach to management to an audit, a process that determines the extent to which this approach fulfills the requirements set in ISO 9001 or ISO 14001.

Audits can be conducted for internal purposes (first party) and serve as a basis for companies to self-declare conformity.[48] They can be conducted also by, or on behalf of, customers (second party),[49] or by an independent external party (third party). Third-party audits can serve as a basis for obtaining assurance that an organization's management system is compliant with the requirements of the standard, for example, in the form of 'certification' or 'registration'. ISO permits companies to get certified (i.e. to undergo a conformity assessment and receive a certificate confirming conformity to the requirements set in the standard) from a third-party certification body that has been accredited.[50] ISO does not require that

[47] Ibid., VII.

[48] Self-declaration is possible for ISO 14001 but not for ISO 9001.

[49] ISO published ISO 19011:2011 'Guidelines for Auditing Management Systems' on 15 November 2011. ISO 19011 provides guidance on internal audits and audits conducted by consumers on their suppliers. It replaced the first edition (19011:2002), which was technically revised. ISO 19011 can be used for self-declaration. ISO 19011:2011 Guidelines for Auditing Management Systems (2011).

[50] Accreditation in this regard entails that a certification body or person has been given formal recognition by an authoritative body of its competence to carry out certification

companies undergo such conformity assessment processes,[51] nor does it certify itself.

Fifth, certification can send strong signals to consumers and other interested parties, as well as regulators and markets. In the case of ISO 9001 and ISO 14001, it can present the company as providing good-quality products and being environmentally responsible. In other words, the implementation of an MSS not only can help companies improve their performance but also communicate results and progress in a manner that is systematic, coherent and, due to verification, assurance and certification, seemingly credible. This is said to assist companies in improving their stakeholder relations and enhancing their corporate image.[52] Certification, in particular to ISO 9001, has been a hugely popular tool to improve supplier-customer relations in the global supply chain. It has also served as a powerful driver in the diffusion of ISO 9001 globally.[53]

2. Missed opportunities

ISO 26000 was not designed to help companies engage in commercial relationships. It does not set any requirements. More generally, the absence of requirements makes assessments to verify the accuracy of claims by companies of adhering to a management system standard difficult if not impossible. It follows that as a tool, ISO 26000 is of limited use in providing credible representations of being compliant with ISO 26000 or in obtaining credible assurance that suppliers and purchasers are socially responsible. In fact, communicating compliance or progress without verification creates the risk that claims amount to an inaccurate representation of company's actual performance. Stakeholders can be mislead into believing that a company is socially responsible when in reality this is not the case. It makes ISO 26000 susceptible to becoming the public relations tool for companies to 'green-wash' their corporate image that many, and

within a particular business sector. ISO Strategic Advisory Group on Social Responsibility, supra note 15, 31.

51 ISO has, however, developed international standards that set criteria and guidance for the certification of management systems. See ISO/IEC 17021:2011. Conformity Assessment: Requirements for Bodies Providing Audit and Certification of Management Systems (2011).

52 Castka and Balzarova, supra note 19, 74–87.

53 By December 2012, 1,101,272 ISO 9001 certificates had been issued by International Accreditation Forum–accredited certification bodies, in 184 countries and economies. See 2012a. ISO, 'ISO Survey of Management System Standard Certifications – 2012' (2012). Available at: http://www.iso.org/iso/iso_survey_executive-summary.pdf. Last retrieved 11 January 2015.

NGOs especially, appall. It also risks causing the reputational damage to ISO brand.

ISO 26000 stipulates that it 'is not intended or appropriate for certification purposes' and that 'any offer to certify, or claims to be certified, to ISO 26000 would be a misrepresentation of the intent and purpose of this International Standard' (Clause 1). Although it is not impossible to get certified to a guidance standard, this provision ruled certification out in the case of ISO 26000. One cannot verify the credibility of claims by companies to be compliant with ISO 26000 through a compliance audit and certification. It follows that compliance to ISO 26000 can, in principle, not serve as a de facto requirement for market access. The fact that companies cannot get certified to ISO 26000 entails that consumers but also other relevant parties in the marketplace (e.g. shareholders, investors and investment funds) cannot rely on ISO 26000 to obtain credible information relating to the SR performance of companies to inform their decision-making. Companies cannot rely on certification to ISO 26000 to assess the performance of purchasers or suppliers in their global supply chains. It follows that ISO 26000 will not offer much commercial benefit to companies seeking to reap the rewards from these market dynamics (e.g. trust, investments, revenues) by proclaiming that they rely on, and are compliant with, ISO 26000 to boost their public image and to distinguish themselves from those companies that are not compliant. Such proclamation amounts to misuse of ISO 26000.

ISO 26000 is moreover not intended for 'regulatory or contractual use' (Clause 1). This reduces its potential value to companies as well as the likelihood that it will enjoy uptake by not only companies seeking to impose contractual social requirements on suppliers and buyers but also governments seeking to require or promote SR among companies through regulatory or policy measures. For instance, governments cannot condition the award of public procurement contracts, technical assistance or subsidies to social responsible performance as demonstrated through certification or to consider compliance to ISO 26000 in the monitoring or enforcement of regulations. ISO 26000 further decouples implementation and compliance from potential legal effects by stipulating that it 'is not intended to provide a basis for legal actions, complaints, defenses or other claims in any international, domestic or other proceedings' (Clause 1). It follows that the implementation of ISO 26000 is of no avail to companies in terms of meeting regulatory or contractual requirements. Such requirements would amount to misuse of ISO 26000 and should not exist in the first place.

The exclusion of ISO 26000 for regulatory or contractual use also largely divorces the utility of ISO 26000 from litigation strategies. It discourages companies to rely on ISO 26000 to avoid corporate liability claims that may arise, for example, under the civil tort liability procedures in Western jurisdictions (see more on this later in the chapter). Such legal liability risks serve as an important driver for CSR because they can inflict grave damage on a company (e.g. as a result of negative publicity and pressure on stock prices).

These aforementioned potential economic benefits to companies (e.g. assistance in implementing management systems, meeting contractual requirements, cost saving, avoiding litigation costs) are mainstream motivating factors for why companies choose to adopt ISO standards and opt for ISO certification, as well as the diffusion of ISO standards more generally.[54] The fact that ISO 26000 is, however, not a management system standard and does not provide the potential commercial benefit to companies that are commonly linked hereto undermines the justification for why obtaining access to ISO 26000 standard is so costly.[55] There are few material incentives for companies to engage with ISO 26000, as is the case for other market actors. The price for accessing the standard risks undermining the usage and dissemination of the guidance standard, and as such, risks jeopardizing the effectiveness of ISO 26000 in encouraging corporations to act on their responsibility.[56]

As outlined earlier, there was extensive discussion before the adoption of the work item proposal in 2004 about what justification an MSS could provide for ISO to get involved in the SR domain. ISO was right to consider the MSS approach at the outset. At least in theory, the evaluation and measurement of performance on the basis of MSS, as demonstrated by ISO 9001 and ISO 14000 standards, can be powerful in driving companies to action, be it for material incentives and for the specific purpose of improving the internal capacity to achieve SR results. For this reason and

[54] Castka and Balzarova, 'ISO26000 and Supply Chains – On the Diffusion of the Social Responsibility Standard' (2008) *International Journal of Production Economics* 111, at 274–286. There is, however, no uncontested evidence that the successful implementation of ISO9001 translates into these benefits. See Castka and Balzarova, supra note 19.

[55] This is not to say that there is no managerial relevance to ISO26000 at all. See R Hahn, 'Standardizing Social Responsibility? New Perspectives on Guidance Documents and Management System Standards for Sustainable Development (2012) 4 *Engineering Management* 59, at 717–727.

[56] 'Best Prices of ISO26000' (2013). Available at: http://www.26k-estimation.com/html/best_prices_for_iso_26000.html. Last retrieved 11 January 2015.

for these particular purposes, an MSS on SR potentially would have been more effective in actually moving companies, a quality that ISO 26000 currently lacks. This does not take away, however, the fact that there are pitfalls to the application of MSS, and market-based measurement tools and practices more generally to the goal of improving SR performance. These are addressed later in the chapter.

What value, then, does ISO 26000 create that justifies ISO 26000 being so expensive? The following section argues that for ISO 26000, in fact, charging money for obtaining access is misplaced because large parts of the text consist of chewed over principles of international environmental and SD law. It is also ill judged, given the societal impact the label seeks to have in developing countries in particular.

III. The added value of ISO 26000: Pushing the development agenda forward?

A conceptual question that the SAG considered at the beginning was under what conditions an ISO standard on SR would be justified. Traditionally, ISO responds to a market need for standardization to harmonize a costly and unnecessary proliferation of standards. From this standardization policy perspective, the value of ISO 26000 could have been to address an unnecessary overlap or proliferation of existing SR standards providing substantive understandings of SR and hence level the playing field. This was not an issue at that time, however. The SAG could not reach a consensus about whether there was indeed an unnecessary proliferation of standards. To ascertain such proliferation, a nuanced analysis had to be undertaken that considered 'the nature of the requirements contained in, the focus of, and the objective of existing SR initiatives' it noted,[57] which the SAG did not do.

When addressing the need for international standardization, justification was grounded on public policy reasons. ISO 26000 phrases its purpose as follows (Clause 1).

> This International Standard is intended to assist organizations in contributing to sustainable development. It is intended to encourage them to go beyond legal compliance, recognizing that compliance with law is a fundamental duty of any organization and an essential part of their social responsibility. It is intended to promote common understanding in the field of social responsibility, and to complement other instruments and initiatives for social responsibility, not to replace them.

[57] ISO Strategic Advisory Group on Social Responsibility, supra note 15, 59.

ISO 26000 defines SD as 'meeting the needs of society while living within the planet's ecological limits and without jeopardizing the ability of future generations to meet their needs' (Clause 3.3.5). SR is characterized by 'the willingness of an organization to incorporate social and environmental consideration in its decision making and be accountable for the impacts of its decisions and activities on society and the environment' (Clause 3.3.1). It follows from the public policy nature of these objectives that the primary criteria for assessing the value of ISO 26000 should be linked to these objectives, that is, to contributing to better and more SR, and ultimately, to SD.

This gives rise to the question of how, and to what extent, ISO 26000 contributes to these public policy objectives. ISO 26000 was developed as an international standard providing guidance to all types of organizations, irrespective of size, sector or area of operation and that can be used as part of public policy activities. To this end, it sets out a common understanding of SR, through agreed terms and definitions, seven substantive principles of SR, stakeholder engagement, core subjects and associated issues, and methods to integrate SR throughout an organization, placing emphasis on 'the importance of results and improvements in performance on social responsibility' (see Figure 12.1).[58] By providing assistance to organizations in improving their SR performance, ISO 26000 can create value for organizations, It is, however, less clear what has been the added value of ISO 26000 as far as complementing existing SR initiatives are concerned, as well as the overarching SD agenda.

1. Clarifying SR confusion?

The added value of ISO 26000 could have been to eliminate any inconsistencies, incompatibilities or incomparability involving substantive SR standards. Such inconsistencies could cause confusion, undermining the public confidence in existing SR standards and, possibly, efforts of companies in improving their SR performance.[59] This did not seem to be the case. In fact, before and during the negotiations of ISO 26000, there were concerns that a new international guidance standard on SR could create such inconsistencies and incompatibilities itself by reinterpreting already existing SR standard presumptuously. It propelled the ILO, followed by the UN Global Compact, the OECD and the GRI to sign

[58] ISO26000:2010: Guidance on social responsibility, vi.

[59] ISO Strategic Advisory Group on Social Responsibility, supra note 15, 56.

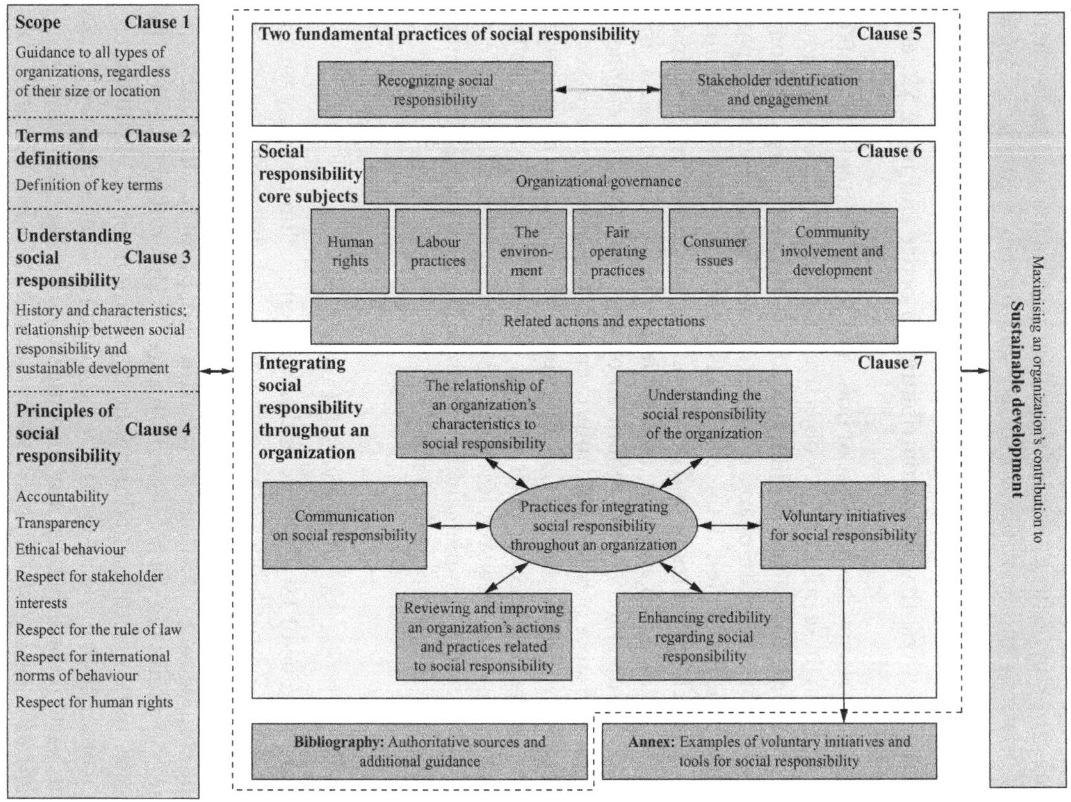

Figure 12.1. Drawing a common understanding of social responsibility

Memoranda of Understanding (MoU) with ISO, and among each other, to ensure formally the consistency of ISO 26000 with the principles and standards already set by these initiatives. The MoUs granted special procedural safeguards to these initiatives to facilitate their proactive engagement with ISO to see to it that the objectives of the MoU would be achieved.

On 4 March 2005, ISO signed a MoU with the ILO to govern their consultations and cooperation in the area of SR. The purpose of the MoU is 'to establish between the Parties cooperation with a view to ensuring that any ISO International Standard in the field of SR, and any ISO activities relating thereto, are consistent with and complement the application of international labour standards world-wide, including fundamental rights at work'.[60] The scope of this cooperation was defined broadly (Art. 1.3):

> The areas of cooperation between the Parties will include the development of any ISO International Standard in the field of SR; any activities and publications for the promotion, support, evaluation and approval of any published ISO International Standard on SR, and any periodic review of such ISO International Standard for confirmation, revision, or withdrawal. This cooperation will encompass any issues involving ILO's mandate, including but not limited to international labour standards and the interpretation and application of ILO instruments.

The MoU provided substantive rights to the ILO regarding the content of the standards. It notes that an ISO SR standard will be 'fully consistent with' the object and purpose of ILO international labour standards and their interpretation by competent bodies of ILO and 'in no way detract from these standards' (Art. 2.1).

On 5 May 2008, ISO signed a MoU with the OECD. The purpose of this MoU was phrased in wording similar to that of the ISO-ILO MoU (Art. 1).[61] Unlike the ISO-ILO MoU, it stipulates that the future ISO standard 'needs to be consistent', rather than 'fully consistent' with the OECD Guidelines and their implementation (Art. 1.1). Also, the scope of the MoU is narrower; the areas of cooperation only encompass the development and periodic review of ISO standard for confirmation,

[60] Memorandum of understanding between the ILO and ISO for standardization in the field of social responsibility 2005, supra note 32.

[61] Memorandum of understanding between the OECD and ISO in the area of Social Responsibility 2008. Available at: http://www.oecd.org/investment/mne/45330481.pdf. Last retrieved 11 January 2015.

revision and withdrawal (Art. 1.2). The MoU also defines the nature of the cooperation in more or less equalized terms, noting how the parties will be 'mutually supportive of each other' (Art. 1).

ISO singed a MoU with the UN Global Compact on 9 November 2006.[62] This MoU also defines its purpose as establishing 'between the parties cooperation', with the view of ensuring that the ISO standard and ISO activities relating thereto 'are consistent with and complement the Global Compact ten principles' (Art. 1). Its scope includes areas of cooperation on the development of the standard, activities and publications relating 'the promotion, support, evaluation and approval' of the standard and any of its periodic reviews (Art. 2.1). In terms of substantive elements, it was agreed that the future standard needed to be consistent with the ten principles of the UN Global Compact (Art. 2.1) and that ISO would address any concerns raised by the UN Global Compact, and in particular by 'the Global Compact Office and the core UN Agencies which form part of the Inter-Agency-Team, other than the ILO, which has its own MOU with ISO' (Art. 2.2).

ISO/ILO went one step further than the other MoUs by not only determining that 'the provisions of ILO instruments serve as the authoritative and definitive source of reference, and minimum base line for any elements which relate to international labour standards' (Art. 6.1), but also that ILO standards would take priority in case of conflict (Art. 2.3) and that ISO activities would 'avoid addressing ILO issues that should only be resolved through representative political or legal processes' (Art. 2.4). This provision applies to conflict that occurs 'in the context of development, and of any promotion, support, evaluation and approval, or periodic review of the SR standard', and 'involving ILO issues with any private initiative with which ISO may collaborate in the context of that Standard' (Art. 2.3). The MoU thus clearly defined a hierarchy between ISO and existing ILO standards codified in ILO instruments and reaffirmed the primacy of the ILO standards over the ISO standard. It also formally established that ISO was to show deference to the relatively greater authority of ILO as the standard-setter in area of labour, as the SAG had called for at the beginning. The MoU extended the scope of this primacy to apply not only to the aforementioned activities relating to the SR standard, but also to ISO's relations with other private initiatives.

[62] Memorandum of understanding between the UN Global Compact Office and ISO (2006). Available at: http://www.globalgovernancewatch.org/docLib/20091230_GGW-_UNGC-ISO_MOU.pdf. Last retrieved 11 January 2015.

2. Clarifying or supplementing national and international SR standards to raise the bar?

Whereas ISO recognized the primacy of ILO standards in relation to ISO standard-setting in SR, ISO was not explicit in recognizing a hierarchy of laws in relation to international and national laws more generally. In fact, research suggests that ISO did not sign the ILO/MoU because it agreed with the rationale put forward by the ILO recognizing the primacy of international labour standards (ILSs). The ILO argued that its instruments were 'adopted, implemented and interpreted on the basis of a representative, tripartite, global process in which all 177 ILO Member States participate, and therefore the ILSs should "serve as the authoritative point of reference and minimum baseline for social responsibility elements related to ILO's mandate, including in cases of conflict with national law or practice"'.[63] It is said that ISO merely recognized and accepted 'the primordial nature of regulations over voluntary standards'.[64] The Working Proposal was equally vague on this issue, noting that the standard should 'not be intended to reduce government's authority to address the social responsibility of organisations'.[65] In other words, ISO gave no clear indication that it would concede its own legitimacy in standard-setting to national sovereignty and international law.

The ambiguities about the possible legal repercussions of ISO standard-setting activities, as it related to the substantive elements of SR, resulted in ISO 26000 processes becoming caught up in the politics of international development law.[66]

Concerns arose about ISO standard-setting trespassing the ambit of laws and regulations, and reshaping states' international legal obligations.[67] As Ward (2012) elaborated on, governments were wary that the behavior and positions of governmental experts on ISO 26000 drafts might signal state practice, and possibly, contribute to the creation of customary international law, despite the fact that these representatives acted in the capacity of independent experts. This resulted in the expert delegates to the WG to no longer act as individual experts, but rather to politically posture themselves as representatives of their governments and to engage in lobbying efforts and maneuvering to ensure that the political

[63] Wood, supra note 30, 13. [64] Ibid, 13.

[65] H Ward, 'ISO 26000 and Global Governance for Sustainable Development', International Institute for Environment and Development (2012), 18. Available at: http://pubs.iied.org/pdfs/16507IIED.pdf. Last retrieved 11 January 2015.

[66] Ibid. [67] Ibid, 65.

positions and legal interests of their governments would not be compromised in the process. Illustrative is the text in ISO 26000 indicating that the standard is not intended 'to be cited as evidence of the evolution of customary international law' (Clause 1).

Political maneuvering put discernable constrains on decision-making within ISO working group, to the effect of ISO 26000 being of little added value in terms of breaking new ground as it relates to the substantive elements of SR.

Controversies in the international law of SD were sucked into ISO 26000. The language on the precautionary principle for instance was clearly controversial. It is standard practice especially for the United States to exercise extreme caution whenever the precautionary principle is discussed anywhere. During the process, critics have been apprehensive about ISO 26000 containing 'problematic proclamations about contested notions of environmental impacts and employee and consumer rights, but no endorsement of shareholder rights'. The 'precautionary principle' figured prominently in the debate because it was said to eschew the cost-benefit analysis.[68] Eventually, the precautionary 'approach' rather than 'principle' was included in the final ISO 26000 text. This excluded 'the only criterion that would specifically recognize that firms could weigh the cost of taking action on principles of social responsibility'.[69] The United States voted against the final ISO 26000 draft.

Opportunities were thus missed to obtain much needed clarity on the precautionary principle. In fact, a large part of the text consists of chewed over principles of international environmental and SD law, written elsewhere by others. The politics also constrained ISO in pushing the status quo in terms of setting new global minimum baselines, for example, to offer guidance to organizations operating in context where legal safeguards may be weak or absent. Ward (2012) comments that it was inevitable that global applicable norms for these circumstances had to derive from international law. Hence, the definition of 'international norms of behavior' in ISO 26000 was framed as: 'Expectations

[68] There also were concerns that ISO 26000 would lead to transnational regulation. Such regulation could threaten national sovereignty, promote protectionism at the expense of innovation, restrict companies' ability to transact business and result in shareholders conflicting with illegitimate NGOs having been granted 'stakeholder status' it was argued. See J Entine, 'ISO 26000: Sustainability as standard?' (2012) Ethical Corporation. Available at: http://www.ethicalcorp.com/business-strategy/iso-26000-sustainability-standard. Last retrieved 11 January 2015.

[69] Ibid.

of socially responsible organizational behavior derived from customary international law, generally accepted principles of international law, or intergovernmental agreements that are universally or nearly universally recognized' (2.11).

Concerned about the inclusion of the principle of respect for international law, China adopted the position that the rule of law at the national level should prevail irrespective of national circumstances. Strong advocacy in favour thereof resulted in ISO 26000 stipulating that 'in applying this International Standard it is advisable that an organisation take into consideration societal, environmental, legal, cultural, political and organizational diversity, as well as differences in economic conditions, while being consistent with international norms of behaviour' (Clause 1).[70] China made various attempts to reject the draft in the balloting process[71] but eventually voted in favor of the final ISO 26000 draft.

During the negotiations certain elements in ISO draft texts were altered to protect deference to national sovereignty. Substantive elements were adjusted to accommodate national legal, social and cultural contexts, even to the effect of lowering the benchmark below those set in other jurisdictions. This was illustrated by the inclusion of 'personal relationships' in the text of ISO, instead of the term 'sexual orientation' after a striking number of Arab and Middle Eastern countries expressed their opposition to the DIS, which included the term 'sexual orientation', through a no-vote on the DIS.[72]

3. Safeguarding the trade interests of developing countries?

ISO 26000 seems to offer limited added value in terms of providing a common standard of SR to advance the development agenda as outlined here. It did not clear any confusions caused by inconsistencies or incompatibilities between existing SR norms and standards or break new ground in terms of raising global benchmarks beyond levels set by national or international law or clarifying vaguely worded principles of international law as established elsewhere. In fact, the WG evidently was careful to avoid any social or legal ripple effects that may undermine the purpose of ISO 26000 as serving as just a 'guidance standard'.

Indeed, ISO had to consider potential negative implications of ISO 26000 on the trade interests of developing countries, which could undermine the standard's legitimacy and credibility. India had expressed a clear concern about the potential misuse of ISO 26000 in trade-related issues.

[70] Ward, supra note 65, 33. [71] Chhabara, supra note 29. [72] Ward, supra note 65, 29.

There was a possibility that ISO 26000 could serve as a basis for technical regulations imposing higher mandatory baselines and hence creating technical barriers to trade that adversely affected developing countries in particular, it argued. Previous to its adoption, India had sought to have Annex A of ISO/DIS26000 removed. The Annex contained the list of voluntary standards, including certifiable standards of private commercial organizations. India sought to avoid the situation in which these initiatives would become interpreted as part of ISO 26000. Such interpretation could provide guidance to the further development of ISO 26000, and risk future misuse of ISO 26000 to obtain trade advantage. This request by India was eventually not followed up by the Working Group. India voted against the final ISO 26000 draft.

Language was adapted to the effect of the standard disqualifying itself as an 'international standard', 'guideline' or 'recommendation' under the WTO TBT Agreement. ISO 26000 also stipulates that it is not 'intended to provide a basis for any assumption or finding that a measure is consistent with WTO obligations' (Clause 1). During the negotiations it was unclear whether clauses of ISO 26000 (e.g. Clause 6.5.2.2) could amount to guidelines on 'products or related production and processing methods' and hence fall within the scope of 'standard' under the TBT agreement,[73] with possible implications for the policy space of states at the national and international level. If this had been the case, ISO 26000 may have indirectly set a mandatory baseline for technical regulations or other kinds of public product policy that WTO members would need to consider under their WTO obligations. ISO delimiting the applicability of an ISO standard in relation to the WTO legal framework in this manner was a first. It raises a number of issues.

4. National certifiable variations of ISO 26000 standards and their effects

The actual societal impact of ISO 26000 seems to be different than ISO intended. In particular, a number of national ISO members have developed national variations of ISO 26000 standards that are certifiable. Such national standardization is permitted by ISO rules, but is problematic for several reasons.[74]

[73] Ibid, 34.

[74] EC Newsdesk, 'ISO 26000's Certification Problems' (2011), Ethical Corporation. Available at: http://www.ethicalcorp.com/communications-reporting/iso-26000%19;s-certification-problems. Last retrieved 11 January 2015.

Most important, such standardizing activities are contrary to the intent and spirit of ISO 26000. ISO 26000 is used for a purpose other than providing guidance – to inform the design of other, national standards that are specifically set out to facilitate and retrieve revenues from certification. Especially for those national member bodies that provide certification services, developing national, certifiable variants of ISO 26000 is lucrative.[75] Moreover, the option for organizations to get certified to ISO 26000–like national standards may create the perception that ISO 26000 is certifiable, which is clearly not the case. This can cause market confusion, especially among stakeholders that are uninformed, and mislead into believing that companies are conformant with ISO 26000.[76]

The standardization activities can result in national certifiable variations of ISO 26000 that impose different, and possibly conflicting, requirements on organizations. Indeed, ISO 26000 was not intended to serve as a global script that can be readily fixed to and inform requirements for management systems. Henriques notes that ISO 26000 does 'not give systematic help with implementing social responsibility practices'.[77] As national standardization bodies set requirements in relation to SR (e.g. to ensure continuant improvement of SR management systems) and incorporate principles, guidelines and recommendations provided by the national ISO 26000 standards, these requirements may be comparable but not perfectly identical to requirements set by other national ISO 26000–like certifiable standards.

The social effects of the diverse requirements being imposed by variations of national ISO 26000–like certifiable standards seem to bear particularly heavy on developing countries. There is already a large number of schemes imposing many CSR 'standards' on suppliers in developing countries. The addition of variations of certifiable national ISO 26000–like standard runs counter to their interest in having greater conformity in SR standards. The standards may be disadvantageous for international trade flowing into developing countries where these standards impose specific requirements and conformity assessment procedures that organizations from developing countries do not wish to – or cannot – comply with.[78]

[75] Ibid. [76] Ibid. [77] Henriques, supra note 2, 15.

[78] Developing countries may not oppose MSS on social responsibility per se. In fact, they expressed themselves in favor of an MSS approach during the ISO International Conference on Social Responsibility in Stockholm, 21–22 June 2004. An ISO MSS standard could create greater convergence in the diversifying field of SR standards. A new ISO MSS standard on SR could become the principal standard in the field, it was stated, and, unlike the case with

Indeed, developing countries may have consented to the substantive guidance provided by ISO 26000, but they may not approve of the substantive elements or procedural requirements imposed by national ISO 26000–like standards. They are often in a worse position to limit the application of these standards within their borders by engaging in the development of these standards. A lack of capacity to participate, especially where numerous procedures are taking place simultaneously, may be one reason. Formal procedures that permit or facilitate such participation may not exist. In the absence of such engagement by developing countries in the development of national standards, requirements imposed may not be appropriate to local priorities, circumstances and challenges. In fact, they may be tailored to benefit organizations in developed countries, resulting in organizations from developing countries, especially small organizations with limited resources, being in a relatively worse position to meet the requirements and reap the economic rewards of conforming to the standard.

The development of national ISO 26000–like standards is problematic for other reasons as well. One is of a more technical nature and relates to the potential pitfalls in the application of measurement standards and tools to the public interest issue of SR.

5. Measuring SR in the interest of sustainable development

National certifiable variations of ISO 26000 appear to conflate the two approaches to standardization: the standardization policy perspective and public policy perspective. The overall purpose of these national standards is defined in light of the public policy perspective, and in similar terms as that of ISO 26000: to promote the SR of all business enterprises and to contribute to the overall objective of SD. However, the nature and design of these standards as MSS are tailored to the logic that serves the traditional standardization policy perspective. They set requirements that provide yardsticks for measuring, evaluating and comparing the efforts and performance of companies in the realization of an SR management system. This conflation is significant. By fixating the SR substantive elements defined in ISO 26000 to management system requirements, the public interest of promoting companies' contribution to SD, as identified by ISO 26000, risks being overrun by the practices and tools of measurement that are commonly applied to MSS.

other standards, developing countries could at least have an influence in its development. Castka and Balzarova, supra note 19, 83.

There is the presumption that by fixating substantive SR elements to internal management systems, improvement in such internal management systems, and hence capacity to fulfill the SR policies, objectives and requirements set, will actually contribute to the objective of better SR performance, and ultimately SD. This is not necessarily the case. As the literature has shown in other social contexts, the application of market-based measurement tools to objectives of a public policy nature can be harmful. SR is of a qualitative, contested and, as pointed out by the SAG, inherently complex and evolving nature.

Measurements that make reference to criteria that are quantitative, definite and generic may not adequately capture reality. First, problems arise in the *creation and design* of national variations of ISO 26000 standards. For instance, standards are designed for a particular purpose and embody a particular logic, which guides the selection and prioritization of components and around which the actions and operations are organized. In the case of national certifiable ISO 26000–like standards, the criteria applied to 'good' requirements are tailored to the purpose of auditing. Auditing can generate information related to the existence or implementation of an SR management system element, which seems objective because it is based on facts that have been obtained through measurement. This implies that technical considerations may take precedence over substantive considerations in defining such SR requirements. In other words, whether improvements can be measured, and if so, how and when, will be more important than how improvements can be best measured in light of promoting better SR.[79]

There is also a risk that in the creation and design of ISO 26000–like national standards, the substantive elements presented by ISO 26000 will be placed within in a different context – that of the management systems processes – and reinterpreted in light of the function of designing requirements that make SR management systems auditable. Such reinterpretations of ISO 26000 guidance in the development of requirements may especially benefit business enterprises, which can rely on the quantitative information provided by such audits to communicate their improvements in internal management in a seemingly accurate, reliable and verifiable manner. However, they risk widening the gap between a

[79] This argument is based on a similar argument developed in A Roga and Satterthwaite, 'UN Indicators in Critical Perspective' in K Davis, A Fisher, B Kingsbury and S Merry (eds), *Governance by Indicators Global Power through Quantification and Rankings* (2012), 305.

business's improvement of internal management and its actual contribution to SR and SD.[80]

Second, problems arise in relation to the practices involved in the *use* of national certifiable ISO 26000–like standards. For instance, there is the risk of measures becoming a target, a problem that is said to apply to all measurement mechanisms that are tied to the goal of improvement.[81] Once SR management system requirements have been set, it is likely that companies will direct their focus toward meeting these requirements. Companies are expected to maximize resources and efforts where measurable results can be obtained. There are clear incentives for companies to do so when they will be judged on the basis of these criteria and can reap market rewards through demonstrating compliance. These practices create the risk of substantive SR objectives flying under the radar.

It follows that audits can shape the strategic policies of companies toward compliance with these requirements. Such strategies may seek to enhance transparency and accountability to relevant market actors; however, they do so to only a limited extent, as far as it relates to progress achieved in meeting those requirements. There is considerable discretion for companies to decide which records and facts to disclose because the evidence of conformity is retrieved from records that are provided by the companies themselves. Hence, companies retain considerable autonomy to provide records that show measurable results indicating improvements, while withholding those that do not, creating a distorted picture of their actual performance. Furthermore, companies can conceal information that they are not required to disclose according to the MSS but that may be essential for stakeholders (i.e. NGOs) to assess and leverage a company's negative impact on the environment or human rights. By shielding such information, a company can effectively evade questions and potential costs inflicted on a company.

The activities of ISO national standardizing bodies may be undesirable, but they are legal; thus ISO cannot put a stop to national standardization activities.[82] Nor can ISO monitor or review such activities by national standardization bodies or control ISO 26000–related activities of accreditation and certification bodies that operate independently from ISO.[83] The potential adverse effects of national standardization activities risk

[80] Ibid, 305. [81] Ibid, 305.

[82] See article 4.2 of ISO Statute; 'Decisions of the Organization in the form of approved International Standards constitute recommendations to the members, each member remaining free either to follow or not to follow them'. ISO, supra note 40.

[83] EC Newsdesk, supra note 74.

undermining the development agenda and turning SR into a mere facade. The fact that ISO cannot contain these effects calls into question whether ISO was the right organization to address the issue.

IV. Some alternatives: using conflict of law and public international law to further the debate

If it is not necessarily compliance with an uncompliable (a nonexistent word, we concede) standard that will drive SR forward, what alternatives might there be?[84]

The use of private and of public international law in the debate on CSR has taken flight recently (but has also taken a hit).[85] In the EU, the debate has focused on the issue of private international law. In the United States, it is public international law that was the subject of recent decisions. Extraterritorial application of national law raises specific difficulties in each of the three steps of private international law: jurisdiction of national courts to hear the case, the choice of law to apply to the facts at issue and the foreign recognition and enforcement of any resulting judgments. Adequate national/international private law scenarios are suggested as offering faster and less complex solutions than interstate disputes involving state responsibility and international public law issues.[86]

Since the groundbreaking *Doe* v. *Unocal* litigation in 1997, more than fifty cases have been brought in the United States against companies under the Alien Tort Statute alleging corporate involvement in human rights and environmental abuse abroad.[87]

Amendments to EU private international law instruments are being suggested to increase the jurisdiction for European courts in cases involving companies without European corporate 'roots' and to expand the application of European law to acts occurring outside the EU.[88]

[84] See also G Van Calster, 'The Role of Private International Law in Corporate Social Responsibility', *Erasmus Law Review* (2014), 125–133.

[85] See also G Van Calster and C Luks, 'Extraterritoriality and Private International Law', in *Recht in Beweging*, 19de VRG Alumni-dag, Antwerpen, Maklu, 2012, 119–135 (also available at ssrn.com).

[86] P Birnie, A Boyle and C Redgwell, *International Law & the Environment*, 3rd ed, Oxford University Press, 2009; 303.

[87] J Ruggie, United Nations Human Rights Council, 22 April 2009, Report of the Special Representative, John Ruggie on the issue of human rights and transnational corporations and other business enterprises, A/HRC/11/13. Available at: http://www2.ohchr.org/english/bodies/hrcouncil/docs/11session/A.HRC.11.13.pdf (p. 26).

[88] D. Augenstein, 'Study of the Legal Framework on Human Rights and the Environment Applicable to European Enterprises Operating Outside the European Union' (2010).

1. United States – Alien Tort Statute Litigation

Reviewing the suitability of employing private international law as a way forward for what are essentially disputes with a high potential for upsetting inter-state relations, is particularly relevant in light of recent developments in case law involving the Alien Tort Statute. In *Kiobel* v. *Royal Dutch Petroleum*, the US Court of Appeals for the Second Circuit held that corporations cannot be sued under the Alien Tort Statute for violations of customary international law because 'the concept of corporate liability . . . has not achieved universal recognition or acceptance of a norm in the relations of states with each other'.[89]

Many viewed Alien Tort legislation as one of the last great hopes for effective enforcement of human rights violations internationally. However, the ultimate decision by the US Supreme Court was less encouraging.[90] The central question in *Kiobel* turned out to be this: whether and under what circumstances US courts may recognize a cause of action under the Alien Torts Statute for violations of the law of nations occurring within the territory of a sovereign other than the United States. In focusing on this question (and replying in the negative), the Supreme Court did not entertain the question that actually led to *certiorari* – namely whether the law of nations recognizes corporate liability.

Extraterritorial application of US law was most recently the issue in *Morrison* v. *National Australia Bank*[91] in the area of securities. In *Kiobel*, the Supreme Court relied on its extensive review of extraterritoriality in *Morrison*. It did so even if in *Morrison* (and other cases before it), the question of extraterritoriality was one of 'merits' (also known as 'jurisdiction to prescribe'): that is, whether an Act of Congress regulating conduct applies abroad. By contrast, in *Kiobel*, the question concerns jurisdiction *pur sang* (also known as 'jurisdiction to adjudicate'). For the Supreme Court, this did not dent precedent value of Morrison: 'we think the principles underlying the canon of interpretation similarly constrain courts considering causes of action that may be brought under the ATS'.

In *Morrison*, the Court held that when a statute gives no clear indication of an extraterritorial application, it has none. In *Kiobel*, the Court

Available at: http://ec.europa.eu/enterprise/policies/sustainable-business/files/business-human-rights/101025_ec_study_final_report_en.pdf. Last retrieved 4 July 2015.

89 Esther Kiobel v. Royal Dutch Petroleum Co. et al., U.S. Court of Appeals for the Second Circuit, Nos. 06-4800-cv, 06-4876-cv (17 September 2010), at 49.

90 Esther Kiobel v. Royal Dutch Petroleum Co. et al., USSC, Certiorari, No. 10–1491 (17 April 2013).

91 Morrison et al. v. National Australia Bank Ltd et al., Certiorari, No. 08–1191 (24 June 2010).

did not find convincing argument in text, history or purpose of the Alien Tort Statute that could rebut the presumption against extraterritoriality. The closest such rebuttal arguably lay in the historic (and more current) examples of employing Alien Tort Statute against piracy: as the SC notes, 'piracy normally occurs on the high seas, beyond the territorial jurisdiction of the United States or any other country[; however,] applying U.S. law to pirates does not typically impose the sovereign will of the United States onto conduct occurring within the territorial jurisdiction of another sovereign, and therefore carries less direct foreign policy consequences'.

The latter of course is where the core of the argument lies and where public and private international law principles of comity come into play: the degree to which in upholding jurisdiction, the courts in ordinary might be obstructing US foreign policy. The Supreme Court concluded as follows:

> On these facts, all the relevant conduct took place outside the United States. And even where the claims touch and concern the territory of the United States, they must do so with sufficient force to displace the presumption against extraterritorial application. See Morrison, 561 U.S. ___ (slip op. at 17–24). Corporations are often present in many countries, and it would reach too far to say that mere corporate presence suffices. If Congress were to determine otherwise, a statute more specific than the ATS would be required.

The Court therefore answers *Kiobel*-type cases (a foreign plaintiff suing a foreign defendant for acts or omissions occurring wholly *outside* of the United States that allegedly violate the law of nations); however, it leaves open many questions that fall outside the factual *Kiobel* box. Does the reference to 'claim' and 'territory' of the United States refer to the tortious *action* (thus requiring that to take place in the United States) or would a US defendant suffice (in all likelihood, no)? What 'link' would be enough for the action to take place in the United States: in particular, lack of corporate oversight over foreign subsidiaries? In all likelihood, further debate will take place in lower courts (and the topic will eventually end up at the Supreme Court again); however, it is clear that the scope for ATS litigation has been severely diminished (as was indeed swiftly confirmed in *Daimler* v. *Baumann*).[92]

[92] Daimler AG v. Bauman et al., USSC, Certiorari, No. 11–965 (14 January 2014).

This means that attention may now be reignited in what has been brewing in the EU for some time: using national courts to apply national law for conduct abroad: in other words, classic private international law/conflict of laws entering the limelight once again.

2. The European Union

In European private international law, as with the Alien Tort Statute, the two main concerns that arise when addressing matters of corporate violation of rights are whether EU Member State courts have jurisdiction and, if so, what laws, national or international, apply.[93] Suing an EU-domiciled company in EU courts is straightforward under the relevant principles of the applicable jurisdiction regulation.[94] Pursuing a holding company that has domicile in the EU may be possible from the jurisdiction point of view; however, it would be more challenging with respect to applicable law. The EU does not operate an Alien Tort–like system, which employs international law to advance the case of plaintiffs seeking 'justice' in environmental or human rights cases. The route that must be followed in the EU is one of *Gleichlauf* (meaning concurrence) between having a court in the EU hear the case and having that court apply the human rights or environmental law of that forum as a benchmark for deciding the merits of the action. The most likely route to pursue a corporation in a court in the EU, is via an action in tort. This generally, under the Rome II Regulation,[95] entails the application of the *lex loci damni*: the law of the place where the damage occurred. In our current hypothesis, that is not the law of an EU Member State, and relevant exceptions in Rome II do not apply; however, they do offer room for maneuver for the judges.

A recent decision by the Court at the Hague (Shell Nigeria/Royal Dutch Shell[96]) confirmed the difficulty in having European law apply to activities that took place outside the EU; however, this judgment applied Dutch law,

93 Augenstein, supra note 88, 16. See also V Van Den Eeckhout, 'Promoting Human Rights within the Union: The Role of European Private International Law', *European Law Journal*, 2008, (105) 127.

94 More detail in G Van Calster, *European Private International Law*, Oxford, Hart, 2013 (2nd ed forthcoming, 2015).

95 Regulation 864/2007 on the law applicable to non-contractual obligations, OJ [2007] L199/40.

96 See 'No Big Surprises in Dutch Shell Nigeria/Royal Dutch Shell Ruling', 3 January 2013. Available at: http://gavclaw.com/2013/01/31/no-big-surprises-in-dutch-shell-nigeria-royal-dutch-shell-ruling/. Unofficial English translation on file with the authors.

because the Rome II Regulation had not yet entered into force at the time the events took place.

D. Conclusion

ISO developed ISO 26000, a common standard providing guidance on SR, to promote better SR by organizations and assist them in contributing to SD; however, ISO's agenda was not public policy. Arguably ISO's involvement was troublesome from the start, for it was so out of sync with its standard *causa operandi.* This could have been remedied had the standard focused on 'how' management can better structure an organization and its operations to make it meet SR standards, rather than on 'what' SR entails in substance, which, as noted, is not what it does.

Through a lengthy, intensive and participatory standard-setting process, ISO sought to build a broad consensus that could give the standard credibility. Interactions with other nongovernmental standard-setting bodies and international organizations were managed through principles of consistency, hierarchy and rules of conflict sanctioned through MoUs. This could not prevent discussions on the substantive elements of SR in the Working Group, which required extensive review of international law, from becoming political, however.

Opportunities were missed by ISO to break new ground; ISO 26000 did not raise, clarify or supplement SR standards established elsewhere. ISO was careful to avert any potential legal or social ripple effects that could disadvantage developing countries. This resulted in ISO 26000 draft achieving a consensus in the fall of 2010; however, it also deprived the document of any real significance.

ISO could not prevent ISO 26000 from taking on a whole new dynamic on its own. ISO 26000 has been recycled into a certifiable MSS at the national level. Rough certificates have been issued. These activities not only go against the intention and spirit of ISO 26000 but also risk creating the adverse effects on developing countries that ISO 26000 sought to avoid in the first place. One cannot help but wonder whether ISO took on more than it could handle.

Given that ISO addresses companies, one must consider the impact of ISO 26000 on these businesses. Are companies really impressed with this scheme – is it this that makes their clock tick? After the emergence of ISO 26000, the world does not seem to look much different to companies. Not engaging with the standard does not impose any discernable cost to businesses. ISO 26000's success in promoting more sustainable companies will depend on the standard's uptake and diffusion. The costs for purchasing access to ISO 26000 may act as a hurdle in this regard, and for this reason, the standard should be available free of charge.

This chapter has argued that ISO 26000 project was an overly costly enterprise for what it delivered. Efforts could have been better spent on procedures that are effective in bringing about actual change in corporate behavior, including selective litigation.

13

The development dimension of the Codex Alimentarius standard-setting process

AWILO OCHIENG PERNET

A. Agricultural trade and the Codex Alimentarius Commission

Agricultural and food exports play an important part in international trade that has increased tremendously over the past thirty years, contributing to an increase in gross domestic product, including that of developing countries (DCs).[1] It is acknowledged that expansion of agricultural trade has a direct impact on poverty reduction and may lead to accelerated economic growth.[2]

Compliance with sanitary and phytosanitary and technical requirements is a prerequisite to successful export trade, and the investments made by DCs to comply with these requirements lead to a better access to export markets and also contribute to upgrading the domestic market food safety level.[3] However, lack of capacity often hinders DCs from realising international trade opportunities.[4] These supply-side constraints

I thank Professor Panagiotis Delimatsis for his excellent support and guidance.

[1] United Nations Conference on Trade and Development (UNCTAD), 'Development and Globalization: Facts and Figures', (2008). Available at: http://www.unctad.org/en/docs/gdscsir20071_en.pdf. Also Food and Agriculture Organisation of the United Nations and World Health Organisation (FAO) (2005), *Understanding the Codex Alimentarius.*

[2] I Foss (2004) *Development of Trade in Africa: Promoting Exports through Quality and Product Safety.* Report commissioned by Sida and Norad. Published by Sida, Department for Infrastructure and Economic Co-operation; also S. John Wilson (2002). *Liberalising Trade in Agriculture: Developing Countries in Asia and the Post-Doha Agenda, World Bank.* Available at: http://www.ppl.nl/bibliographies/wto/files/1280.pdf.

[3] S Henson and S Jaffee, 'A Strategic Perspective on the Impact of Food Safety Standards on Developing Countries' (2006). Invited paper prepared for presentation at the International Association of Agricultural Economists Conference, Gold Coast, Australia, August 12–18, 2006.

[4] S Henson and W Mitullah, 'Kenyan Exports of Nile Perch: The Impact of Food Safety Standards on an Export-Oriented Supply Chain' (2004), World Bank Policy Research Working Paper 3349, June 2004.

include lack of compliance resources – for example, limited access to information on SPS requirements, weak food safety and quality management capacities, financial constraints and poor infrastructure. Nevertheless, despite what could be considered as extremely high costs of compliance,[5] non-compliance could have dramatic consequences for exporting countries as demonstrated by the example of the European Union's (EU) ban on fish exports from East Africa in the 1990s.

Although tariffs and quota restrictions have gradually been removed, it is recognised that technical measures (e.g. SPS requirements, labelling, quality and compositional standards) can act, either explicitly or implicitly, as barriers to trade in a manner similar to tariffs and quantitative restrictions.[6] Data from the EU Rapid Alert System for Food and Feed (RASFF) regarding food import notifications suggest that DC food exports often do not meet the SPS and technical requirements of importing countries. DCs face an additional challenge posed by the plethora of private standards set by retailers in developed country markets with which they have to comply to access the high value international food markets.

The Codex Alimentarius Commission (CAC) is the intergovernmental food standards-setting body that was established in 1963 by the Food and Agriculture Organisation of the United Nations (FAO) and the World Health Organisation (WHO) with the mandate to develop food safety and quality standards aimed at protecting the health of consumers and ensuring fair practices in the food trade. Currently, the CAC has 186 Members (185 Member Countries and one Member Organisation, the European Union) from seven geographic regions: Africa, Asia, Europe, Latin America and the Caribbean, Near East, North America and the South West Pacific. There are also 229 intergovernmental organisations (IGOs) and non-governmental organisations (NGOs) that enjoy observer status in Codex. Observer IGOs and NGOs participate in the work of the

[5] J C Cato and C A L Santos, 'Costs to Upgrade the Bangladesh Frozen Shrimp Processing Sector to Adequate Technical and Sanitary Standards and to Maintain a HACCP Programme', in L Unnevehr (ed), *The Economics of HACCP: New Studies of Costs and Benefits* (2000), pp. 385–402.

[6] Ministerie van Landbouw, Natuur en Voedselkwaliteit, *Proceedings of the International Seminar on Setting Food Safety Standards, Effects on International Trade, Particularly for Developing Countries*, The Hague, The Netherlands, 9–12 June 2008 (2008). Available at: http://www.minlnv.nl/portal/page?_pageid=116,1640360&_dad=portal&_schema=PORTAL&p_file_id=29284. World Trade Organisation (2012), *World Trade Report 2012, Trade and Public Policies: A Closer Look at Non-Tariff Measures in the 21st Century*.

CAC in line with the Guidelines and Principles developed by the CAC.[7] The CAC is the highest decision-making body and also reviews and adopts Codex standards, guidelines, codes of practice, recommendations, maximum residue limits (MRLs) for veterinary drugs and pesticides, and so forth. This body of texts is generally referred to as Codex standards. Codex standards are developed by the subsidiary bodies (Codex committees) of the CAC, which are Codex general subject committees, commodity committees, Ad hoc Intergovernmental Task Forces or FAO/WHO Regional Coordinating Committees. The Codex Procedural Manual lays down the statutes and rules of procedure of the CAC and also contains General decisions of the CAC. The CAC relies on several independent FAO/WHO expert bodies to provide risk assessments and scientific advice, for example, the Joint FAO/WHO Expert Committee on Food Additives (JECFA), Joint FAO/WHO Expert Meetings on Microbiological Risk Assessment (JEMRA) and the Joint FAO/WHO Meetings on Pesticides Residues (JMPR).

Codex develops standards according to an eight-step procedure that is generally initiated at the technical committee level by the submission of a new work proposal included in a Project document that is developed by a Codex Member according to the procedure laid down in the Codex Procedural Manual. Once the Committee approves the proposal for new work, it submits a request for New Work to the CAC. Before its consideration by the Commission, the New Work Proposal is critically reviewed by the Codex Executive Committee (CCEXEC). The Commission takes due account of the views expressed by the CCEXEC before giving its final opinion. When the Commission decides to undertake new work on the proposed subject, the text officially enters the Codex eight-step procedure as a Proposed Draft text. All along the eight-step procedure, Codex Members and Observers have the opportunity to give inputs into the Draft at several steps at Committee level and finally before the adoption by the Commission at Step 8. Codex Observers contribute to the decision shaping without the right to vote, whereas only Codex Members take the decisions.

[7] FAO/WHO (2013). 'Guidelines on Cooperation between the Codex Alimentarius Commission and International Intergovernmental Organisations in the Elaboration of Standards and Related Texts' and 'Principles Concerning the Participation of International Non-Governmental Organisations in the Work of the Codex Alimentarius Commission', Codex Alimentarius Commission, *Procedural Manual*, 21st ed, Rome, 2013.

The elaboration process varies in length with extreme cases of draft standards being held at Step 8 at the Commission level for several years. Nevertheless, the official average period for the elaboration of a Codex Standard is generally considered to last four years with a few standards having been elaborated in one year (e.g. the revision of the health-related substances in the Standard for Natural Mineral Waters or the setting of the ML for melamine in infant formula).

The WTO SPS Agreement explicitly recognises the competence of the CAC in setting international food safety standards, while the food quality, nutrition and labelling aspects of Codex standards are covered by the Agreement on Technical Barriers to Trade (TBT Agreement). The SPS Agreement also recognises the standards set by the World Organisation for Animal Health (OIE) and the International Plant Protection Convention (IPPC). Codex standards are thus the benchmark used to measure national food standards in international trade. Furthermore, both the SPS and the TBT Agreements acknowledge the importance of harmonising standards internationally to prevent WTO Members from adopting SPS or other technical measures that could constitute unjustified barriers to trade. Article 3.1 of the SPS Agreement encourages governments to establish national SPS measures that are consistent with international standards, guidelines and recommendations (harmonisation). Article 2.6 of the TBT Agreement states that 'with a view to harmonising technical regulations on as wide a basis as possible, Members shall play a full part, within the limits of their resources, in the preparation by appropriate international standardising bodies of international standards for products for which they have either adopted or expect to adopt, technical regulations'. Health and safety aspects of Codex decisions and recommendations should be based on a risk assessment, as appropriate to the circumstances. The food safety risk assessment should incorporate the four steps of the risk assessment process and should be documented in a transparent manner. The SPS Agreement considers Codex standards to be scientifically justified. Indeed, Article 3.2 of the SPS Agreement lays down the presumption of conformity of a national food standard if it conforms to a Codex standard. National standards that are more stringent than Codex standards have to be based on a proper risk assessment demonstrating that the Codex standard does not provide a sufficient level of protection. This means that WTO Members who implement SPS measures based on Codex standards do not need to provide justification in case of a trade dispute.

B. The development dimension of the CAC standard-setting

The adoption of Codex standards is considered normally to have overwhelming positive economic impacts because Codex standards allow countries to trade their food products on international markets.[8] In general, whenever the CAC develops standards and in line with the general criterion, the main objectives of this new work proposal should be consumer health protection, food safety and ensuring fair practices in the food trade. In addition, the new work proposal should take into account the identified needs of DCs. There are criteria that apply to general subjects, such as food labelling, food additives and food hygiene, and others that apply to commodities. For the latter, in addition to information regarding the volume of production and consumption of the commodity that is up for standardisation and the pattern of trade between countries, data should also be provided regarding the international or regional market potential. In this respect, DCs can aim for the standardisation of their local/regional commodities, which could lead to the setting of international/regional food safety and quality standards for the given commodity, and this in turn could lead to an increase in intraregional (between or among countries of the same region) and/or international trade in the given commodity.

Development of regional standards enables the standardisation of food products that are of importance to the particular region. Currently more than twenty Codex regional standards have been developed by the Codex regional coordinating committees. The CAC has also adopted global standards for commodities that are primarily traded and consumed in DCs, such as the Codex standard for bitter cassava. It is worth noting that many commodity standards cover products that are produced in DCs, for example, the standards developed by the Codex Committee on Fresh Fruits and Vegetables (Host country: Mexico), the Codex Committee on Spices and Culinary Herbs (Host country: India) or the Codex Committee on Fats and Oils (Host country: Malaysia). In addition, many commodity standards have also been developed for commodities for which the essential raw materials come from DCs such as cocoa beans, which are necessary for the production of cocoa products and chocolate. Nevertheless, horizontal or general subject committees also develop standards that are of

[8] Report of the 28th session of the Codex Committee on General Principles, Paris, France, 7–11 April 2014. Available at: http://www.codexalimentarius.org/meetings-reports/en/. Accessed 23 April 2014.

great relevance to DCs, such as the setting of maximum limits (MLs) for aflatoxins in peanuts or pistachios. Exporting countries thus have to comply with the given Codex standards for their commodities to access international food markets. The effective participation of DCs in Codex technical meetings where the proposals are drafted as well as in the CAC sessions where the draft standards are adopted as international/regional Codex standards is thus of great importance.

Most Codex Members including DCs produce food both for domestic consumption and for international trade. Codex standards, guidelines and recommendations can be applied at various stages of the value chain to ensure food quality and safety, from primary production to consumption. Good Agricultural Practices (GAP), Good Manufacturing Practices (GMP) and Good Hygienic Practices (GHP), including the Hazard Analysis and the Critical Control Point (HACCP), can be applied at various stages, among which are primary production, harvesting, transportation, processing and distribution. On the other hand, Codex guidelines on food import and export inspection and certification systems may be applied at the relevant stages in view of export and import trade. Codex standards serve as a reference in the international food trade negotiations as they are the recognised benchmarks for food safety within the WTO SPS system.

Many DCs adopt Codex standards in their national legislation, and they are implemented on the domestic market to ensure food safety and quality. Thus Codex standards provide the legal requirements for food hygiene, food additives, food labelling, contaminants in foods, pesticide residues and residues of veterinary drugs, for example. Codex standards can also be used for consumer information about food safety and quality.

I. Codex standards, reference points for DCs

Codex standards serve as reference points for DCs when the latter are developing food standards. The East African Community (Burundi, Kenya, Rwanda, Uganda and Tanzania) explicitly refers to Codex standards for various commodities as clearly indicated in the Catalogue of East African Standards,[9] and in addition, it is foreseen that Codex standards could be used to resolve food-related trade disputes that may arise

[9] East African Community Catalogue of East African Standards 2010. Available at: http://www.trade.eac.int/. Accessed on 23 April 2014.

within the East African Community. Furthermore, many DCs adopt Codex standards in their national legislation. For example, Kenya has adopted more than 100 Codex standards as the national standards. In fact, in the country food and agriculture profile for Kenya, it is stated that 'national food standards do not differ from Codex standards'.[10] DCs thus benefit directly from Codex Standards by implementing internationally harmonised standards that are based on sound risk assessment carried out by the international independent FAO/WHO expert bodies, thereby saving the resources that they would have had to invest to carry out the necessary risk assessments for the food safety standards. In addition, DCs can directly implement Codex food quality standards both for domestic and international trade. Finally, DCs use Codex standards not only to ensure that their products comply with food requirements in the international markets but also to control imported foods to ensure that they comply with the national food standards, which are based on Codex standards.

II. How does Codex accommodate development concerns?

The 28th session of the Codex Committee on General Principles[11] reviewed the mechanisms for examining economic impact statements submitted by governments concerning possible implications for their economies of some of the individual draft standards or some of the provisions thereof, including the development of a template or harmonised format for the submission of such economic impact statements.

The authors of the proposal[12] argued that the consideration of economic impact should be seen as an important aspect of a standard development process. Although protecting health of consumers is a priority for Codex, the role of Codex in ensuring fair practices in the food trade should also be given adequate attention; thus, economic considerations should also be examined. It was argued that food standards affect producers, distributors and consumers and that a new measure or standard may present

[10] World Health Organisation Regional Office for Africa, Food and Nutrition. *Food and Nutrition Country Profiles – Kenya.* Available at: http://www.afro.who.int/en/clusters-a-programmes/hpr/food-safety-and-nutrition-fan/fan-country-profiles.html. Accessed on 23rd April 2014.

[11] CCGP, Paris, 7–11.4.2014 Agenda Item 5, document CX/GP 14/28/5.

[12] Malaysia and Australia. Agenda Item 5: Issues Related to Economic Impact Statements. Document CX/GP 14/28/5. Available at: http://www.codexalimentarius.org/meetings-reports/en/. Accessed on 23 April 2014.

challenges to some countries, especially DCs, in ensuring compliance. In some cases, it may also lead to major trade barriers for DCs and adversely affect their economies. The authors went on to note that although Codex standards are voluntary, the WTO uses Codex standards as the benchmark in the adjudication of international trade disputes involving foods. In their view, it was thus important to ensure that new measures or standards being developed do not unnecessary obstruct trade. Consequently, they were of the view that the development of standards should be based on principles such as openness for other countries' comments, transparency in setting rules and regulations and inclusiveness of others' comments and suggestions. Finally, they argued that by giving serious consideration to the potential economic impact statement of an intended measure, there could be better acceptance of the proposed standard.

Although the committee noted the view, among others, that the proposed Codex standard could create requirements that would be costly for DCs to meet, it indicated that 'the current provisions in the Codex Procedural Manual were sufficient to take into consideration the economic implication of a new or revised Codex text in the Codex standards development process'.[13] Therefore, the committee did not approve the proposal to revise the current provisions as had been proposed by the proponents of this initiative.

The foregoing discussion clearly illustrates the fact that Codex standards may have an impact on the economies of DCs. The latter should therefore voice their concerns at the appropriate standards development procedures and ensure that these are addressed and duly taken into consideration during the various steps of the development of the Codex standards. Thus, DCs have to ensure an effective participation in Codex work and also contribute data for the risk analysis process. DCs should participate actively and make contributions in both the Codex decision-shaping and decision-making process.

C. DC participation in Codex and development

DC participation in Codex is important because it 'results in a more inclusive, equitable and participatory global standards-setting body and a "triple-win" for participating countries as they benefit from Codex standards through: better protection of public health on both ends

[13] Paragraph 35, Report of the 28th session of the Codex Committee on General Principles.

of the trade chain; enhanced trade opportunities and enhanced economic development'.[14] Indeed, many DCs adopt Codex standards in their national legislations by reference as has been seen earlier with the case of Kenya and the East African Community.

I. The Codex Trust Fund

In 2003, the FAO/WHO Project and Fund for Enhanced Participation in Codex (Codex Trust Fund [CTF]) was launched with the objective of increasing DC participation in Codex standard-setting activities to enhance their capacity to develop and implement effective national food safety systems and thereby develop their opportunities to take advantage of international trade in food. The main objective of the CTF is to help Codex Members that are DCs or those with economies in transition to enhance their level of effective participation in the development of global food safety and quality standards by the CAC. The CTF also aims at strengthening participation (objective 2) and enhancing scientific and technical participation of DCs in Codex (objective 3).

II. DC eligibility for the CTF

Eligibility for the CTF is based on the four official classifications produced within the United Nations system:

1. the list of Least Developed Countries established by the United Nations Economic and Social Council,
2. the World Bank classification of economies,
3. the UNDP Human Development Index, and
4. the list of Small Island Developing States established by the United Nations Department of Economic and Social Affairs, Division for Sustainable Development.[15]

According to the information on the CTF website,[16] in 2012, there were ninety-seven Codex Members eligible for support from the CTF. The eligible Members were divided into four groups:

[14] World Health Organisation, supra note 1.

[15] Available at: http://www.who.int/foodsafety/codex/trustfundeligible/en/. Accessed 24 April 2014.

[16] CX/CAC 13/36/14, Part 1, FAO/WHO Fund and Project for Enhanced Participation in Codex Annual Report of the CTF 2012. Available at: http://www.who.int/foodsafety/codex/AR-cac36_14e.pdf?ua=1. Accessed on 25 April 2014.

Group 1a countries were the least developed countries (LDCs), which were entitled to five sponsored participations
Group 1b countries were the low-income and the low or medium human development countries, which were also entitled to five sponsored participations
Group 2 countries were the lower middle-income and the medium or high human development countries, which were entitled to four sponsored participations
Group 3a countries were upper middle-income and the medium human development countries, which were entitled to two participations
Group 3b countries were upper middle-income and the high human development countries, which were entitled to two participations
Group 4 countries were the LDCs and the Small Island Development States, which have graduated from regular support and were entitled to two sponsored participations

III. Effective participation of DCs in Codex

As a prerequisite to an effective DC participation in the Codex standard-setting procedure, the latter should be represented in the relevant fora to ensure that their views and positions are taken into account during the decision-shaping process, which would enable them to contribute directly to and influence the standard-setting process. DCs should also have the opportunity to participate in the decision-making process when the standards are actually adopted by the CAC in line with the recommendations of the Codex technical committees.[17] Effective participation also includes the opportunity for DCs to submit relevant information and data for the risk assessments that are carried out by the FAO/WHO expert bodies and the results of which form the basis for Codex food safety standards. Unfortunately, many DCs lack the capacity to generate scientific data and thus have no possibility to influence the outcome of the risk assessments, which as can be expected will not take into account their specific concerns because of the lack of data. It is therefore believed that Codex Members who have a stronger capacity to generate

[17] A Engler, L Nahuelhual, G Cofré and J Barrena, 'How Far from Harmonization Are Sanitary, Phytosanitary and Quality-Related Standards? An Exporter's Perception Approach', *Food Policy*, 37 (2012), 162–170.

data may also have a greater ability to influence the outcomes in the CAC.[18]

The implementation of the CTF has led to an increase in the DC participation in Codex by providing financial assistance to DCs so that they may attend Codex meetings, which take place all over the world. Through the various workshops and training seminars organised in all Codex regions, the CTF has also contributed to strengthening DC participation in Codex. In addition, the CTF financed a pilot initiative using a 'mentoring approach' so that less experienced DCs (mentees) could learn from the more experienced lead countries and/or Observer Organisations (mentors).[19] Through funds provided by the European Union since 2012, the CTF has contributed to enhancing scientific/technical input to Codex by enabling the designated four major sorghum-producing and/or -exporting DCs to generate data for mycotoxins in sorghum. These data will enable the relevant expert body JECFA to carry out a detailed risk assessment, with a view to the establishment of MLs for selected mycotoxins in sorghum.[20]

The CTF Mid-Term Review[21] examined the lessons learned from the first six years of the CTF. The positive impact that the CTF has had on increasing DC participation in Codex sessions was noted, and it was indicated that the CTF had reached its first objective of widening DC participation in Codex work. However, it was stated that more attention needed to be given to the second and third objectives of the CTF: to strengthen overall participation in Codex and to enhance DC technical and scientific participation in Codex.[22] The CTF will come to an end in December 2015, and steps have been taken to ensure a successor initiative after that date. The final project evaluation of the CTF started in January 2014 and the recommendations and conclusions of the evaluation were

[18] LA Jackson and M Jansen, 'Risk Assessment in the International Food Safety Policy Arena. Can the Multilateral Institutions Encourage Unbiased Outcomes?', *Food Policy* 35 (2010), 538–547.

[19] CTF Report, Elaboration of practical examples for the establishment and application of microbiological criteria for foods as part of the revision of the Principles for the establishment and application microbiological criteria for foods (CAC/GL 21–1997).

[20] Part 2, 17th Progress Report (January–June 2013), FAO/WHO Project and Fund for Enhanced Participation in Codex, CX/CAC 13/36/14, pp. 20–25.

[21] K Forss, J Andersson and E Kasirye-Alemu, 'Codex Trust Fund Mid-Term Review', Final Report, Andante – Tools for Thinking AB, 30 April 2010.

[22] The Codex Alimentarius Commission, Joint FAO/WHO Food Standards Programme: CX/CAC 10/33/14 Add.1, p. 9. Available at: http://www.codexalimentarius.net.

submitted to the 70th session of the CCEXEC and the 38th session of the CAC in July 2015.

It is worth highlighting the long-term goal of the CTF, which is to improve global public health and food security by promoting the provision of safer and more nutritious food and also contributing to a reduction in food-borne diseases. It is hoped that increased participation in Codex activities will help to enhance international food trade opportunities for DCs and those with economies in transition. The three immediate outputs of the CTF were expected to be as follows: (1) widening participation in Codex by increasing the number of countries routinely sending delegations to CAC sessions and to its committees/task forces, and by increasing the number of countries who address issues of priority health and economic concern for their specific countries; (2) strengthening overall participation in Codex, whereby the number of countries routinely developing and presenting national considerations in the Codex standard-setting process will have increased along with their participation in the Codex committees/task forces; and (3) enhancing scientific/technical participation in Codex, whereby the number of countries that are actively providing scientific/technical advice in support of the Codex standard-setting process will have increased.

Connor[23] explored the linkages between increased participation in Codex and enhanced international food trade opportunities. Although recalling the long-term goal of the CTF that increased participation in Codex activities should help enhance international food trade opportunities for DCs, he noted that overall, the CTF may have contributed to improving the uptake of Codex standards in beneficiary countries. He also noted that the application of Codex standards may have enabled exporters in more countries to trade internationally. However, his analysis showed that many countries failed, for various reasons, to apply the standards that are required in international trade. He further noted the difficulty of clearly defining the impact of the CTF activities on the ability of national food safety systems to enhance international food trade opportunities due to the complex interaction of a large number of trade-related development activities. Consequently, he recommended to international

[23] RJ Connor, 'Initiative to Explore Linkages between Increased Participation in Codex and Enhanced International Food Trade Opportunities', Final Report (2007). Funded by the Department of International Development (DFID), United Kingdom, Project Number RNRAT 012 (AG0396).

capacity-building actors to collaborate and coordinate their activities to help DCs meet this long-term CTF objective.

D. SPS requirements: non-tariff measures

SPS requirements can impede trade[24] and significantly affect the ability of DCs to access international food markets.[25] In addition to increasing the overall production costs, SPS requirements can have other negative impacts such as prohibiting trade by imposing an import ban, prohibitively increasing production and marketing costs, diverting trade from one trading partner to another and adopting food safety and quality regulations that result in discrimination among potential suppliers, for example.[26] In addition, the need to gather information about foreign regulations, hiring lawyers and adjusting product designs to make them consistent with import requirements may also constitute barriers to the food trade.[27] According to the World Trade Report 2012, 'evidence suggests that procedural obstacles are the main source of difficulties for exporting firms from DC'. These include time constraints and unusually high fees or 'informal' payments.

The WTO SPS Agreement lays down obligations regarding the application of food safety measures. Article 2 of the SPS Agreement acknowledges WTO Members' basic rights and obligations. Indeed, they are allowed to

[24] UNCTAD has been actively involved in research and programmatic activities on issues related to non-tariff measures since the early 1980s. In 1994, it began to collect and classify non-tariff barriers (NTBs) according to a customised Coding System of Trade Control Measures that classified tariffs and para-tariffs and non-tariff measures into more than one hundred subcategories. UNCTAD (2010), 'Non-tariff Measures: Evidence from Selected Developing Countries and Future Research Agenda', Developing countries in international trade studies, United Nations, New York and Geneva, 2010. ITC International Trade Centre (2010). 'ITC's programme on non-tariff measures (NTMs). Understanding the non-tariff obstacles to trade exporters and importers experience when trading their goods.' Presentation to STDF Working Group by Mondher Mimouni, 2 July 2010.

[25] S Henson, 'Food Safety Issues in International Trade', Food Safety in Food Security, Focus 10. Brief 5 of 17. September 2003. International Food Policy Research Unit, Washington, DC. Available at: http://www.ifpri.org/sites/default/files/publications/focus10_05.pdf. A Wiig and I Kolstad, 'Lowering Barriers to Agricultural Exports through Technical Assistance', *Food Policy 30* (2005), 185–204.

[26] S Henson and R Loader (2000), 'Barriers to Agricultural Exports from Developing Countries: The Role of Sanitary and Phytosanitary Requirements', in S Henson and R Loader (eds), *The WTO and Technical Barriers to Trade, Critical Perspectives on the Global Trading System and the WTO.*

[27] JE Anderson and E van Wincoop, 'Borders, Trade and Welfare', in S Collins and D Rodrick (eds), Brookings Trade Forum (2001), pp. 207–243.

take SPS measures that are necessary to protect human, animal or plant life or health provided that such measures are not inconsistent with the provisions of the SPS Agreement. In addition, Members shall ensure that any sanitary or phytosanitary measure is applied only to the extent necessary to protect human, animal or plant life or health, is based on scientific principles and not maintained without sufficient scientific evidence, except as provided for in paragraph 7 of Article 5. Also, Members shall ensure that their sanitary and phytosanitary measures do not arbitrarily or unjustifiably discriminate between Members where identical or similar conditions prevail, including between their own territory and that of other Members. Sanitary and phytosanitary measures shall not be applied in a manner that would constitute a disguised restriction on international trade. Finally, WTO Members are encouraged to apply standards that are consistent with those developed by the relevant international standard-setting bodies: in the field of food safety, the CAC. Nevertheless, WTO Members may adopt measures which result in higher levels of health protection than would be achieved by measures based on relevant international standards provided that these measures are based on an appropriate risk assessment and that the approach is consistent and not arbitrary.

According to UNCTAD[28] and the World Trade Organisation,[29] available evidence indicates that NTMs[30] are often applied as alternative trade policy instruments because multilateral trade agreements impose limits on the use of traditional trade policy instruments, such as tariffs.[31] These measures can create challenges to producers because of the costs that result from various requirements. According to Foss,[32] there is still dramatic tariff escalation on many processed African agricultural products, and he notes that current tariff rates for some sorts of jams and preserved fruit are 200 to 600 percent.

Reports of the EU Food and Veterinary Office (FVO) and the RASFF show that the most affected products in the international food trade are those that are subject to microbiological and animal health requirements, pests and plant diseases, fish and fishery products, meat and other animal products and unprocessed fruits and vegetables. As the biggest global food importer, the EU market is of paramount importance to DCs. The

[28] UNCTAD (2010), supra note 24. [29] World Trade Organisation, supra note 6.

[30] UNCTAD defines Non-Tariff Measures as follows: 'Non-Tariff Measures (NTMs) are policy measures, other than ordinary customs tariffs, that can potentially have an economic effect on international trade in goods, changing quantities traded, or prices or both' (UNCTAD supra note 21).

[31] Supra note 6. [32] I Foss, supra note 2.

loss arising from food rejection is way beyond the value of the rejected product because it also includes transportation and other export costs, which are incurred by the exporter. Otsuki et al. (2001) estimated that the EU standard on aflatoxin, which was higher than the Codex ML, would result in losses for African products by a decrease in exports worth US$670 million.[33] In 1997, the EU imposed an SPS-related five-month-import ban on shrimp from Bangladesh, and it is estimated that the adverse effect of this ban mounted to $65.1 million.[34]

It is argued that some SPS requirements set by developed countries are based on technological limits rather than on public health considerations. Otsuki et al. (2000)[35] illustrate this with the example of the 1998 EC Regulation that raised the maximum limit for aflatoxin in foodstuffs and animal feed to a level higher than the Codex level. The results suggest that the EU standard, which would reduce the risk by approximately 1.4 deaths per billion a year, would affect exports from nine affected countries by a reduction of more than 60 percent. It is worth noting that the SPS Agreement acknowledges the problems that DCs might face in complying with SPS requirements, and it thus calls upon Members to exercise special and differential treatment with DCs when adopting food safety and other SPS measures.[36]

In their study on the impact of sanitary measures on export of fishery products from India (the case of Kerala), Henson et al. (2004)[37] noted the challenges that DC fish exporters encountered in trying to comply with strict SPS requirements imposed by major industrialised countries. Noting the prohibitively high costs of compliance, the authors also highlighted some positive returns in terms of continued and/or expanded

[33] T Otsuki, JS Wilson and M Sewadeh, 'What Price Precaution? European Harmonisation of Aflatoxin Regulations and African Groundnut Exports', *European Review of Agricultural Economics* 28 (2001), 263–283.

[34] Cato et al., supra note 8.

[35] T Otsuki, JS Wilson and M Sewadeh, 'Saving Two in a Billion: Quantifying the Trade Effect of European Food Safety Standards on African Exports' in *Food Policy*, 26 (2001): 495–514.

[36] Article 10, SPS Agreement: Special and Differential Treatment. '1. In the preparation and application of sanitary or phytosanitary measures, Members shall take account of the special needs of developing country Members, and in particular of the least-developed country Members.' WTO, 1998.

[37] SJ Henson, M Saqib and D Rajasenan, 'Impact of Sanitary and Phytosanitary Measures on Fishery Products from India: The Case of Kerala' (2004), Agriculture and Rural Development Discussion Paper 17, Cost of Compliance with SPS Standards, The International Bank for Reconstruction and Development/The World Bank, Washington, DC. Available at: http://siteresources.worldbank.org/INTARD/Resources/India_Fish_final.pdf.

access to high-value markets for the exporters that are able to comply. Compliance with standards for niche products can be a 'successful experience' as illustrated by the increase in Peruvian asparagus exports, making Peru the 'world leader in asparagus'.[38]

E. Private standards: challenges and opportunities

Although they fall outside the governance structures established by the WTO because they inter alia do not have to follow the transparency principles, private standards are playing an increasing role in the international food trade,[39] and they are becoming the predominant drivers of agri-food systems.[40] Private standards have evolved in response to regulatory developments and consumer concerns, and as a means of competitive positioning in markets for high-value agricultural and food products. Private standards can be applied at various stages of the value chain: primary production, processing, trading, retailing and consumption. Nevertheless, private standards can become de facto trade barriers depending on the buying power of the retailers.[41] The International Trade Centre has developed a Standards Map that provides information on more than 130 standards, codes of conduct and audit protocols addressing sustainability hotspots in global value chains.[42] It is acknowledged that agri-food businesses use private standards as a tool for product differentiation to highlight specific characteristics of their products, to segment the market and to ensure quality-based competition.[43] In addition, private standards in the agri-food trade may entail requirements that go beyond food safety and quality considerations such as environmental

[38] A Mello, 'The Use of Standards and Its Impact in the Agribusiness Sector: The Peruvian Experience'. (2009) TBT Committee Workshop on the role of international standards in economic development, 16–17 March 2009. The author enumerates several key factors of change and success among these: standards development and implementation; food safety and quality assurance; investment and private innovation (logistics, new product presentations and packaging), use of modern technology as well as strategic alliances between the public and private sectors.

[39] S Henson and S Jaffee, supra note 3.

[40] S Henson and NH Hooker, 'Private Sector Management of Food Safety: Public Regulation and the Role of Private Controls', *International Food and Agribusiness Management Review*, 4 (2001), 7–17.

[41] WTO (2005), *World Trade Report 2005: Trade, Standards and the WTO.*

[42] Available at: http://www.standardsmap.org.

[43] S Henson and T. Reardon, 'Private Agri-food Standards: Implications for Food Policy and the Agri-food System', *Food Policy* 30 (2005), 241–253.

and ethical requirements.[44] Private standards can therefore act as barriers to trade for food products, especially from DCs.[45] DCs are considered 'standard-takers' rather than 'standard-makers' because developing their own standards is generally considered to be more costly than adopting the standards of their major markets.[46]

The issue of private standards has been raised in several international fora including the WTO SPS Committee, where it was first raised in June 2005 by St Vincent and the Grenadines (G/SPS/GEN/766; specific trade concern no. 219) who complained about the requirements for EUREGAP's (now called GLOBALGAP) certification for bananas destined for sale in the United Kingdom.[47] Since then, the issue of private standards has been on the agenda of every SPS Committee meetings.

The concerns raised by WTO Members in the SPS committee[48] regarding private standards include, inter alia, the fact that private standards are not always based on science; that they deviate from international standards or from official governmental requirements (e.g. MRLs), that they are not harmonised; that certification and compliance with private standards is costly; that there is a lack of transparency and that there are no possibilities for consultation or systems of appealing; that private standards prescribe how measures should be applied rather than what the outcome should be; that they ignore the principle according to which equivalent outcomes achieved by different means should be recognised; that they pose disproportionate burdens on small and medium-size enterprises including producers and exporters in DCs, and so on. However, some WTO Members have also highlighted the benefits of private standards because they help suppliers comply with national and international

[44] JFM Swinnen and M Maertems (2009), 'Standards, Trade and Developing Countries'. Note prepared for the World Bank, Trade and Research Departments, for the Project on Adjustment Costs to Trade.

[45] S Henson and J Humphrey (2009), 'The Impacts of Private Food Safety Standards on the Food Chain and on Public Standard-Setting Processes', Rome: Food and Agriculture Organisation and World Health Organisation, Codex Alimentarius Commission, ALINORM 09/32/9D-Part II.

[46] SM Stephenson (1997), 'Standards, Conformity Assessment and Developing Countries', World Bank, Policy Research Working Paper 1826.

[47] G/SPS/GEN/766; specific trade concern no. 219 in which they complained about the requirements for EUREGAP (now called GLOBALGAP) certification for bananas destined for sale in the United Kingdom (WTO, 2005; Chia-Hui Lee, 2006). Chia-Hui Lee, G. (2006). Private Food Standards and Their Impacts on Developing Countries (DG Trade Unit G2, European Commission, Brussels). Available at: http://trade.ec.europa.eu/doclib/docs/2006/november/tradoc_127969.pdf

[48] Available at: http://www.wto.org/english/news_e/news11_e/sps_30mar11_e.htm.

standards when they prescribe how those standards should be met; they promote best practices and improved productivity; they give brands a better reputation and help suppliers have access to markets and credit and they address emerging risks in a rapid manner, filling gaps; and that private standards make it easier for international standards to eventually be adopted.

The SPS Committee's discussions on private standards initially focused on three themes: market access, development and WTO law. As regards market access, WTO Members have differing views as to whether private standards are an opportunity or a threat to exporters. Many Members are concerned that the cost of certification, sometimes for multiples sets of standards for different buyers, can be a problem, especially for small-scale producers and particularly (but not exclusively) in DCs. However, WTO Members have differing views as to whether private standards fall under the jurisdiction of the SPS Agreement. The concern that the proliferation of private standards could undermine some of the progress made in regulating SPS measures through the adoption and implementation of the SPS Agreement is at the root of these divergent views.[49] In 2008, the SPS Committee set up an ad hoc working group on SPS-related private standards to bring more structure to discussions on private standards within the SPS Committee. This group discussed the subject of SPS-related private standards and possible future work of the committee based on a questionnaire survey on the effects of SPS-related private standards, a descriptive report summarizing the responses received, and a report on possible actions for the SPS Committee regarding private standards.

At the SPS Committee meeting that took place in Geneva, 30–31 March 2011, the committee adopted five of the six actions put forward by a working group for endorsement.[50] Although WTO Members remain divided as to whether private standards legally fall within the scope of the SPS Agreement, the SPS Committee Members had agreed to develop a working definition for private standards related to SPS measures.[51] This work has been pursued since then and was reviewed at the SPS Committee meeting held 16–17 October 2013 when WTO encouraged the 'three

[49] WTO, World Trade Report 2012.

[50] G/SPS/55 and G/SPS/R/62. Also refer to 'WTO/Members Take First Steps on Private Standards in Food Safety, Animal and Plant Health'. Available at: http://www.wto.org/english/news_e/news11_e/sps_30mar11_e.htm.

[51] G/SPS/55.

sisters' (Codex, OIE and IPPC) to continue their collaboration with the WTO on this subject.[52] However, at the SPS Committee meeting that took place in Geneva 25–26 March 2014, three years after Members had agreed to develop a definition for private standards, there was a total disagreement on a definition for private standards for food safety and animal and plant health. The SPS Committee therefore agreed to look at definitions for private standards, used in other international forums and to try to adapt these to food safety, animal and plant health. At its meeting held 9–10 July 2014, the SPS Committee examined the list of existing definitions of private standards that had been compiled by the Secretariat. A revised version of this document was later issued as document G/SPS/GEN/1334/Rev. 1 and it includes a list of 'existing definitions of private standards in other international organisations' (FAO/WHO Codex Alimentarius Commission, International Trade Centre, OECD, OIE, UNCTAD and the World Bank) and 'existing definitions of a private standard or a standard from additional sources'.[53] However, several Members had noted that none of the definitions of private standards referred to in the Secretariat's list were helpful in defining SPS-related private standards, and thus they proposed that the working definition tabled by the Electronic Working Group (e-WG) co-stewards, China and New Zealand (document G/SPS/W/276), form the basis of any further discussion of a definition. Therefore, Members were requested to submit any additional comments regarding elements of other existing definitions that could be incorporated into the co-stewards' definition by 5 September 2014. Furthermore, it was agreed that the e-WG would circulate its report on a compromise working definition of an SPS-related private standard to the Committee for consideration at the October 2014 meeting.[54]

In a communication from Belize dated 15 April 2014 titled 'SPS-related private standards',[55] Belize informed the Committee about the discussions that the Belize Agricultural Health Authority held with major papaya and citrus exporters whose products were being affected by private standards. They cited specific requirements that were not guarding against any actual

[52] Supra note 50.

[53] WTO, SPS Committee (2014), 'Members to Try New Approach for Defining Private Sanitary-Phytosanitary Standards'. Available at: http://www.wto.org/english/news_e/news14_e/sps_25mar14_e.htm.

[54] WTO, SPS Committee (2014), G/SPS/R/75 Summary of the Meeting of 9–10.7.2014. Note by the Secretariat, 18 September 2014. Available at: http://www.wto.org/english/tratop_e/sps_e/work_and_doc_e.htm.

[55] WTO (2004) G/SPS/GEN 1327. Available at: http://www.wto.org/english/res_e/res_e.htm.

risk of transmission of pests or diseases but that could be described as measures to prevent 'theoretical risk of transmission'. In addition, the exporters indicated that the requirements were not supported by a valid risk assessment. Belize went on to inform the committee that because several exporters from that country lacked the volume of trade that would allow them to negotiate with the buyers and because they lacked the appropriate technical expertise, several exporters from Belize had ceased to exist while others raised concerns regarding the high costs related to certification and annual audits. Belize called on the SPS Committee to advance work on private standards 'as developing countries were affected on a daily basis by the requirements imposed by these nongovernmental entities'. Belize concluded by recommending that 'the sanitary and phytosanitary requirements for trade should be applied in compliance with the provisions of the SPS Agreement'.

I. Private standards and other relevant international standard-setting organisations

The OECD noted the additional constraints that DCs face due to the increase in private standards that could restrict the potential suppliers in particular by increasing compliance costs, among others.[56] The OIE undertook work on assessing the implications of private standards that might conflict with OIE standards. In June 2009, an OIE expert ad hoc group on private standards was convened and a questionnaire addressed to all countries. The results of the questionnaire showed a significant divergence of views between developed countries and DCs. In May 2010, the OIE general session adopted a resolution recommending that the OIE maintain and strengthen links and dialogue with relevant global private standard-setting bodies and global private industry organisations to encourage the compatibility of private standards with OIE standards.[57] ISO has also undertaken work regarding private standards. The representative from ISO informed the 33rd session of the CAC that 'ISO had

[56] OECD (2007). *Private Standard Schemes and Developing Country Access to Global Value Chains: Challenges and Opportunities Emerging from Four Case Studies.* AGR/CA/APM (2006)20/FINAL, Directorate for Food, Agriculture and Fisheries, Committee for Agriculture, Paris, 3 August 2007. Available at: http://www.olis.oecd.org/olis/2006doc.nsf/NEWRMSENGREF/NT0000942A/$FILE/JT03230765.PDF.

[57] Available at: http://www.oie.int/en/international-standard-setting/implications-of-private-standards/.

recently published a special brochure explaining the difference between ISO Standards and private standards and schemes'.[58]

II. *The discussion on private standards in the CAC*

The CAC has also examined the issue of private standards since 2009 by considering their impact and has also sought ways of addressing the challenges that private standards pose both to Codex and its Members. At its 33rd session (2010),[59] the CAC noted that private standards were more stringent and more prescriptive than Codex standards – for example, in stating how food hygiene requirements should be met and that the prescriptions they contained were often inappropriate in DC contexts and difficult or impossible to apply in small-scale food businesses in DCs. In addition, the cost of certification disproportionately penalised small-scale producers. Nevertheless, to get regional views on the issue, the CAC requested the FAO/WHO Coordinating Committees (Regional Coordinating Committees) to conduct a further analysis of the problems that their Members encountered with private standards and to make recommendations for follow-up at the 34th session of the CAC (July 2011).[60]

Over the period of 2010–2011, all six Regional Coordinating Committees[61] examined the issue of private standards at their respective committee sessions, and they came up with recommendations that could be summed up as follows.

The Regional Coordinating Committees generally noted that private standards represented serious burdens especially for small-scale farmers and small and medium-sized enterprises due to their more stringent and prescriptive nature and that the additional burden related to their implementation could prevent the latter from accessing international food markets. It was further noted that the financial burden resulted from

[58] Paragraph 216, Report of the 33rd session of the Codex Alimentarius Commission. ALINORM 10/33/REP. Available at: http://www.codexalimentarius.net/web/archives.jsp?year=10. Also refer to http://www.iso.org/iso/private_standards.pdf.

[59] The Codex Alimentarius Commission, 'Impact of Private Standards', 33rd session of the CAC, Geneva, Switzerland (5–9 July 2010), Agenda Item 13. Available at: ftp://ftp.fao.org/codex/cac/cac33/cac33_13e.pdf.

[60] Report of the 33rd session of the Codex Alimentarius Commission. ALINORM 10/33/REP, paragraphs 218–243. Available at: http://www.codexalimentarius.net/web/archives.jsp?year=10.

[61] Codex Alimentarius Commission (2011). REP11/AFRICA; REP11/ASIA; REP11/EURO; REP11/LAC; REP11/NASWP and REP11/NEA.

increased costs for inspection and certification, training personnel and capacity building to comply with the requirements to upgrade equipment and laboratory facilities, the infrastructure and so forth. Codex Members also indicated that the development of private standards lacked in transparency and that there was no harmonisation among the various private standards. Nevertheless, they acknowledged that, in general, large-scale food producers were in a better position to meet private standards' requirements, whereas small-scale producers, who could not meet the requirements set in private standards were excluded from the market and had to seek alternative market opportunities.

Regional Coordinating Committee Members were of the view that all food safety private standards should be harmonised with Codex standards, and they encouraged private standard-setting organisations to seek observer status with the CAC and to participate actively in the work of the CAC as observers. They also encouraged the CAC to establish a dialogue with global private standard-setting bodies, engage with them and also exchange information with them as this dialogue could facilitate the harmonisation of private standards with Codex standards. In addition, the CAC was encouraged to invite private standard-setting bodies to participate in Codex as observers. Regional Coordinating Committees also encouraged the CAC to accelerate the Codex standard-setting procedure, and they recommended that relevant Codex committees give appropriate consideration to the development of more 'end-user relevant' standards. In addition, the CAC should promote greater awareness of its standards and their status as international benchmarks and that Codex Members should identify gaps in Codex standards and channel suggestions for new work through the appropriate Codex committees. Finally, Regional Coordinating Committees also called on the CAC to work in close cooperation with OIE, IPPC and the SPS and TBT committees on this subject.

Furthermore, the Regional Coordinating Committees also recommended to private standard-setting bodies to address the legitimate concerns expressed by increasing transparency in the development of private standards, dialogue with DCs during the private standard-setting process and seek to harmonise private food safety standards with Codex standards.

At its 34th session held in July 2011, the CAC discussed the challenge posed by private standards and an opinion was expressed that 'in the face of the challenge of private standards, Codex had to show that it was capable to fulfil its role to set science-based international food standards, otherwise private standards would quickly fill the gap left by Codex'. The

CAC concluded that Codex would continue to work closely with WTO, IPPC and OIE on the issue of private standards and that private standard-setting organisations would be invited to participate as observers in Codex meetings especially in view of the fact that the mandate of the CAC included 'promoting coordination of all food standards work undertaken by international governmental and non-governmental organisations'.

III. *Follow-up and future perspectives regarding private standards and Codex*

The CAC has proactively encouraged collaboration and cooperation with private standard-setting bodies and at its June 2013 session, the Codex Executive Committee agreed to 'recommend to the Directors-general of FAO and WHO to grant observer status to . . . and CGF'.[62] The CGF-GFSI has thus been formally recognized as the 221st Observer of the CAC.[63] In its submission requesting observer status with the CAC, CGF-GFSI indicated under its expected contribution to the Joint FAO/WHO Food Standards Programme that 'One of the key activities within CGF/GFSI includes the definition of food safety requirements for food safety schemes through a benchmarking process. Food safety schemes are mainly based on Codex standards however there have been divergences in the past. The participation of CGF/GFSI would allow them to signal issues related to the implementation of Codex standards and also adapt the benchmarks to Codex standards'.[64]

With these recent developments, we would expect more collaboration between the CAC and the relevant private standard-setting organisations, especially as these contribute actively to the Codex decision-shaping process in their capacity as observers. The CAC should also heed to the recommendation made by the Regional Coordinating Committees and accelerate the Codex standards development procedure so that Codex standards remain relevant and up-to-date and that Codex responds proactively to address current and emerging food-related issues.[65]

[62] CX/EXEC 13/68/5; CRD 4 (CGF-GFSI), REP13/EXEC Paragraph 139.

[63] GFSI Newsletter, Winter 2013, p. 17, presentation by Mr Tom Heilandt, Senior Officer, Codex Alimentarius Secretariat Available at: http://www.mygfsi.com/gfsifiles/Newsletter/GFSI_NEWSLETTER_WINTER_2013_FINAL.pdf.

[64] CX/EXEC 13/68/5; CRD 4 (CGF-GFSI).

[65] Codex Strategic Plan 2014–2019 (2013). Available at: ftp://ftp.fao.org/codex/Publications/StrategicFrame/Strategic_plan_2014_2019_EN.pdf.

F. Implementation of Codex standards and trade opportunities: Achieving compliance through product quality and safety

DCs have to devise strategic development plans to promote economic development and trade while ensuring food safety and the protection of public health on the domestic market through the implementation of national food safety management strategies. Thus, complying with SPS requirements should be viewed not merely as a barrier to international trade but also as an opportunity to upgrade domestic quality and food safety standards.[66]

DC products have to be competitive on the target market, and their products should meet all SPS and other technical requirements of the target market, otherwise they risk being excluded from these markets in case of noncompliance. NTM and stringent and prescriptive private standards are an additional threat to DCs because they could result in their exclusion from international food markets. Nevertheless, it is acknowledged that new institutional arrangements such as public-private partnerships could play a key role in enabling compliance with market requirements. Jaffee and Masakure[67] examined the impact of the strategic decision made by Kenya to comply with private standards. Indeed, this strategic decision enabled Kenya to enhance international competitiveness in the vegetable trade to Europe. This strategy entailed substantial investments to improve production and procurement systems, upgrade pack-house facilities and adopt quality assurance and food safety management systems. With examples from Kenya and India, Narrod et al.[68] demonstrate that given adequate institutional support, smallholders can cope with stringent SPS requirements.

The identification and prioritisation of capacity building needs is an important step in the process of strengthening national food control systems[69] that will enable DCs to overcome the constraints that limit

[66] J Jongwanich, 'The Impact of Food Safety Standards on Processed Food Exports from Developing Countries', *Food Policy* 34 (2009), 447–457.

[67] S Jaffee, and O Masakure, 'Strategic Use of Private Standards to Enhance International Competitiveness: Vegetable Exports from Kenya and Elsewhere', *Food Policy* 30 (2005), 316–333.

[68] C Narrod, D Roy, J Okello, B Avendaño, K Rich and A Thorat, 'Public-Private Partnerships and Collective Action in High Value Fruit and Vegetable Supply Chains', *Food Policy* 34 (2009), 8–15.

[69] FAO, 'Strengthening National Food Control Systems: Guidelines to Assess Capacity Building Needs'. (2006). Available at: ftp://ftp.fao.org/docrep/fao/009/a0601e/a0601e00.pdf.

their access to international markets. Article 9 of the SPS Agreement calls on developed countries to provide technical assistance to DCs to help them meet SPS requirements. In their study on technical assistance providers, Wiig and Kolstad[70] concluded that technical assistance was provided in an ad hoc manner. The authors therefore argued for a more systematic allocation of technical assistance to DCs based on relevant data and comparison of benefits and costs of different kinds of capacity building. The Global Trade-Related Technical Assistance Database[71] was set up to provide information on trade-related technical assistance and capacity-building projects.[72]

In 2002, FAO, WHO, WTO, the World Bank and the OIE established the Standards and Trade Development Facility (STDF), which is a global programme for capacity-building and technical assistance in SPS matters including issues covered by Codex.[73] The STDF is a coordination and resource mobilisation mechanism that acts as a forum for information sharing on SPS-related technical cooperation activities. It also aims to mobilise resources to address SPS issues that limit the trading possibilities of DCs and thereby strives to improve the protection of human and animal life and health as well as plant health from SPS risks. The strategic aims of the STDF are to assist DCs to enhance their expertise and capacity to analyse and to implement international SPS standards; to improve DC human, animal and plant health situations and their ability to gain and maintain market access; and to act as a vehicle for raising awareness about the importance of SPS issues. In addition, the STDF ensures coordination among technical cooperation providers and also assists in the mobilisation of funds, the exchange of experience and the dissemination of good practices in relation to the provision and receipt of SPS-related technical cooperation.

G. Conclusion

The scientific literature examined seems to agree that compliance with SPS and technical requirements in the international food trade is still a challenge for DCs. It is also acknowledged that DCs struggle to comply

[70] A Wiig and I Kolstad, supra note 25. [71] Available at: http://gtad.wto.org/.

[72] A list of current projects is available at: http://gtad.wto.org/entity_project.aspx?EntityID=304.

[73] Available at: http://www.standardsfacility.org.

with public Codex SPS requirements,[74] so NTM and the requirements set by private standards hinder their ultimate access to international food markets. Nevertheless, there is a clear, strong and logical linkage between DCs increased and effective participation in Codex and enhanced international food trade opportunities and development.[75] The CTF has led to an increase in DC participation in Codex and to an improved uptake of Codex standards in DCs by contributing to an increased level of awareness and understanding of the benefits of setting and applying Codex standards. However, various supply-side constraints including the lack of administrative capacity to implement international standards and the lack of technical and scientific capacities still hinder DCs from complying with SPS and technical requirements in the international food trade.

DCs should participate actively and effectively in Codex, OIE, IPPC and the WTO SPS and TBT Committees to contribute effectively to the process of standardisation and frame the agenda by also raising issues and priorities of importance to them. In addition, an increased DC participation in the CAC will help ensure that Codex standards take into account the widest interests possible thus providing greater legitimacy and global relevance, acceptance and implementation of the standards themselves and that the CAC is a truly global and inclusive international organisation. The technical assistance initiatives implemented by intergovernmental agencies such as FAO, WHO, STDF, UNCTAD and World Bank and national donors should target the major impediments to exports caused by SPS requirements in collaboration with DCs.

SPS requirements should not only be considered as 'barriers' because they could also serve as 'catalysts'[76] enabling the adoption of efficient strategic approaches by both DC governments and the private sector to improve food safety management both for the domestic and international markets. In this case, the costs of complying with food safety standards could provide a powerful incentive for the modernisation of export supply chains in DCs.[77] Consequently, capacity-building efforts should not solely address the development of technical infrastructure but also provide the

[74] REP11/ASIA, reference to Indian small-scale producers.

[75] Connor, supra note 2. [76] S Henson and S Jaffee, supra note 3.

[77] S Henson and S Jaffee, 'Understanding Developing Country Strategic Responses to the Enhancement of Food Safety Standards', *The World Economy* (2008). Available at: http://onlinelibrary.wiley.com/doi/10.1111/j.1467–9701.2007.01034.x/pdf.

'tools' to both governments and the private sector, including public-private partnerships.[78] This will enable DCs to develop and maximise the strategic options available to them and thereby allow them to address future stringent SPS and technical requirements adequately and ensure that their products continue to have access to the international food trade.

[78] C Narrod et al., supra note 68.

14

Standards of responsible investment and international investment law

ANGELOS DIMOPOULOS

A. The development of international standards on socially responsible investments

Over the past two decades, public opinion has increasingly focused its attention on the broader economic, social and environmental consequences of the investment activities of multinational enterprises (MNEs). Rather than accepting that foreign investment contributes ipso facto to the development of investment recipient countries and their societies, the impact of the business activities of MNEs has been the subject of controversial debates[1] and spurred various initiatives aiming to promote socially responsible investments and guarantee their positive effect on local societies.

Indeed, despite the existence of conflicting empirical evidence,[2] a common conclusion reached in the vast majority of scholarly work on this topic is that MNEs can significantly contribute to the economic development of the countries where they invest, adding to their economic growth.[3] Of course, MNEs' activities *may* result in such positive outcomes.

I thank the participants at the International Standardisation Conference held in Tilburg on 7–8 November 2013 for their useful comments.

[1] On the impact of MNEs activities to the development of investment recipient countries indicatively, see O de Schutter, J Swinnen and J Wouters (eds), *Foreign Direct Investment and Human Development* (2013).

[2] For an overview of the questions raised and the empirical studies conducted, see United Nations Conference on Trade and Development (UNCTAD), *World Investment Report 2006: FDI from Developing and Transition Economies: Implications for Development* (2006), 184–194; UNCTAD, *Economic Development in Africa, Rethinking the Role of Foreign Direct Investment* (2005), 26–36, 76–82.

[3] For an overview of the literature, see indicatively J Dunning and S Lundan, *Multinational Enterprises and the Global Economy* (2008); T Moran, *Harnessing Foreign Direct Investment*

The positive impact of the business activities of MNEs depends on various factors, such as host-country economic environment, their institutional infrastructure and regulatory framework.[4] More important, MNEs may actually have a negative impact on the growth of the recipient country, for example, in countries with imperfect competition conditions, because it can lead to the creation of foreign-owned monopolies, crowding out domestic firms and eventually generating unemployment. In addition, foreign investment may potentially cause significant social and environmental harm, leading to a 'race to the bottom', as recipient countries, in their effort to attract foreign investment, may lower, or tolerate the violation of, their environmental, labour and other social standards.[5]

In that respect, an important tool that has been used for the promotion of public policy goals by private entities has been corporate social responsibility (CSR). The rise of MNEs with a global reach of activities and a complicated structure across different jurisdictions has indicated the limitations of traditional governance mechanisms and facilitated the introduction of new tools, which engage private actors in their development, monitoring and enforcement. CSR programmes and tools have become commonplace and are effective in influencing the ways in which MNEs and their suppliers conduct business around the world.

The development of standards of socially responsible investments has been proven significant in practice, given the inadequacies of regulatory instruments to ensure that MNEs effectively comply with social policy goals. Although there is now a significant amount of domestic law governing the wider relations between corporations and society, 'domestic' regulatory solutions cannot always provide an appropriate solution to combat socially irresponsible behaviour of MNEs. Considering that investment recipient developing countries often face problems of corruption and lack institutional foundations that can effectively guarantee good governance and rule of law, they enable MNEs investing in such countries to become 'polluters, abusers of market power, corruptors of State officials, exploiters of workers, and potential accomplices to violations

for Development: Policies for Developed and Developing Countries (2006); T Moran et al. (eds), *Does Foreign Investment Promote Development?* (2005).

4 L Alfaro, *Foreign Direct Investment and Growth: Does the Sector Matter* (2003).

5 For a brief overview of the potential negative impact of foreign investment, see indicatively: Dunning and Lundan, supra note 3, chapters 10 and 17; R Lipsey, 'Home and host country effects of FDI' in Baldwin & Winters (eds), *Challenges to Globalization* (2004), 333–382.

of fundamental human rights'.[6] Besides, even when national laws impose social responsibility obligations on MNEs, their ability in practice to contribute successfully to global compliance with CSR is doubtful, even if they apply extraterritorially. Notwithstanding any legitimacy questions that such 'internationalisation' of national rules on what constitutes social responsibility raises,[7] the problems that arise with regard to their monitoring exemplify their limited practical effectiveness.[8]

Nevertheless, the development of standards of socially responsible investment presents significant challenges. The different processes involved in their adoption, implementation and compliance affect not only their binding force but more importantly their normative value, their legitimacy and ultimately their effectiveness. Considering the nature of the goals that CSR standards serve and the incentives that MNEs have to comply and engage actively with socially responsible investments, it is highly controversial whether there should be more regulatory intervention in the field of CSR or whether it should be left entirely voluntary to the discretion of MNEs.

Within this framework, the purpose of this contribution is to examine whether and how international investment law can successfully address the key challenges that the development of socially responsible investment standards raise. In that respect, this chapter examines first the different types and categories of standards of socially responsible investments. It explores the key characteristics of responsible investment standards to identify how they can contribute to the achievement of public policy goals. Afterward, the chapter looks at the key challenges that arise from their voluntary nature and the role of regulation in dealing with them. Bearing in mind the 'public' nature of the objectives such standards aim to achieve, this part assesses the main concerns that different types of international CSR standards present and discusses how international law tools may be used to tackle them. Finally, this chapter examines whether and how international investment law can contribute to promoting CSR standards and address their deficiencies. By examining the different options available for addressing CSR concerns through International Investment Agreements (IIAs), this part identifies new ways through which IIAs can

[6] J Hepburn *and* V Kuuya, 'Corporate Social Responsibility and Investment Treaties' in M.-C. Cordonier Segger, M. Gehring and A. Newcombe (eds), *Sustainable Development in World Investment Law* (2011), 590–591.

[7] See in particular J Zerk, *Multinationals and Corporate Social Responsibility* (2006), chapter 3; P Muchlinski, *Multinational Enterprises & the Law*, 2nd ed (2006), chapter 4.

[8] Zerk, supra note 7, chapters 4 and 5.

enhance the effectiveness of CSR standards without jeopardising their voluntary nature as standards.

B. The categories of international standards on socially responsible investments

Corporate Social Responsibility tools have long been used for the promotion of public policy objectives. Indeed, since the 1950s businesses have been viewed as responsible for respecting and promoting important social policy objectives.[9] Alongside regulation, CSR presented a new form of governance, which acknowledged the power held by private actors in a market economy and the potential of public-private cooperation toward achieving non-economic and social policy objectives. Without entering into a detailed analysis of how voluntary CSR standards contribute to the achievement of social policy goals,[10] for the purposes of this chapter, it is necessary to identify the different categories of standards on socially responsible investments, and explain their key characteristics. Although it is difficult to provide an exhaustive list of different standards on socially responsible investments, two broad categories of standards can be distinguished based on their origins and the normative weight they carry.

I. Industry-driven standards

There are more than twenty key international initiatives providing standards for social and environmental practices of firms at home and abroad. Following the recognition of the special responsibility that MNEs carry for promoting the materialisation of social policy objectives, MNEs were motivated to create and adopt CSR standards. Considering the variety and multiplicity of MNEs activities, such initiatives have been originally established at firm level and have also gradually contributed to the development of industry-wide standards.

More specifically, more than 90 percent of MNEs have gradually developed their own voluntary company CSR codes, which are most closely

[9] On the development of the CSR concept, see A Carroll, 'A History of Corporate Social Responsibility: Concepts and Practices' in Crane et al. (eds), *The Oxford Handbook on Corporate Social Responsibility* (2008), 19–46.

[10] For an insightful discussion of this topic, see D McBarnet, 'Corporate social responsibility beyond law, through law, for law: the new corporate accountability', in D McBarnet, A Voiculescu and P Campbell (eds), *The New Corporate Accountability: Corporate Social Responsibility and the Law* (2007), 9–58.

aligned with their values and operations.[11] Such codes consist typically of a set of principles and priorities that guide MNEs through their actions. Although such company codes rely on broad principles, they tend to address the specific social policy concerns that arise from the activities of the MNEs, providing for example detailed rules as to how standards are incorporated into an MNE's operations and how their implementation is being monitored and guaranteed.[12] Although the adoption and implementation of such voluntary standards may be costly for an MNE, by adopting standards of socially responsible investment, MNEs in reality can benefit immensely: abiding by internal social responsibility standards can enhance their reputation, ensure compliance with consumer protection laws, or mitigate any damages, thus creating a competitive advantage.[13]

To strengthen the value and facilitate the adoption and implementation of company CSR instruments, there have been many initiatives in different economic sectors to develop industry-wide standards of socially responsible investments. Today there are hundreds of industry association standards in existence covering major areas of national and international commercial activity, both industry-specific and general. Such codes are usually jointly developed by the leading firms within an industry after consultations with the civil society and other institutional actors. They offer additional competitive advantages because compliance with their provisions may result in preferential treatment. For example, the financial sector has developed socially responsible investment schemes, which enable MNEs that abide by their standards to gain access to beneficial treatment, such as being listed separately on the stock exchange.[14] As a result, the development of and compliance with such standards presents a huge incentive for companies, as they present an additional tool to have better access and treatment in capital markets.

A unique example of a well-recognised, industry-driven code on socially responsible investments presents the ISO 26000 standard on social responsibility.[15] As explained in detail by Stephanie Bijlmakers and Geert Van Caster,[16] the International Organisation for Standardisation

[11] UNCTAD, *Investment and Enterprise Responsibility Review* (2010), 12.

[12] OECD, *Overview of Selected Initiatives and Instruments Relevant to Corporate Social Responsibility* (2009).

[13] D Vogel, *The Market for Virtue: The Potential and Limits of Corporate Social Responsibility* (2005).

[14] For example, the FTSE4Good is a separate stock-exchange index using CSR standards. Ibid, 61–63.

[15] ISO 26000 (2010) Guidance on Social Responsibility.

[16] See Chapter 12 in this volume.

(ISO), which is a non-governmental organisation whose members are the national standard-setting bodies in countries around the world, has developed an international standard on social responsibility. Bridging the interests of MNEs, national governments and other stakeholders, the development of the ISO standard was developed because it was considered to present an excellent opportunity to raise awareness and 'facilitate the establishment, implementation, maintenance and improvement of social responsibility frameworks' and 'contribute to increasing the confidence and satisfaction' from their application.[17] In that respect, ISO 26000 presents a significant landmark for standardisation of socially responsible investments.

II. International Organisations standards

In addition to private initiatives, socially responsible investment standards have attracted the attention of a number of international organisations. Recognising the normative dimension of CSR standards, a number of international organisations have been particularly involved in the establishment of international CSR instruments. In fact, the most successful international initiatives on CSR have been developed by international organisations, including the Organisation for Economic Co-Operation and Development's Guidelines for Multinational Enterprises (OECD Guidelines),[18] the International Labour Organisation's Tripartite Declaration of Principles concerning Multinational Enterprises and Social Policy (ILO MNE Declaration),[19] (ILO) and the United Nations' (UN) Global Compact.[20]

The OECD Guidelines for Multinational Enterprises comprise a set of voluntary recommendations in all major areas of corporate activity, including employment and industrial relations, human rights, environment, information disclosure, combating bribery, consumer interests, science and technology, competition and taxation.[21] Following the 2011

[17] Ibid.

[18] OECD Guidelines for Multinational Enterprises, latest version adopted on 25 May 2011, available online at http://mneguidelines.oecd.org/text/.

[19] ILO, Tripartite Declaration Concerning Multinational Enterprises and Social Policy, (2000), LXXXIII OB, Ser A, No (November 1977, amended November 2000).

[20] UN Global Compact, available at http://www.unglobalcompact.org/ (accessed on 25 January 2015).

[21] For an analysis of the CSR protection offered by the OECD Guidelines, see E Morgera, 'OECD Guidelines for Multinational Enterprises' in T Hale and D Held (eds.), Handbook of Transnational Governance: Institutions and Innovations (2011), 314–322; A McBeth, International Economic Actors and Human Rights (2011), 261–263.

amendments, the Guidelines include new recommendations on human rights and company responsibility for their supply chains, making them the first inter-governmental agreement in this area. Similar to industry-based CSR standards, the Guidelines are voluntary in their application because the OECD invites MNEs to adopt them and comply by their standards. Despite their voluntary nature, their implementation involves a unique combination of binding and voluntary elements. In a nutshell, the OECD Guidelines establish a dispute settlement system, as they require from the countries that have adopted them to set up National Contact Points, which can hear complaints made against MNEs, conduct investigations and produce non–legally binding recommendations concerning possible remedies.[22]

A similar set of labour-specific standards that promote socially responsible investments are included in the ILO MNE Declaration. The Declaration provides detailed guidance on how MNEs that invest abroad can maximise their positive contributions to economic and social development and minimise the negative impacts of their operations in fields such as employment promotion, skills development, conditions of work and life, and industrial relations.[23] Moreover, the Declaration includes a dispute settlement mechanism, which produces non-binding recommendations to the parties, as to how the declaration should be interpreted. Going beyond the OECD Guidelines, the ILO Declaration includes as well a surveillance mechanism, which, through the use of periodic surveys, offers useful data as to the popularity of the instrument with MNEs and their compliance record.[24]

Aiming to gain catholic acceptance across developed and developing countries, the UN were also proactively involved in the establishment of another set of socially responsible investment standards. The UN Global Compact was a UN initiative that was developed by the at the time UN Secretary-General, with support from different UN agencies,

[22] OECD, National Contact Points for the OECD Guidelines for Multinational Enterprises, available online at http://www.oecd.org/document/60/0,3343,en_2649_34889_1933116_1_1_1_1,00.html. For a description of the OECD enforcement mechanism, see S Robinson, 'International Obligations, State Responsibility and Judicial Review under the OECD Guidelines for Multinational Enterprises Regime' (2014) 38 *Utrecht Journal of International & European Law* 68.

[23] McBeth, supra note 21, p. 264.

[24] J Cernic, 'Corporate Responsibility for Human Rights: Analyzing the ILO Tripartite Declaration of Principles Concerning Multinational Enterprises and Social Policy' (2009) 6 *Miskolc J Int'l Law* 24.

governments and representatives of business, labour and other civil society bodies. The UN Global Compact is based on ten principles in the areas of human rights, labour standards, the environment and anti-corruption that are derived from the Universal Declaration of Human Rights, the ILO's Core Labour Standards, the United Nations Convention Against Corruption and – in the field of environmental protection – the Rio Declaration. Similar to the OECD Guidelines and the ILO Declaration, the Global Compact is a voluntary system, which invites MNEs to abide by its principles. Compliance relies on a public reporting mechanism, as well as a complaints process, which helps identifying MNEs that fail to comply with the principles.[25]

All these instruments have been well received by MNEs as positive steps toward promoting socially responsible investments. An increasing number of MNEs use these instruments to a certain extent as guiding their operations and activities overseas. Like private standards, the benefits of MNEs abiding by these principles in most cases can outweigh the costs linked to their implementation and monitoring. More important, the credibility and international recognition of the OECD Guidelines, the ILO Declaration and the UN Global Compact help MNEs obtain reputational gains, thus providing a powerful incentive to MNEs to abide by their rules.[26]

C. The challenges of standards on socially responsible investments and the role of regulation

Notwithstanding the significant role that CSR standards play in achieving widely accepted social goals, the private and voluntary nature of the processes involved in their adoption, implementation and compliance raises concerns about their legitimacy and effectiveness. Given that CSR standards aim to serve social policy goals, including the protection of human rights or environmental protection, the fact that their application relies primarily on MNEs appears problematic. The current debate over CSR focuses on whether corporations should be trusted to discharge their moral obligations to society and the environment by voluntarily

25 For a critical appraisal of the success of the Global Compact, see E Oshionebo, 'The UN Global Compact and Accountability of Transnational Corporations: Separating Myth from Realities' (2007) 19 *Florida Journal of International Law* 2.

26 McBeth, supra note 21, 260; A Litovsky et al. *Investing in Standards for Sustainable Development: The Role of International Development Agencies in Supporting Collaborative Standards Initiatives* (2007).

implementing CSR policies, or whether they should legally be compelled to do so. On the one hand, these societal concerns are considered to be too important for their adoption, monitoring and implementation to be left in the hands of profit-seeking MNEs.[27] At the same time, the proponents of the voluntary approach argue that markets are self-correcting, thus rewarding MNEs that develop and abide by socially responsible standards whilst complying with local laws and regulations.[28] Combining elements of the 'regulatory' and 'business' case for CSR, there is a growing number of voices in favour of a mixed approach, arguing that regulation can play a facilitative role, offering incentives to abide by voluntary standards and maximise their benefits.[29]

Indeed, the nature and normative value of the objectives that CSR standards aim to protect advocates in favour of a more regulatory approach to CSR. Without entering into a discussion as to the normativity of social goals and the purposes of allowing, promoting and encouraging MNEs' activities,[30] suffice it to raise the example of corporate human rights responsibilities to showcase why voluntary standards are not deemed adequate to respond to the challenges that the protection of human rights raise.[31] Nevertheless, the significance of the objectives pursued by CSR does not render them obsolete. Given the limitations of international law as a tool for imposing positive law obligations on MNEs and ensuring appropriate remedies, CSR standards remain a significant instrument that can be used to guarantee and promote compliance with human rights concerns. As the UN Special Representative on the issue of Human Rights and Transnational Corporations and other Business Enterprises,

[27] P Muchlinski, 'Corporate Social Responsibility and International Law: The Case of Human Rights and Multinational Enterprises, in D McBarnet, A Voiculescu and P Campbell (eds), *The New Corporate Accountability: Corporate Social Responsibility and the Law* (2007), 431–441.

[28] E Kurucz, B Colbert *and* D Wheeler, 'The Business Case for CSR' in A Crane et al (eds), *The Oxford Handbook on Corporate Social Responsibility* (2008), 83–112.

[29] McBarnet, supra note 10, 24–25; S McLeod, 'Reconciling Regulatory Approaches to Corporate Social Responsibility: The European Union, OECD and United Nations Compared' (2007) 13 *European Public Law* 671.

[30] A. Voiculescu, 'Human Rights and the Normative Ordering of Global Capitalism' in A Voiculescu and H Yanacopulos (eds.), *The Business of Human Rights: An Evolving Agenda for Corporate Responsibility* (2011), chapter 2.

[31] A Clapham, *Human Rights Obligations of Non-State Actors* (2006), 195–270; P Alston, 'The 'Not-A-Cat' Syndrome: Can the International Human Rights Regime Accommodate Non-State Actors?' in Alston et al. (eds), *Non-State Actors and Human Rights* (2005), 1, at 13–17; McBeth, supra note 21, 293–294.

John Ruggie, concluded in his report,[32] because MNEs bear limited international law obligations to protect human rights, voluntary standards can play a significant role: they can ensure not only compliance of MNEs with human rights norms but also proactively contribute toward the realisation of human rights objectives.[33]

But even if the development of voluntary instruments on CSR is indispensable, the nature of the social objectives pursued raises concerns with regard to the processes leading to their adoption and their inclusiveness. As CSR standards are developed by non-governmental actors aiming to spell out what is reasonable to expect of MNEs, inclusiveness is crucial for their success. However, it is difficult to identify CSR standards with broad inclusiveness. On the other hand, inclusiveness in the private standard-setting process is less broad geographically and does not typically extend beyond the industry and limited affected stakeholders.[34] The fact that national governments are not always involved in the procedures raises concerns regarding whether such CSR standards can actually reflect the social objectives and needs of the civil society.[35]

Nevertheless, there are an increasing number of quasi-normative CSR standards, especially those adopted by International Organisations that offer broad legitimacy coverage. For example, the recent renegotiation of the OECD Guidelines, illustrates the attempts to render this set of CSR standards more authoritative, as they reflect international consensus, including the views of developing countries, NGOs, as well as those of MNEs.[36] The participation of the latter is equally important since, the success of voluntary CSR standards depends on whether they become

[32] J Ruggie, 'Protect, Respect and Remedy: A Framework for Business and Human Rights Report of the Special Representative of the Secretary-General on the issue of human rights and transnational corporations and other business enterprises', UN Doc A/HRC/8/5, 2008; and Guiding Principles on Business and Human Rights: Implementing the United Nations 'Protect, Respect and Remedy' Framework, UN Doc A/HRC/17/31, 2011.

[33] Ibid., paras. 30–31.

[34] On current practices of private standard-setting, the International Social and Environmental Accreditation and Labelling Alliance (ISEAL Alliance), which is an association of international organisations that develops best practice guidelines for social and environmental standard-setting bodies, has set up a Code of Good Practice. ISEAL, Setting Social and Environmental Standards v5.0: Code of Good Practice (latest revision in 2010).

[35] U Mueckenberger and S Jastram, 'Transnational Norm-Building Networks and the Legitimacy of Corporate Social Responsibility' (2010) 97 *Journal of Business Ethics* 223.

[36] On the debates leading to the adoption of the 2011 reform of the OECD Guidelines see OECD, *Annual Report on the OECD Guidelines for Multinational Enterprises 2010*. Available at: http://www.oecd.org/daf/inv/mne/corporateresponsibilityreinforcingauniqueinstrument-2010annualreportontheoecdguidelinesformultinationalenterprises.htm.

a mainstream part of an industry. If an initiative is too unrealistic in its expectations or too costly to comply with, few MNEs will be able to follow it.

Moreover, the nature of the objectives pursued by CSR standards renders the adoption of effective standards challenging. Unlike other areas where international standards have been developed, socially responsible standards are by their nature more ambiguous and open-ended. The inherent difficulty in measuring social objectives with an uncontested, scientifically proven method presents a major hurdle in the adoption of universal, industry-wide standards that are meaningful and effective. Indeed, CSR standards intended to be applicable to all sectors in all countries are necessarily general, leaving most MNEs to identify themselves exactly how to implement the standards, and in some cases monitor compliance with them.[37] As a result, low precision may undermine the overarching aim of CSR standards to get MNEs to embrace their core values. In such cases, voluntary codes are frequently amounting to little more than 'vague statements of principle that cannot provide reliable guidelines for behaviour in concrete situations'. Unfortunately, most attempts to establish general CSR standards suffer from lack of specificity, undermining their value. For example, the UN Global Compact illustrates perfectly how lack of specificity undermines the value of CSR standards. Given the vague nature of its ten principles, which MNEs have to abide by, while allowing them to decide how to implement them and report compliance with them, the Global Compact can been seen as an ineffective system that can be used by MNEs to promote their CSR profile.[38]

These difficulties are manifest in the processes regarding the implementation of CSR standards. CSR standards aim to replicate the implementation methods found in other standard-setting frameworks. They often have proactive certification or accreditation programmes, which may involve a higher degree of compliance assurance and typically include periodic inspections, corrective action programmes, reporting and

[37] Of course there have been efforts to provide industry or issue specific guidance, such as the OECD 2011 Due Diligence Guidance for Responsible Supply Chains of Minerals from Conflict-Affected and High-Risk Areas and the ILO Codes of Practice on Safety and Health for specific industries. For a more detailed analysis of the specific standards developed for specific industries, see ISEAL's work, supra note 34.

[38] R Ghafele and A Mercer, 'Not Starting in Sixth Gear: An Assessment of the UN Global Compact's Use of Soft Law as a Global Governance Structure for Corporate Social Responsibility' (2010) 17 *UC Davis Journal of International Law and Policy* 41, at 53–55.

consumer labelling schemes.[39] Nevertheless, given the inability to translate all social objectives into measurable standards, most international CSR standards lack the certification guarantees that are typical of widely accepted voluntary standards. For example, it is worth pointing out that the ISO 26000 standard on social responsibility presents one of few instances in which no certification is provided for proving compliance with an ISO standard.[40]

Last but not least, the voluntary nature of CSR standards casts doubts on the effectiveness of their enforcement because monitoring and compliance with them cannot rely on the remedies available for legally binding obligations. Indeed, MNEs can only be made accountable through legally enforceable obligations. Voluntary CSR initiatives, are 'no substitute for the legislative actions of a recognised political authority'[41] given the lack of any basis for legal claims or redress under such initiatives. However, this does not make them ineffective. Soft law mechanisms often include other sticks and carrots besides the threat of legal action, and at least in some cases, CSR 'soft law' norms have succeeded in enforcing higher standards of social responsibility. Many CSR standards include mechanisms that promote reporting on compliance with CSR, and others avail themselves of quasi-judicial mechanisms that offer the opportunity to interested parties to challenge compliance of MNEs with specific standards.[42] The dispute settlement system set up in the OECD Guidelines is for example noteworthy, as National Contact Points can hear complaints made against MNEs, conduct investigations and produce non–legally binding recommendations concerning possible remedies and sanctions.[43] Nevertheless, much of current reporting is highly misleading or incomplete, while many CSR initiatives lack any compliance mechanism. It is worth pointing that this was considered to be one of the main disadvantages of the UN Global Compact, which was criticised as being toothless, even after the introduction of the complaints system, which has partly addressed these concerns.[44]

[39] The most comprehensive CSR reporting framework is that of the Global Reporting Initiative (GRI; https://www.globalreporting.org/Pages/default.aspx). GRI, which has been extensively used by MNEs, has developed a Sustainability Reporting Framework, which gains constantly popularity among MNEs.

[40] See Bijlmakers and van Calster, Chapter 12, this volume.

[41] S Hughes et al., 'The Global Compact: Promoting Corporate Responsibility' (2001) 10 *Environmental Policy* 155, at 157.

[42] Vogel, supra note 13, 46–74. [43] See supra note 22.

[44] Ghafele and Mercer, supra note 38, 48–53.

Aiming to compensate for the drawbacks of voluntary systems of CSR, there is an increasing tendency toward using the legal framework to provide incentives for developing appropriate CSR standards and ensuring effective compliance with their demands. Without creating too much international regulation imposing CSR principles, which can stifle innovation and damage competition, appropriate regulatory measures can offer significant support to the 'business case' for CSR, contributing toward their success. The supportive role of regulation toward enhancing CSR initiatives has in fact gained political support in a number of jurisdictions. Without entering into details, suffice it to refer to the European Union (EU) CSR agenda,[45] where the EU institutions aim to utilise different regulatory instruments to promote the adoption of CSR standards and enhance their credibility, as well as identifying avenues for redress and sanctions in cases of noncompliance.[46]

Next to national and regional attempts, international law presents a unique regulatory tool in which facilitative rules that help enhance the effectiveness of CSR standards can emerge. Indeed, responding to the emergence of integrated international production systems and the globalising world economy,[47] the establishment of a comprehensive and efficient regulatory framework that extends beyond national borders is necessary. Following the global character of MNE activity, the internationalisation of rules facilitative of CSR standards presents an effective means for addressing the problems and the concerns that their voluntary nature raises. Setting common rules for enhancing the credibility of CSR standards and incentivising compliance with them, alongside establishing an impartial and effective mechanism for their enforcement, an international regulatory framework would succeed in promoting CSR goals globally. Besides, it could offer an opportunity to national policy makers to review

[45] European Commission, Communication from the Commission to the European Parliament, the Council, the European Economic and Social Committee and the Committee of the Regions 'A renewed EU strategy 2011–14 for Corporate Social Responsibility', COM 2011/681/ FINAL; European Parliament, Report on Corporate Social Responsibility: promoting society's interests and a route to sustainable and inclusive recovery, EP 2012/2097(INI).

[46] J Wouters and L Chanet, 'Corporate Human Rights Responsibility: A European Perspective' (2008) 6 *Northwestern Journal of International Human rights* 262, at 272–283; D Kinderman, 'Corporate Social Responsibility in the EU, 1993–2013: Institutional Ambiguity, Economic Crises, Business Legitimacy and Bureaucratic Politics', (2013) 51 *Journal of Common Market Studies* 701, at 711–715.

[47] T Agmon, 'Who Gets What: The MNE, the National State and the Distributional Effects of Globalization' (2003) 34 *Journal of International Business Studies* 416.

and modernise national legislation, in particular in developing countries, which, despite needing such regulatory frameworks the most, are the ones lacking essential capital and the tools to set up and administer them.

D. Promoting responsible investment through international investment law

Within the range of available international law tools that can be facilitative to CSR standards, international investment law plays a prominent role. The increasing significance that international investment law has obtained, becoming one of the most dynamic fields of international law, presents a unique opportunity for establishing international norms facilitative of CSR standards.

I. The emergence of a new generation of IIAs

IIAs have become the main international law tools for promoting and protecting international investments. Indeed, there is today a nexus of more than 2500 IIAs that contain provisions regarding the promotion and protection of international investments.[48] The vast majority of IIAs are bilateral investment treaties (BITs), which primarily aim to provide investment protection. Recalling past nationalisations that occurred in many developing countries in their attempt to gain control over their vital resources and industries, developed countries were primarily interested in establishing international law norms granting protection to foreign investors. Building on customary international law on diplomatic protection of aliens, which granted 'prompt, adequate and effective' compensation to foreign investors in cases of expropriation of their property, BITs offer substantive rights of protection to foreign investors, as well as guarantee an impartial avenue for investors to raise their claims via investor-state arbitration.[49] Next to protection-oriented BITs, in recent years the international community has witnessed a rise in the number of new generation IIAs that go beyond traditional investment protection norms. Combining elements of investment protection and liberalisation

[48] UNCTAD, *IIA Databases*. Available at http://unctad.org/en/pages/DIAE/International%20Investment%20Agreements%20%28IIA%29/IIA-Tools.aspx.

[49] On the history and development of international investment law, see M Sornarajah, *The International Law on Foreign Investment* (2010), 19–47; K Miles, *The origins of International Investment Law* (2013), 19–61.

and linking them with international trade, a growing number of bilateral and regional preferential trade agreements include comprehensive investment chapters.

This remarkable law-making activity of states has been premised not only on the desire of investment-exporting nations to protect the property of their nationals abroad but more importantly on the belief of investment-recipient countries that international investment norms help attract foreign investment and secure benefits resulting from foreign investment activity.[50] However, over the past couple of years, BITs have been the subject of greater scepticism: their ability to attract foreign investment is questioned because empirical evidence is controversial as to whether and to what extent BITs advance foreign investment flows.[51] Moreover, the vague and potentially broad scope of BITs has given rise to challenges of national measures implementing national public policies, such as environmental protection, public health and (inter)national security.[52]

Within this framework, an increasing number of countries have reconsidered the scope and content of their BITs in light of the foregoing criticism and their own negative experiences with the international investment regime. Moreover, the tremendous increase of investment flows from developing countries and emerging economies to traditional investment-exporting countries has raised awareness of BITs' potential impact on national public policies in Western, developed countries and added new international actors in investment treaty law-making, such as China.[53] This trend is evidenced by the steadily growing number of new IIAs that depart from traditional model-BITs and experiment in different ways with innovative rules.

50 Supra note 3.

51 L Colen, M Maertens and J Swinnen, 'Determinants of Foreign Direct Investment Flows to Developing Countries: The Role of International Investment Agreements' in O deSchutter, J Swinnen and J Wouters (eds), *Foreign direct Investment and Human Development* (2013), 116–137; A Chaudhuri and H Benchekroun, 'The Costs and Benefits of IIAs to Developing Countries: An Economic Perspective' in A deMestal and C Levesque (eds), *Improvising International Investment Agreements* (2013), 93–112.

52 See, for instance, *CMS Gas Transmission Company* v. *The Argentine Republic*, Award, at 274–5 and 280; *LG&E Energy Corp., LG&E Capital Corp., and LG&E International, Inc* v. *The Argentine Republic*, at 121–139; *Sempra Energy International* v. *The Argentine Republic*, Award, at 298 and 303–304; *Enron Corporation and Ponderosa Assets, L.P* v. *The Argentine Republic*, Award, at 259–268.

53 C Liu, 'The evolution of Chinese approaches to IIAs' in A deMestal and C Levesque (eds) *Improvising International Investment Agreements* (2013), 59–64.

A growing number of IIAs are experimenting with new provisions, aiming to balance investors' protection and the pursuance of public policy objectives. A number of IIAs have witnessed the introduction of provisions that limit investors' rights or impose (indirect) obligations on them to respect or promote public policy objectives. Examples of such provisions would be broad exceptions from investment protection, detailed definitions of expropriation or clauses that demand adherence to international labour standards. Moreover, a growing number of IIAs introduce procedural checks and balances to the system of investor-state arbitration. To minimise uncertainty and enhance the legitimacy of the international investment regime, they promote procedural fairness and transparency.[54]

The reshaping of IIAs has also provided fertile ground for developing CSR-related norms and making linkages to voluntary CSR standards. Although CSR-specific clauses do not feature prominently in IIAs, a small but growing number of agreements, especially recent IIAs with trade and investment chapters, include CSR-specific provisions. The most common tools that have been used so far include the transformation of specific CSR standards into hard law and the adoption of preambular references to CSR standards in IIAs as well as best endeavour clauses addressed to the investors. Although these efforts present significant positive steps toward creating an appropriate international regulatory framework supportive of CSR standards, IIAs offer additional, yet unexplored opportunities for enhancing the international regulatory framework on CSR.

II. *Transforming CSR standards into hard law*

The most traditional option for incorporating CSR standards in IIAs has been their transformation to hard law, through the insertion of legally binding provisions that impose obligations on the signatory states. Adhering to the 'regulatory' approach to CSR, many countries have been eager to introduce in their IIAs provisions that establish minimum standards regarding environmental or labour protection and forbid the lowering of

[54] On recent trends in investment treaty making, see A deMestal and C Levesque (eds), *Improvising International Investment Agreements* (2013), especially chapters 1, 11 and 14; F Ortino, 'Refining the Content and Role of Investment 'Rules' and 'Standards': A New Approach to International Investment Treaty Making' (2013) 28 *ICSID Review* 152; P Muchlinski, 'Trends in International Investment Agreements, 2008/2009: Review of the Model Bilateral Investment Treaties of Norway, South Africa and the United States' (2010) 2 *Yearbook of International Investment Law & Policy* 35.

existing standards. Nevertheless, these IIAs stop short of imposing obligations on MNEs, opting instead to strengthen the domestic regulatory framework under which MNEs operate.

More specifically, CSR standards are indirectly transformed into hard law via the insertion of provisions requiring the maintenance of existing national environmental, labour or other social standards. A milestone provision of this type is Article 1114 of the North American Free Trade Agreement (NAFTA), which provides that each party should not waive or derogate from national health, safety or environmental measures in order to encourage investment. Although the wording of this provision is hortatory, a number of subsequent agreements have introduced mandatory language, demanding non-lowering of standards. For example, the free trade agreements (FTAs) that both the EU and the United States have concluded with Peru provide that the Parties shall ensure that foreign direct investment is not encouraged by lowering domestic environmental standards or laws.[55] These provisions present an innovation in comparison to Article 1114 NAFTA, as they adopt clearer and much stronger wording and do not appeal to the Parties' 'best efforts', but oblige them to avoid lowering their national standards.[56]

Moreover, the transformation of CSR standards into legally binding rules has been pursued through the adoption of provisions on minimum standards. For example, the EU is pioneering the introduction of such norms in its FTAs, as it has introduced binding norms on minimum environmental and labour standards in its EPA with the CARIFORUM States and the FTAs it concluded with Korea, Peru and Colombia. These agreements establish legally binding commitments for the parties to incorporate minimum international standards and oblige the parties to take all appropriate measures to ensure that foreign investment activity conforms to environmental standards. Without imposing direct obligations on private individuals, which would be controversial under investment law,[57] these EU FTAs are carefully drafted to introduce basic limits on foreign investment activity. For example, Article 72 of the EPA with the CARIFORUM States obliges the Parties to take 'measures as may be necessary

[55] Articles 277 of the EU-Peru and Colombia FTA and Article 18.3 of the US-Peru FTA.

[56] For example, contrary to Art. 1114 NAFTA, which uses words such as 'it is inappropriate' and 'should not', the FTAs include the term 'shall'. See also J Vinuales, *Foreign Investment and the Environment in International Law* (2012), 40–44.

[57] O DeSchutter, 'The Challenge of Imposing Human Rights Norms on Corporate Actors' in O DeSchutter (ed.), Transnational Corporations and Human Rights (2006), 1–43.

to ensure' that international labour and environmental obligations arising from agreements signed by the parties are not breached by investors' activities. The use of imperative language emphasises that the EPA not only recognises the right of the parties to pursue policies that ensure these standards, but it also imposes an obligation on them to take the appropriate and necessary measures. The FTAs with Korea, Colombia and Peru adopt a more elaborate approach, placing similar obligations on the parties under the broader framework of the relation between trade and sustainable development. Recognising that 'economic development, social development and environmental protection are interdependent',[58] they stress the *obligation* of both parties to adopt high levels for protecting labour and environmental standards. Articles 13.4 and 13.5 of the FTA with Korea and Article 279 of the FTA with Peru and Colombia reaffirm the commitments of the parties under international environmental agreements and the obligation of the parties to effectively implement them and express their willingness to cooperate further to promote internationally set environmental goals.

The legalisation of CSR standards, which these provisions introduce, places the responsibility for monitoring good corporate practices on the host State, one step removed from the MNE itself. Such provisions can contribute significantly to the pursuance of social policy objectives because they put pressure on investment recipient countries to adopt and implement high international standards and avoid a race to the bottom, maintaining a high level of protection of the environment and labour rights domestically.

Nevertheless, the effectiveness of such provisions is in most cases questionable. They rely on the ability and willingness of investment-recipient countries to adopt, implement and monitor compliance of MNEs with international sustainable development standards. However, as was already discussed, it is exactly the lack of institutional capacity in developing countries that makes reliance on domestic mechanisms ineffective.[59] In that regard, the insertion of legally binding provisions on labour and environmental standards in IIAs does not render the need for adoption, monitoring and compliance with voluntary CSR standards obsolete, which retain their significance.

Moreover, minimum standards provisions cannot fully address the public policy concerns of investment-recipient states and their societies.

[58] Art. 13.1. (2). of the EU-Korea FTA. [59] See supra notes 7 and 8.

They do not include the proper mechanisms for ensuring that civil society can effectively raise development concerns because they do not include any mechanisms for redressing or sanctioning MNEs that do not comply with such standards.[60] If the provisions on maintenance of standards are breached, the persons who are directly affected are not provided with any remedies for demanding the enforcement of these provisions. Indeed, dispute settlement under IIAs is not designed for complaints to be raised against private persons, as investor-state arbitration enables MNEs to bring complaints against states rather than the opposite. But even if counter-claims or defences could be raised against MNEs in dispute settlement, this does not necessarily mean that host states will always decide to pursue them. Considering that only investors and states can be parties to dispute settlement under IIAs, neither party may wish to rely on a violation of minimum standards because adherence to such standards may be economically burdensome for certain developing countries and, instead of contributing to their development, may actually impede economic progress.

Yet strengthening the domestic regulatory framework provides indirectly incentives for MNEs to behave responsibly because a jurisdiction with weak legal guarantees of social policies can no longer be a determinant of where their investment operations will be located.

III. Best endeavours provisions and CSR standards

Given the drawbacks of hard law approaches to CSR, provisions aiming to facilitate the voluntary adoption and compliance with CSR standards have been introduced in IIAs. The first type of IIAs provisions that could be facilitative to CSR standards are preambular references in IIAs. There is a growing tendency in investment treaty–making to contain references to CSR standards in the preambles of IIAs. For example, the Canada-Peru FTA provides in the Preamble that wish of the Contracting Parties is to 'encourage enterprises operating within their territory or subject to their jurisdiction, to respect internationally recognised [CSR] standards and principles'.[61] Similar hortatory language has also been used in the

[60] On the limits of IIA mechanisms for sanctioning investors' behaviour, see A Newcombe, 'Investor Misconduct', in A deMestal and C Levesque (eds), *Improvising International Investment Agreements* (2013), 195–211.

[61] Preamble of the Canada-Peru FTA.

Norwegian Model BIT,[62] while the European Parliament strongly favours the inclusion of similar provisions in future EU investment treaties.[63]

Such a recognition of adherence to CSR standards as a goal of IIAs could contribute significantly, albeit indirectly, toward the achievement of the social policy objectives pursued by CSR standards. According to public international law, the objectives of international treaties establish a relevant context for the interpretation of the other provisions of the treaty.[64] In that respect, the recognition of CSR objectives as a general preambular objective of IIAs may influence the interpretation of value-open norms, such as fair and equitable treatment and indirect expropriation.[65]

More important, introducing CSR standards as an indispensable objective of IIAs could enhance the attractiveness of CSR standards and boost their validity and credibility. By rendering compliance with CSR standards a goal of investment regulation, MNEs are encouraged to publicly demonstrate their commitment to CSR by adopting and complying with relevant CSR standards. Moreover, the recognition of CSR standards as a key element of the investment regime exemplifies the commitment of states to enhance the credibility and transparency of CSR compliance mechanisms in their domestic legal systems.

Of course, such preambular provisions are not by themselves suitable to address the public policy goals that CSR standards strive to achieve. In practice, the value of preambular references to CSR may be limited, since the interpretation of IIAs provisions, including references to CSR, arises in investor-state dispute settlement only *ex post facto*, namely when investors challenge the legality of states' measures. Hence, any preambular references to CSR standards cannot act preventively, so as to guarantee that MNEs will adopt relevant CSR standards or comply fully with them.

62 The Preamble to the Norwegian BIT stresses the importance of CSR by including clauses such as '*Emphasising* the importance of corporate social responsibility; *Recognising* that the development of economic and business ties can promote respect for internationally recognised labour rights'.

63 European Parliament, 'Resolution of 6 April 2011 on the future European international investment policy, 2010/2203(INI)' (Investment Policy Resolution), OJ C 296E. Available at: http://www.europarl.europa.eu/sides/getDoc.do/type=TA&reference=P7-TA-2011-0141&language=EN

64 Art. 31(1) of the Vienna Convention on the Law of Treaties (VCLT) states that '[a] treaty shall be interpreted in good faith in accordance with the ordinary meaning to be given to the terms of the treaty in their context and in the light of its object and purpose'.

65 M Gehring and A Kent, 'Sustainable Development and IIAs: Mixed Objectives, Mixed Results' in A deMestal and C Levesque (eds), *Improvising International Investment Agreements* (2013), 299.

Moreover, similar to hard law obligations, preambular references to CSR do not add any remedy to the toolbox available for redress and sanctioning MNEs that violate CSR standards.

In addition to preambular references, an increasing number of IIAs include 'best endeavours' clauses on CSR. Such provisions do not directly impose obligations on MNEs, nor do they create any systems for accountability of MNEs for social or environmental harm that they may cause. On the contrary, they include hortatory language, encouraging MNEs to adopt and abide by CSR standards, making in certain instances specific references to CSR standards, such as the OECD guidelines. Following UNCTAD's recommendations to encourage the adoption and compliance with CSR standards via IIAs,[66] a number of states have experimented with introducing such provisions in their IIAs, in different formations.

Canadian IIA practice presents an excellent example of how adoption and compliance with CSR standards can be promoted via IIAs.[67] More specifically, encouragement of CSR standards can appear as a general, stand-alone provision, which constitutes part of the core provisions of a BIT. For example, the Canadian Model BIT requires that 'Each Party should encourage enterprises operating within its territory or subject to its jurisdiction to voluntarily incorporate internationally recognised standards of corporate social responsibility in their practices and internal policies, such as statements of principle that have been endorsed or are supported by the Parties'.[68] Moreover, provisions encouraging the adoption of CSR standards could be placed in more specific chapters or accompanying documents, which focus on the relation between investment and sustainable development, the environment, labour rights, and the like. For example, CSR clauses have been introduced in the agreement on labour cooperation and the environment that accompany the Canada-Peru FTA.[69] Although these specific provisions are referring in general to CSR standards, they illustrate the potential of such clauses to encourage the adoption of specific CSR standards. For example, MNEs that are active in polluting or labour-intensive industries can be encouraged to adopt

[66] UNCTAD, *Social Responsibility* (2001), Doc. UNCTAD/ITE/IIT/22. Available at http://unctad.org/en/docs/psiteiitd22.en.pdf.

[67] On the Canadian approach to CSR references see *Hepburn and Kuuya*, supra note 6, 601–603.

[68] Art. 16 of the 2012 Canadian Model BIT.

[69] Preamble of the agreement on labour cooperation and Article 6 of the agreement on the environment accompanying the Canada-Peru FTA.

and comply with specific CSR standards that address environmental protection or labour rights, such as ISO 14001.

The introduction of best endeavours provisions in IIAs can play a significant role in facilitating adoption and compliance with CSR standards by MNEs. Indeed, the hortatory language used allows MNEs to have the benefits of a voluntary approach to CSR, thus avoiding overregulation, which may not always be fruitful or productive. Moreover, similar to introducing CSR language in the preambles of IIAs, the declaratory function of such provisions can present an important incentive for MNEs to abide by CSR standards and practices.

More important, the use of CSR language in IIAs can have a positive indirect effect on the development of new forms of CSR standards and mechanisms for ensuring compliance with them. The growing tendency to make references to specific CSR standards in IIAs may offer an incentive for MNEs, states and other stakeholders to develop and improve industry-specific standards, for example, by increasing their transparency, developing new methods for their implementation, establishing standards suitable for certification, and promoting appropriate monitoring mechanisms that can ensure continuous compliance of MNEs activities with CSR standards. By introducing CSR language in international treaties, even if it is nonbinding, states clearly indicate the value they attribute to responsible corporate behaviour, which could advance new international initiatives on developing and improving existing CSR standards and their implementation and compliance mechanisms.

On the other hand, reliance on voluntary mechanisms does not necessarily create sufficient incentives for all MNEs to comply with CSR standards. Given that CSR commitments in IIAs are phrased in hortatory language, compliance remains entirely voluntary. In that respect, best endeavours clauses cannot deter MNEs from harmful conduct, when they wilfully operate without respecting CSR standards. Although best endeavours provisions may guarantee that MNEs that already adhere to CSR principles strive to improve their social impact and monitor better their compliance record, it is questionable whether such provisions could equally provide an incentive for MNEs for which reputational costs are minimal and thus have less economic incentives to comply with CSR.[70]

[70] Vogel, supra note 13, 29–33.

IV. Compliance with CSR standards as a condition for obtaining IIAs benefits

Considering the drawbacks of both voluntary and regulatory approaches to CSR, other approaches toward encouraging adoption and compliance with CSR standards could be also used in IIAs. An effective means for promoting compliance with international CSR standards would be to subject the conferral of benefits under IIAs to the compliance of MNEs with relevant CSR standards, a sort of 'CSR conditionality'.[71] Rather than imposing obligations on MNEs or relying entirely on economic incentives that MNEs may have so as to adopt CSR standards, such an approach would mean that MNEs would not benefit from rights accrued to them under IIAs, unless they behaved socially responsible and complied with relevant CSR standards. The establishment of specific linkages between conferral of benefits under IIAs and compliance with CSRs could present a more effective way to promote adoption of CSR standards without facing the legal and practical problems mentioned earlier and provide new incentives for MNEs to comply with CSR standards without creating gaps in MNEs' protection.

Provisions offering incentives for MNEs to comply with CSR standards have been rarely used in IIAs so far and can be found mainly in investment liberalisation agreements. For example, article 13.6(2) of the EU-Korea FTA enables states to grant preferential treatment to investors that comply with CSR obligations.[72] However, even this provision does not condition FTA benefits to compliance with CSR standards but only acknowledges the possibility that investment recipient countries may impose CSR conditionality. Another interesting example presents the new Indian model BIT, which provides for conditional accrual of BIT benefits. The model BIT provides that accrual of benefits under the BIT is subject to the investors' compliance with national laws on corruption, taxation, labour and environmental law and human rights. Yet, it stops short of referring to CSR standards.[73]

[71] On a similar proposal see R McCormick and D Thompson, 'The reverse and responsible MAI' (2004) 14 *Journal of Corporate Citizenship* 27.

[72] Article 13.6.2 of the EU-Korea FTA provides that 'The Parties shall strive to facilitate and promote trade in goods that contribute to sustainable development, including goods that are the subject of schemes such as fair and ethical trade and those involving corporate social responsibility and accountability'.

[73] Article 8.3 of the Indian Model BIT provides that 'Investors and their Investments must comply with the obligations in Articles 9, 10, 11, and 12 to benefit from the provisions

CSR conditionality in IIAs could be designed on the successful use of similar mechanisms under international and domestic law. Similar techniques have already been used effectively both in international as well in domestic law to boost compliance with CSR and provide incentives for MNEs to adopt appropriate CSR standards. For example, compliance with CSR standards is used as a tool by World Bank institutions to offer financing or insure foreign investment operations abroad. It is common practice of the Multilateral Investment Guarantee Agency (MIGA) to insert CSR provisions in insurance contracts with potential investors in developing countries.[74] Moreover, in domestic legal orders, compliance with CSR standards has been used as a condition for eligibility of bidders and/or award of public procurement contracts. For example, EU legislation requires that compliance with relevant CSR standards is taken into consideration as a relevant factor when concluding public contracts.[75]

In that respect, IIAs could equally present another regulatory framework where access to benefits is subject to compliance with relevant CSR standards. Thus, they can provide a true economic incentive for MNEs to adopt socially responsible behaviour. Indeed, one could envisage the inclusion of a provision in an IIA, which would subject benefits regarding market access and/or investment protection to compliance with relevant CSR standards.

Of course, such a provision subjecting IIA benefits to CSR compliance should not become a tool for protectionism. CSR conditionality seems to be more of an appropriate tool for limiting market access benefits, rather than benefits concerning the lawful protection of investors' property. Although CSR conditionality could be used for screening market access, it should be viewed with extra caution when access to investment protection rights under IIAs is at stake. Indeed, arbitral tribunals have been cautious to exclude investors from the scope of investment protection only in

of this Treaty', while Articles 9–12 spell out the national law rules on anti-corruption, disclosure, taxation, and labour, environment and human rights respectively.

74 To offer insurance to investment projects, MIGA expects projects to have a positive development outcome with regard to social and environmental sustainability. To assess such outcomes, MIGA has developed a specific policy, applying a comprehensive set of social and environmental performance standards. On MIGA's Policy on Social and Environmental Sustainability, see http://www.miga.org/documents/environ_social_review_021507.pdf.

75 European Commission, Interpretative Communication on the Community Law Applicable to Public Procurement and the Possibilities for Integrating Social Considerations into Public Procurement, COM (2001) 566 final, 2001; European Commission, Buying social – A Guide to Taking Account of Social Considerations in Public Procurement, 2010; Wouters and Chanet, supra note 46, 289–294.

rare circumstances, such as when investors violated international public policy or domestic laws and regulations related to corruption of local authorities.[76] In that respect, lack of compliance with CSR standards should not become a tool in the hands of recipient states to evade their obligations under investment treaties to provide compensation when state measures violate IIA standards such as fair and equitable treatment or expropriation.

To overcome such difficulties, CSR conditionality should be devised carefully. In that regard, were CSR conditionality to be introduced in IIAs, it should avoid references to vague and ambiguous CSR principles and set of standards. Such references would only increase legal uncertainty, which would possibly outweigh any potential benefits that the conclusion of IIAs have on increasing investment flows and attracting FDI. CSR conditionality would be successful only if IIAs made references to rigorous CSR standards, which on the one hand satisfy minimum legitimacy requirements and thus promote sustainable development objectives, but on the other are specific enough to provide appropriate guidance to different types of industries and rely on a robust institutional framework that can guarantee monitoring of their implementation.

In that respect, not all CSR standards would be suitable for CSR conditionality. Given the concerns about inclusiveness, specificity, certification and monitoring of compliance that were discussed earlier in the chapter,[77] it seems that only certain types of CSR standards would be suitable. For example, a field in which CSR conditionality could be successful would be for investments that may have an impact on environmental protection, where well-respected international standards exist.[78]

In such cases, CSR conditionality can complement significantly the role that IIAs can play in promoting socially responsible investments. It can offer incentives to MNEs to comply with CSR standards. Considering that market access and investment protection are important determinants of FDI, MNEs would have strong incentives to comply with relevant CSR standards to benefit from IIAs. Besides, such promotion of CSR standards would not amount to a transformation of standards into legally

[76] For example *World Duty Free Company Ltd* v. *Kenya*, ICSID case No ARB(AF)/00/7, Award of 4 October 2006; *Plama Consortium Ltd v Bulgaria*, ICSID Case No ARB(AF)/03/24, Award of 27 August 2008; *Inceysa Vallisoletana S.L. v El Salvador*, ICSID case No ARB/03/26, Award of 2 August 2006; *Anderson and others v Costa Rica*, ICSID Case No ARB (AF)/07/03, Award of 19 May 2010.

[77] See section IV of this chapter.

[78] For example the ISO standards, as well as specific standards recognised by ISEAL.

binding rules, as compliance with relevant CSR would only be relevant for determining access to IIAs benefits, without any legal obligations being imposed on investors. Moreover, CSR conditionality would strengthen the role of sustainable development in international investment law, which has been the subject of significant criticism of IIAs.

Last but not least, the use of CSR conditionality could contribute tremendously towards strengthening CSR standards. By subjecting investment benefits to compliance with CSR standards, the debate regarding IIAs can have a positive spill-over on CSR. It could provide the necessary incentive for MNEs as well as states to invest in developing inclusive and specific CSR standards for different industries and establish appropriate implementation and monitoring mechanisms.

E. Concluding remarks

The growing significance of CSR standards lies in the ever increasing influence of corporate activities on the achievement of social policy goals. States have long recognised that by relying solely on traditional regulatory tools, social policy goals could not be always achieved, while the corporate world has been eager to accept that it bears social responsibility. Albeit not premised on legal obligations, CSR has developed into a significant governance tool that is instrumental towards promoting social policy objectives globally, especially in countries where regulation is usually failing to provide effective solutions.

To help boost the effectiveness of CSR, standardisation mechanisms are indispensable. In order to be able to contribute towards social policy goals in a credible and legitimate manner, CSR standards should overcome the inherent ambiguities of their scope and result in the development of inclusive, specific and certifiable standards that are appealing to the business community and states alike. Moreover, the development of credible monitoring and compliance mechanisms are indispensable for the success of CSR standards. There is already a number of international initiatives and sets of CSR standards, the evolution of which testifies their adaptation to the needs of modern investment opportunities, yet no initiative appears to satisfy all these conditions.

Within this context, international investment law can play a key role in assisting towards the development of comprehensive CSR standards and strengthening compliance with them. Considering the constraints of traditional international law mechanisms, which can impose only limited obligations on MNEs, IIAs can be much more effective by providing incentives to MNEs to adopt and comply with CSR standards. The growing number of best endeavours provisions in IIAs, encouraging MNEs to adopt CSR standards is a testament to that goal. Yet, IIAs have a tremendous potential. By introducing CSR conditionality and subject the conferral of IIA benefits to compliance with relevant CSR standards, they can provide real economic incentives for different types of MNEs in different sectors of the economy to abide by CSR standards and principles. Whether this potential will be exploited remains to be seen.

PART V

International standardisation and innovation

15

Interoperability standards, patents and competition policy

PIERRE LAROUCHE AND GEERTRUI VAN OVERWALLE

A. Introduction

GSM, 3G, 4G, LTE, WiFi, mp3, USB, CD, DVD, PDF, HTML, HTTP, FireWire – these terms have almost invisibly slipped into our daily lives and become household names.[1] Common and self-evident as they sound nowadays, these stylised abbreviations conceal a deeper layer of meaning because they refer to technical standards that enable us to connect and communicate smoothly. Indeed, today, the technologies embodied in many consumer and business products are governed by hundreds of standards.

Generally speaking, standards are regarded as voluntary quality or technical specifications with which current or future products, production processes or services may comply.[2] Standards may be categorised as de facto or de jure. De facto (or informal) standards are standards that have acquired recognition as such, by the relevant industry through widespread use, even though they have not been officially endorsed by a standard-setting organisation (SSO).[3] They attain market approval even though they have not been officially defined, researched and prescribed by an SSO. De jure standards are developed and ratified by a recognised SSO. They are developed through a detailed process that emphasises consensus

[1] See J. Contreras, 'Implementing Procedural Safeguards for the Development of Bioinformatics Interoperability Standards' (2012) 39 *Northern Kentucky Law Review* 87.

[2] See Regulation 1025/2012 of 25 October 2012 on European standardisation [2012] OJ L 316/12, Rec. (1) and (3). For a somewhat older and wider typology of standards based on aims and functions, see W. Hesser and A. Inklaar, *An Introduction to Standards and Standardisation* (1998).

[3] A. Fitzgerald and K. Pappalardo, 'Moving towards Open Standards' (2009) 6:2 *SCRIPT-ed* 467, 468 and the references cited there.

and vendor neutrality.[4] European standardisation legislation focuses on de jure standards, where it defines a standard as 'a technical specification, adopted by a recognised standardisation body, for repeated or continuous application, with which compliance is not compulsory'.[5]

Over time, standards have been developed in multiple technical areas. They appear in the field of agriculture, foodstuffs and medicines, where they assure minimum safety and health levels. They are established in the sector of electricity generation and the car industry, where they improve air quality. They are equally employed in the bank sector, where they safeguard financial stability. Last but not least, standards are applied in information technology, communications, media and consumer electronics, where they enable components and products designed and produced by different firms to operate and communicate with one another, thus establishing interoperability. It is in this category that WiFi, mp3, USB et al. fit in. For example, the DVD standard relates to a technology that was developed by a consortium, called the DVD Forum.[6] Core developers include Hitachi, Matsushita, Mitsubishi, Philips, Sony, Time Warner, Toshiba and JVC. Later, other firms such as LG Electronics, IBM, Pioneer and Sanyo joined the Forum. The present chapter focuses on such interoperability or compatibility[7] standards, which have become increasingly important[8] and have engendered regulatory complexity.[9]

The present contribution aims to bring together the various strands of literature and policy-making concerning technical compatibility standards, with specific focus on the EU. The starting point is the traditional European literature on standardisation, which generally takes a favourable view of standards (Section B). Over time, however, standardisation has given rise to issues both under intellectual property law (Section C) and competition law (Section D). More specifically, three major issues have arisen in conjunction with the work of SSOs, namely deceptive practices ('patent ambush'), disputes regarding licensing ('patent holdup')

4 Ibid, 468–469 and the references cited there.

5 Regulation 1025/2012, supra note 2, Art. 2(1).

6 See http://www.dvdforum.org.

7 The term 'compatibility' standards is applied by J. Farrell, 'Standardisation and Intellectual Property' (1989) 30 *Jurimetrics Journal* 35.

8 It seems reasonable to assume that compatibility standards are the main activity of a large majority of SSOs, see N. Gandal and P. Régibeau's contribution in this volume (Chapter 16).

9 For the concept of 'regulatory complexity', see P. Drahos and I. Maher, 'Innovation, Competition, Standards and Intellectual Property: Policy Perspectives from Economics and Law' (2004) 16 *Information Economics and Policy* 1.

and the accumulation of patent rights ('patent thicket'). Both intellectual property and competition law have sought to provide solutions to these issues (Section E). There is a risk, however, that standardisation becomes reduced to these issues, so in closing, a broader research agenda is proposed (Section F).

B. Standards and standardisation literature

In Europe, there is a long history of standardisation via official channels, dating back to the first national standardisation bodies in the early twentieth century.[10] After the Second World War, worldwide cooperation between national bodies was intensified with the creation of the International Organisation for Standardisation (ISO) in 1947.[11] Following the entry into force of the Treaty of Rome in 1958, a specific organisation, the European Committee for Standardisation (CEN), was put in place to handle the specific needs of the then European Economic Community (now the EU) with regard to standards. In two particular product subsets – electro-technical products and telecommunications – the same evolution took place, albeit at a different pace and with different institutions, namely the International Electrotechnical Commission (IEC) and the European Committee for Electrotechnical Standardisation (CENELEC), and the International Telecommunications Union (ITU) and the European Telecommunications Standards Institute (ETSI), respectively.

The historical centrepiece of the EU, the internal market, can be significantly hampered by diverging national standards, especially when they relate to product safety, public health, the environment or consumer protection. Yet to overcome such divergence, Member States must be confident that standards issued in other Member States provide the same level of protection as their own. This trade-off between the internal market and competing public policy objectives was at the root of the mutual recognition approach put forward by the Court of Justice of the European Union (CJEU) in its historic *Cassis de Dijon* judgment of 1978.[12] In 1985, following on the footsteps of the Court, the Council then endorsed the Commission proposal for a 'new approach' to harmonisation,[13] which

[10] What is now the British Standards Institute (BSI) was formed in 1901, and the ancestor of the Deutsche Institut für Normung (DIN) was created in 1917.

[11] See ISO, *Friendship among Equals* (1997) for the history of the organisation.

[12] Case 120/78, *Rewe-Zentral AG* v. *Bundesmonopolverwaltung für Branntwein* [1979] ECR 649.

[13] See Council Resolution of 7 May 1985 on a new approach to technical harmonisation and standards [1985] OJ C 136/1.

prevails to this day. One of the hallmarks of the 'new approach' is that EU legislation is kept at a general level, that is, the definition of 'essential requirements' that products must meet to move freely through the EU. Of course, more detailed technical specifications are needed if these requirements are to be implemented and applied in practice. These technical specifications, however, are drawn up not via the EU legislative process but rather through a European Standardisation Organisation (ESO) – CEN, CENELEC or ETSI – acting on a mandate given by the Commission. The primary objective of these European 'harmonised standards' is to define specifications for current or future products or services.[14] Member States must then recognise products conforming to these standards as meeting the 'essential requirements' set out in the applicable Directive and accordingly allow them to be marketed freely. The 'New Approach' is therefore centred on a limited number of more general legislative enactments, which form the basis for a much larger set of European standards elaborated by CEN, CENELEC or ETSI.

Enshrined as they are at the core of the New Approach, CEN, CENELEC and ETSI are quasi-official bodies. Their membership – except for ETSI – is made up of national standardisation bodies, which in turn try to bring together all stakeholders in their respective Member States. Typically, the European standardisation organisations and national standardisation bodies are nonprofit organisations, which conceive of themselves as serving the general interest.[15]

This long-standing practice of standardisation, seen from the angle of the EU internal market, has spawned a strand of literature that is largely positive about the impact of standardisation.[16] Indeed, as reflected in European policy documents,[17] the traditional narrative is that standards

[14] See Regulation 1025/2012, supra note 2, Rec. (1) and (3). The Regulation aims at modernising and improving the European standards setting to make it faster and at the same time more inclusive.

[15] In some cases, they have even expressly been entrusted with a mission of general interest, as is the case with AFNOR, for instance, through Decree 2009–697 of 16 June 2009, JORF 2009, 9860.

[16] See, for instance, the work of Jacques Pelkmans: 'The New Approach to Technical Harmonization and Standardisation' (1987) 25 J Common Market Studies 249; 'The GSM Standard: Explaining a Success Story' (2001) 8 J Eur Public Policy 432; 'Mutual Recognition in Goods. On Promises and Disillusions' (2007) 14 J Eur Public Policy 699.

[17] See, for instance, the recent Commission Communication, 'A Strategic Vision for European Standards: Moving Forward to Enhance and Accelerate the Sustainable Growth of the European Economy by 2020', COM (2011) 311 (1 June 2011), especially pp. 1–8, from which the following discussion is inspired.

are *beneficial*. Of course, the ultimate objective of standardisation is to enhance economic and consumer welfare.[18] A number of more specific positive effects have been compiled, and they are rehearsed briefly hereafter. To begin with, standards lead to static efficiencies (cost savings) through economies of scale in production and supply,[19] reduction of transaction costs, and in many cases, network effects.[20] They enhance trade and growth by reducing information asymmetries between demand and supply. They facilitate compatibility and interoperability,[21] thereby also improving market access across borders. They increase competition at product level around the standardised solution[22] by creating a paradigm (ecosystem) around which incremental innovation can take place. Standardisation can also contribute to a better transfer of research results to markets by creating a platform for basic or fundamental research results to be applied by various firms.

Despite all these advantages, through the interplay with patents and competition policy, a more *sceptical* view has arisen in other corners of the literature and policy-making, resulting in critiques on the traditional narrative.

C. Standardisation and patent law

It is widely accepted that there is an *inherent* tension – even in the absence of any patents or any standards – between the interest of technology makers/inventors seeking economic returns on their R&D investments and technology users wanting access on affordable terms. The tension

18 See Regulation 1025/2012, supra note 2, Rec. (3).

19 See ibid. Also see K. Maskus and S. A. Merrill (eds), *Patent Challenges for Standard-Setting in the Global Economy: Lessons from Information and Communication Technology*, National Research Council (2013), 16 (available at http://www.nap.edu/catalog.php?record_id=18510).

20 See P.D. Curran, 'Standard-Setting Organizations: Patents, Price Fixing, and Per Se Legality' (2003) 70 *University of Chicago Law Review* 983, 986; P.A. David, 'Clio and the Economics of QWERTY' (1985) 75 *American Economic Review* 332.

21 See Regulation 1025/2012, supra note 2, Rec. (1). Also see Curran, supra note 20; Fitzgerald and Pappalardo, supra note 3, 470 ff; U. Gasser and J. Palfrey, 'Breaking Down Digital Barriers: When and How ICT Interoperability Drives Innovation' (2007), Berkman Center Research Publication No. 2007–8 (available at http://papers.ssrn.com/sol3/papers.cfm?abstract_id=1033226).

22 See Regulation 1025/2012, supra note 2, Rec. (3). See also J. Farrell, 'Standardisation and Intellectual Property' (1989) 30 *Jurimetrics Journal* 35, who gives the example of a computer: 'the more people use a given computer operating system, the more software is likely to be written for that system, and so the more choice and competition those users will have available'.

between the individual and the collective, between the development of technology and its diffusion, is by no means new.[23]

However, this tension aggravates when the technology is part of a standard that involves patents. In the relationship between standards and patents, the potential for conflict may emerge when the implementation of a standard, by its essence, necessitates the application of proprietary technology.[24] The private interest of the patent holder then collides with the collective interest of the users of the standard: patents are predestined for private, exclusive use, whereas standards are intended for collective use.[25] In other words, patents stimulate the development of the new and different by creating exclusivity in the market, whereas standards foster harmonisation and compatibility through shared and common systems.[26] Even though standards protected with intellectual property rights (IPRs) may not be all that common,[27] the scope of conflict has increased over time,[28] especially when essential patents are at stake. A Standard-Essential Patent (SEP) is such that it (its claims) would necessarily be infringed by any implementation or adoption of that standard.[29]

[23] See E.J. Iversen, E. Øversjøen and H.T. Lie, 'Standardisation, Innovation and IPR' (2004) 100 *Telektronikk* 65. Also see Maskus and Merrill, supra note 19, 16.

[24] J. Verbruggen and A. Lorincz, 'Patents and Technical Standards' (2000) 33 *Int'l Rev. of Indus. Prop. And Copyright Law* 125.

[25] Ibid. For a wider – and also schematic – analysis of the characteristics of patent law and standardisation, see K.-S. Park, Y.-T. Kim and H. Sohn, 'Disputes over the Intellectual Property Rights of Telecommunications Standardisation' (1998) 20 *ETRI Journal*, 74, 80.

[26] Drahos and Maher, supra note 9.

[27] Worldwide some 100,000 standards apply; less than a third of those standards relate to interoperability (the other standards relate to quality and safety); of all interoperability standards, only some 5000 involve intellectual property having high economic value. Some of these standards, however, like those relating to mobile communications, govern significant activities: See Henk de Vries's contribution in this volume (Chapter 1).

[28] Iversen, Øversjøen and Lie, supra note 23. The International Patent Classification (IPC) classes containing most claimed essential patents are the classes related to telecommunication and audio/video coding techniques, according to K. Blind, R. Bekkers, Y. Dietrich, E. Iversen, F. Köhler, B. Müller, T. Pohlmann, S. Smeets and J. Verweijen, *Study on the Interplay Between Standards and Intellectual Property Rights (IPRs)* (2011), 41 ff.

[29] The common definition relating to standard essential patents can be found at K. Blind and T. Pohlman, 'Trends in the Interplay of IPR and Standards, FRAND Commitments and SEP Litigation' [2013] *Les Nouvelles* 177. The more nuanced definition, referring to the essential *claims* of such patents, is applied in Maskus and Merrill, supra note 19, 35: 'Standard-essential patent (SEP) refers to both patents that are essential to the use of a standard and the essential claims of such patents. Some IPR policies focus on essential patent claims rather than SEPs and, indeed, a properly crafted rule rarely imposes licensing obligations to claims other than the essential ones. However, the analytical concepts involved are largely common to both terms and SEP appears to be in wider use in the literature' (also see p. 38). The word 'necessary' is sometimes used in place of 'essential'. Although some SSOs

An analysis of the IPR databases of eleven of the most important SSOs worldwide by Blind et al. revealed that approximately 250 distinct standards include technologies that are covered by one or more declared IPR and that many of these standards are successful and widely employed.[30] Because there are several hundred thousands of standards available worldwide, this is a quite small group. By far, patents are the most relevant type of essential IPR, whereas copyrights and other types of IPR are virtually not claimed. Most patents in standards relate to telecommunications standards and to standards for consumer electronics products. Standards and IPR traditionally occur prominently in consumer electronics and mobile information and communication technology (ICT) because of the interoperability character of these markets.[31] However, Blind et al. also observed that patents in standards extend to other fields and sectors, such as transport, logistics, energy and health. This extension seems to be driven by ICT-type enabling technologies, though, and less by 'genuine' IPR in those areas. Blind and his team observed that the distribution of patents in standards is skewed, both in terms of standards and in terms of owners. In particular, a few standards cover a large number of patents, whereas most standards include only a few patents, or no patents at all. And a relatively small group of companies own a large number of essential patents in standards, whereas most companies own only a few or none of these patents. Most IPR-owning companies are from the United States, Japan and Europe. Essential patents were mostly found among some small and medium-sized enterprises (SMEs) including a number of non-producing entities (NPEs). However, due to convergence of technologies and globalisation, more and more companies also in emerging economies, research organisations and NPEs have entered the game.[32]

D. Standardisation and competition policy

At first sight, standardisation does not give rise to significant competition law issues in the EU as long as it is carried out within 'official'

define essentiality to mean there are no 'technological' alternatives, others also mention the concept of 'commercial' essentiality (see p. 38).

30 See Blind et al., supra note 28, p. 11.

31 See Blind, supra note 28, 112: 'IPRS, especially patents, in standards are a reality in particular in information and communication technologies driven industry sectors'. Also see Blind and Pohlman, supra note 29, 177 especially Figure 1: Share of SEPs Holding Firms' Main Sector of Operation'.

32 Blind et al., supra note 28.

standardisation bodies, as described earlier. From a competition law perspective, the ESOs (CEN, CENELEC and ETSI) and their work do not represent a major concern, to the extent that they come very close to public authorities in their mission and operations. And indeed, in the first major policy statement by the Commission to discuss the application of competition policy to standardisation, the Horizontal Guidelines of 2000, the Commission begins by stating summarily that the work of standardisation bodies and organisations falls outside of Article 101 TFEU, if and as long as that work takes place according to non-discriminatory, open and transparent procedures.[33]

Starting in the 1980s, however, the Commission has been confronted with a different form of standardisation, namely that which arises from the efforts of firms in an industry to make their products compatible with one another. Although they are not uniformly present, the following features distinguish private-driven standardisation from the traditional public-driven standardisation outlined earlier.

	Public-driven standardisation	Private-driven standardisation
Parties involved	Public entities, i.e. national standardisation bodies, with private firms associated	Private firms, often cooperating within a standard-setting organisation
Purpose	Wide range of purposes (including compatibility); usually compliance with public interest goals relating to safety, health, environmental or consumer protection	Narrower range of purposes; usually product compatibility across manufacturers
Timing	*Ex post* (after products are on the market)	*Ex ante* (before products are developed)
Nature	Harmonisation of pre-existing national standards	Production of a new standard
Membership	Open	Closed
Publicity	Transparent	Closed
Time perspective	Static (ensure compatibility of existing products)	Dynamic (enable introduction of new products)

[33] Guidelines on the applicability of Article [101 TFEU] to horizontal cooperation agreements [2001] OJ C 3/2, paras. 162–163.

The Commission cut its teeth on this new private-driven standardisation in *X/Open Group*, a case concerning a private effort to develop standards for applications based on the Unix operating system.[34] There the Commission laid the basis for its analysis of standardisation under Article 101 TFEU: standard-setting involves agreements between participating firms, which can lead to restrictions of competition between the firms, as well as exclusionary effects toward competitors not taking part in the standardisation effort. The solution to the first concern depends on whether the standard is compulsory, and the solution to the second concern depends on the rules for participation in standard-setting and for access to the standard. That approach was set out more fully in the 2000 Horizontal Guidelines.[35]

Throughout the 2000s, the Commission expanded its decision practice relating to private-driven standardisation by adding an intellectual property dimension and analysing cases under Article 102 TFEU as well. The starting point for these new developments can be traced back to the creation of ETSI in 1988.[36] ETSI differs from other ESOs in that its membership is more open to firms from the industry, and the standards it develops – starting with GSM and now covering a large swathe of ICT – come closer to private-driven than public-driven standardisation. In particular, as explained earlier, the standards relating to mobile communications are known to involve components or elements protected by essential patents (SEPs), resulting from the R&D efforts of the firms taking part in the development of the standard.[37] Implementers of the standards must obtain a license from each SEP holder.[38] The unilateral

[34] Decision of 15 December 1986, *X/Open Group* [1986] OJ L 35/36. This decision is an individual exemption decision pursuant to the old Regulation 17 [1962] OJ 13/204. There were ten notifying parties, comprising most significant European computer manufacturers of the time. The X/Open Group merged into The Open Group in 1996.

[35] 2001 Guidelines, supra note 32, now replaced by the 2010 Guidelines, discussed later in the chapter.

[36] The youngest of the three European standardisation organisations, ETSI was born out of the need to unbundle the sector-specific European telecommunications cooperation structure, in anticipation of a liberalised environment. Following a proposal of the Commission in its Green Paper on the Development of the Common Market for Telecommunications Services and Equipment, COM(87)290 (30 June 1987), the Conference of European Post and Telecommunications administrations (CEPT) entrusted its standardisation activities to ETSI, let go of operational issues (now discussed within the European Telecommunications Network Operators association – ETNO, to the limited extent allowable in a liberalised environment) and concentrated on regulatory matters only.

[37] See supra, Section B.

[38] These can be individual licenses, but also licenses granted through a patent pool or generic licenses granted by reference to an established General Public License (GPL) or other similar model.

behaviour of SEP holders in the course of their licensing practice can, under certain circumstances, fall under Article 102 TFEU, as an abuse of a dominant position. This was at the core of some of the momentous cases of the 2000s, such as *Rambus*[39] or *Qualcomm*.[40] The experience of the Commission was distilled in the new 2010 Horizontal Guidelines.[41] It is discussed in greater detail later in the chapter.

Given that most of the cases involving SEPs arose out of standards elaborated under the supervision of ETSI, the Commission was also able to develop a set of principles on the conduct of SSOs, based on the workings of ETSI. It will be recalled, however, that ETSI is one of the three ESOs (with CEN and CENELEC), and that as such its structure and functioning derive from public institutional models more than anything else.

Since the 2010 Horizontal Guidelines, the Commission has embarked on a new generation of cases relating to disputes between SEP holders and implementers, arising out of the worldwide litigation battle between, on the one hand, Apple, and on the other hand, Motorola Mobility (now part of Google) and Samsung. These cases allowed the Commission to delve deeper into the application of competition law to dealings between SEP holders and implementers. At the same time, these dealings have to be seen against the background of competition between SEP holders and implementers on the product markets. This cluster of cases is dealt with later in the chapter.

E. Three issues concerning SEPs

Once a standard involves patented technology, so that some firms become SEP holders, a number of *problems* can potentially arise. Most of these problems have arisen in one or a few cases – all of which have achieved great visibility in academic and policy circles – but the precise severity and prevalence of those problems remains an open empirical question.

[39] Commission decision of 9 December 2009, Case COMP/38.636, *Rambus* [2010] OJ C 30/17.

[40] There is no official record of this large multi-year case beyond Commission press releases: 'Commission closes formal proceedings against Qualcomm', Press Release MEMO/09/516 (24 November 2009). It has however led to much academic writing: for an overview, see D. Geradin, 'The European Commission Policy Towards the Licensing of Standard-Essential Patents: Where Do We Stand?' (2013) 9 *J Comp L Econ* 1125.

[41] Guidelines on the applicability of Article 101 TFEU to horizontal co-operation agreements [2011] OJ C 11/1.

After all, as mentioned earlier, some 250 standardisation processes involve SEPs, with a much larger number of associated licensing dealings between SEP holders and implementers. The appropriate legal response to these problems depends on their severity and prevalence. Available empirical evidence so far would suggest that the problems discussed in this section do not arise in a significant number of cases,[42] therefore justifying that any legislative or regulatory intervention to address these problems would have to be carefully targeted. Otherwise, the better could become the enemy of the good, if, for instance, a broad intervention would upset the balance between negotiated and litigated cases in favour of the latter.

For each issue, we examine actual or potential *solutions*. The first and most immediate response arises out of the interaction between the parties, in a form of self-regulation. This often takes place under the auspices of the SSO, which seeks to prevent problems by establishing a policy on the IP aspects of standard-setting (IP Policy), the most famous one coming from ETSI.[43] Beyond that, competition law could also play a role. Finally, legal reform could also provide solutions.

I. Deceptive behaviour on the part of the SEP holder ('patent ambush')

First of all, the SEP holder could be engaging in deceptive behaviour, a so-called patent ambush. A patent ambush occurs when a firm withholds information about a patent that is relevant to the standard from other SSO members, at the development and standard-setting stage, where that patent could be sidestepped. When the standard is subsequently set such that the patent in question becomes an SEP, the firm in question then asserts its patent against implementers.[44]

To prevent patent ambushes, SSOs have elaborated IP policies. The vast majority of SSOs nowadays require that if a firm holds patents that could read onto an industry standard, that firm must disclose such patents to all SSO members before the patented technology is considered for inclusion in a potential standard.[45] Critics argue, however, that *ex ante* disclosure

42 See Gandal and Regibeau's contribution in this book (Chapter 16).

43 ETSI Intellectual Property Rights Policy, being Annex 6 of the ETSI Rules of Procedure of 20 March 2013 (available at www.etsi.org). The Policy is explained in more detail in the ETSI Guide on Intellectual Property Rights (also available at www.etsi.org).

44 Blind et al., supra note 28, 24 and the references cited there.

45 For a further discussion, see Blind et al., supra note 28, 25 and the references cited there; Curran, supra note 20, 991; Maskus and Merrill, supra note 19, 71, more in particular p. 73.

will impede standardisation processes and create additional legal risks for participants. A recent empirical study partially rebuts this thesis in a nuanced manner.[46]

Furthermore, patent ambush caught the attention of competition authorities early on, with the *Rambus* case, which played out on both sides of the Atlantic. The crux of that case is that, in the first half of the 1990s, Rambus deceived other members of JEDEC (Joint Electron Device Engineering Council), an SSO developing standards in the area of dynamic random-access memory (DRAM), a type of memory chip present in every computer. In apparent contradiction with JEDEC's IP policy, Rambus failed to disclose *ex ante* that it held patents over technology relevant to the establishment of a JEDEC standard. The standard ended up incorporating that technology. Once the standard was established and the industry had committed to it, Rambus turned around and asserted its patents. Both the US Federal Trade Commission (FTC) and the European Commission found breaches of their respective antitrust or competition law. The European Commission found that, had Rambus disclosed its patents, JEDEC members would have designed the standard to avoid impinging on those patents. The Commission closed the case when Rambus entered into commitments concerning licensing terms and royalties for its IP.[47] In the United States, however, the FTC decision was annulled by the court of appeal, which found that the FTC had not proven that Rambus's conduct breached the Sherman Act.[48] For the court of appeal, had Rambus disclosed its patents on time, it was equally possible that JEDEC, instead of designing around those patents, would instead have requested Rambus to enter into a FRAND commitment regarding those patents. In that latter case, Rambus's deception merely led it to obtain higher royalties than it might otherwise have obtained, an outcome that does not breach US antitrust law.[49]

[46] J.L. Contreras, 'Technical Standards and Ex Ante Disclosure: Results and Analysis of an Empirical Study' (2013) 53 *Jurimetrics* 163.

[47] *Rambus*, supra note 39.

[48] An Act to Protect Trade and Commerce against Unlawful Restraints and Monopolies, 26 Stat. 209 (1890), 15 USC § 1. The *Sherman Act* (also known as the *Sherman Antitrust Act*) is the core of US antitrust law.

[49] *Rambus* v. *FTC*, 522 F.3d 456 (D.C.C.A., 2008). It will be recalled that exploitative pricing (using monopoly power to obtain supra-competitive prices) does not run afoul of US antitrust law (see Verizon v. Trinko 540 US 398 (2004)), whereas it can constitute a breach of Article 102 TFEU.

II. Unwillingness of SEP holder to license or dispute regarding license terms ('patent holdup')

Beyond cases of deception, SEP holders may have respected the IP policy of the relevant SSO and disclosed their patents *ex ante*; yet later on, once their patents have become SEP, disputes may ensue around SEP licensing. Indeed, once a patent has become an SEP, the SEP holder gains market power because by definition a license on that SEP becomes essential to implement the standard. In some cases, the SEP holder might outright refuse to grant a license.[50] In other cases, the SEP holder is willing to license, but it differs in opinion with the implementer on the level of royalties.[51] Both situations have been brought under the header of 'patent holdup' in the literature.[52]

After some nudging by public authorities,[53] ETSI and other SSOs have added to their IP policies some requirements that patent holders enter into commitments regarding the licensing of their patents ahead of any standardisation exercise, failing which the patented content can be excluded from the process. In some cases, that policy is to require an upfront commitment by participating firms to license eventual SEPs on a royalty-free basis (or even to put SEPs under an open license).[54] Not every firm is willing to agree to this, however. Another option would be for patent holders to pre-negotiate license agreements with prospective standard implementers *before* standard-setting begins; that solution is both impracticable (not all implementers are known) and inefficient (massive upfront transaction costs, with a significant number of licenses ending up having been negotiated for nothing).

The most popular form of *ex ante* requirement, however, has been a FRAND commitment, that is, a prior commitment by eventual SEP

[50] Maskus and Merrill, supra note 19, 16; Verbruggen and Lorincz, supra note 24, 322.

[51] See Blind et al., supra note 28, 23 ff and the references cited there. Also see Maskus and Merrill, supra note 19, 16, 60.

[52] See the seminal article of M. Lemley and C. Shapiro, 'Patent Holdup and Royalty Stacking' (2007) 85 *Texas Law Review* 1991. For an overview of the state of the art in economic literature, see B. Ganglmair and E. Tarantino, 'Patent Disclosure in Standard Setting', Working Paper, October 2011 (available at http://www.academia.edu/2841055/Patent_Disclosure_in_Standard_Setting).

[53] See the position taken by the European Commission, *Intellectual Property Rights and Standardisation*, COM(92) 445 final (27 October 1992).

[54] For instance, W3C, an SSO responsible for a number of key Internet standards, requires participants to give royalty-free licenses of any patents on which W3C standards might read: see the W3C Patent Policy (Version of 5 February 2004).

holders to grant licenses to implementers of the standard on a fair, reasonable and non-discriminatory (FRAND) basis.[55] In the meantime, the requirement to commit to FRAND licensing upfront is enshrined in European legislation: 'Intellectual property rights *essential*[56] to the implementation of specifications are licensed to applicants on a (fair) reasonable and non-discriminatory basis ((F)RAND), which includes, at the discretion of the intellectual property right-holder, licensing essential intellectual property without compensation'.[57]

FRAND commitments are now widespread across the numerous SSOs, and as could be expected, some disputes have arisen regarding their operation. In general, SSOs require FRAND commitments yet do not get involved in the implementation and enforcement of such commitments.

A first issue relates to the level of royalties that would qualify as FRAND. Unsurprisingly, SEP holders and implementers do not spontaneously agree on how to translate general ideas of fairness and reasonableness into a concrete royalty rate. Nevertheless, in most cases a negotiated outcome is reached. When negotiations break down, parties can turn to courts or arbitrators, depending on what might have been provided in the FRAND commitment or what they can agree on. Alternatively, some parties have turned to competition authorities with complaints about the implementation of FRAND commitments. In *Broadcom* v. *Qualcomm*, the Court of Appeals for the Third Circuit found that a breach of a FRAND commitment could amount to a violation of s. 2 of the Sherman Act,[58] without, however, specifying how to tell whether the royalty rate requested by the SEP holder is not fair and reasonable.[59] Similarly, the complaint filed against Qualcomm before the European Commission did not result in any official pronouncement.[60] As a starting point, the royalties should reflect the value of the patent, but case law on this point rapidly descends into a checklist approach, which offers limited guidance.[61] The relevant literature suggests that royalties should be at the level they would have

55 FRAND commitments are sometimes called RAND, leaving out the F for 'fair', but by all accounts the two concepts are identical. For a further discussion and some examples, see Curran, supra note 20, 991 ff.

56 Our italics.

57 See item 4 (c) of Annex II (*Requirements for the Identification of ICT Technical Specifications*) of Regulation 1025/2012.

58 Supra note 47, 15 USC § 2.

59 *Broadcom Corp* v. *Qualcomm Inc.*, 501 F. 3d 297, 314 (2007).

60 The Commission abandoned the case after all complainants withdrew: supra note 39.

61 See the test put forward in *Georgia-Pacific Corp.* v. *United States Plywood Corp.*, 318 F. Supp. 1116, 1120 (S.D.N.Y. 1970).

been in a hypothetical *ex ante* negotiation, carried out before standard-setting took place (i.e. before the patent became an SEP).[62] The value of the patent is then its incremental value when compared with the next best alternative. That approach aims at depriving the SEP holder of any windfall rent accruing because of standardisation. It has been taken up in the European Commission's 2010 Horizontal Guidelines.[63]

The *Qualcomm* cases mentioned in the previous paragraphs also featured a difficult issue relating to business models. Some SSO participants are technology owners and *users*, with vertically integrated business models that are based on the sales of products that implement standards and employ patented technologies. They are not interested in asserting their own SEPs and desire low – or no – royalties for the SEPs they license from others. Other SSO participants are technology *sellers* whose business models are based on earning royalties from licensing their SEPs to implementers and who want higher royalties. Still other participants are both technology users and sellers.[64] As between vertically integrated firms, reciprocity can play a role, leading to royalty-free cross-licenses; such an option is not attractive for pure patent holders or pure implementers.

Over the years, scholars have come up with various alternative ideas to solve licensing and royalty setting issues,[65] ranging from *ex ante* disclosure of licensing terms,[66] to a more controversial obligation of the patent owner to enter into binding 'baseball-style' arbitration.[67]

Second, because the patent policies of many SSOs are quite vague, conflicts often arise over the precise rights and obligations of a SEP

62 Lemley and Shapiro, supra note 52.

63 2011 Guidelines, supra note 40 at para. 287–291. The Commission also accepts alternative methods.

64 Maskus and Merrill, supra note 19, 52.

65 For an overview, see Maskus and Merrill, supra note 19, 44 ff. In an attempt to better understand the dynamics of licensing, Kesan looked into the interplay between open source software licensing and FRAND licensing, see J.P. Kesan, 'The Fallacy of OSS Discrimination by FRAND Licensing: An Empirical Analysis', *Illinois Public Law Research Paper 10–14* (2011) (available at http://papers.ssrn.com/sol3/papers.cfm?abstract_id=1767083).

66 See D. Geradin and A. Layne-Farrar, 'The Logic and Limits of Ex Ante Competition in a Standard-Setting Environment' [2007] 3:1 *Competition Policy International* 3.

67 See M.A. Lemley and C. Shapiro, 'A Simple Approach to Setting Reasonable Royalties for Standard-Essential Patents', 28 *Berkeley Tech L J* 1135 (2013). For a reply, see P. Larouche, J. Padilla and R. Taffet, 'Settling FRAND Disputes: Is Mandatory Arbitration a Reasonable and Non-Discriminatory Alternative?' (2014) 10 *J Comp Law & Econ* 581 and G. Sidak, 'Mandating Final-Offer Arbitration of FRAND Royalties for Standard-Essential Patents' (2015) 18 *Stanford Tech. L. Rev.* 1.

holder.[68] In particular, the availability of injunctive relief to the SEP holder in the course of license negotiations has attracted much attention in recent cases.[69] High-profile patent disputes involving SEPs from 3G and 4G standards are being fought worldwide between Apple, on the one hand, and Motorola (now Google) or Samsung, on the other hand. As a starting point, such litigation should be discouraged, so the companies can focus (again) on research, development and innovation.[70]

On the US side, the FTC set out its views in the course of ruling in *Google/Motorola Mobility*.[71] According to the FTC, the SEP holder's right to injunctive relief is narrowly circumscribed. The SEP holder must first offer a negotiation process to any implementer, coupled with judicial or arbitral settlement. Only if that process fails, and the SEP holder has given notice to the implementer (with a final offer), may the SEP holder proceed with injunctive relief; in that case, the implementer is considered unwilling to enter into a license agreement.

In the EU, two approaches are competing. On the one hand, in *Orange Book*, the German Supreme Court ruled that a SEP holder does not breach competition law if and when it seeks injunctive relief against an unwilling licensee.[72] A licensee is deemed unwilling if it has not made a firm proposal for a license agreement (including refraining from challenges to patent validity or infringement) and has not unilaterally implemented it by holding royalties at the proposed level in escrow for the SEP holder. That ruling did not bear on a FRAND dispute as such, but it has been influential in FRAND cases in Germany and across Europe.

On the other hand, the European Commission took a position much closer to the FTC in *Motorola*[73] and *Samsung*.[74] In *Motorola*, the

[68] See Curran, supra note 20, 992 ff. For a discussion of this topic through the lens of comparative law and economics, see T.F. Cotter, 'The Comparative Law and Economics of Standard-Essential Patents and FRAND Royalties' (2014) 22 Texas Intellectual Prop. L. J. 311.

[69] See P. Camesasca, G. Langus, D. Neven & P. Treacy, 'Injunctions for Standard-Essential Patents: Justice Is Not Blind' (2013) 9 *J Comp Law & Econ*, 285–311; Cotter, supra note 67. Also see Maskus and Merrill, supra note 19, 95 ff.

[70] H. Chia, 'Fighting the Smartphone Patent War with RAND-Encumbered Patents' (2012) 27 *Berkeley Tech. L. J.* 209, 240.

[71] FTC, *Motorola Mobility/Google*, Docket C-4410, Decision of 23 July 2013 (available at www.ftc.gov).

[72] BGH, 6 May 2009, *Orange Book*, BGHZ 180, 312.

[73] Case AT.39985, *Motorola*, Decision of 29 April 2014, nyr (available at ec.europa.eu/competition/index˙en.html). This is a prohibition decision finding an infringement of Article 102 TFEU on the part of Motorola.

[74] Case AT.39939, *Samsung*, Decision of 29 April 2014, nyr (available at ec.europa.eu/competition/index˙en.html). This is a decision endorsing commitments offered by Samsung, and as such the Commission reasoning is not as elaborated as in *Motorola*.

Commission finds that once an SEP holder has given a FRAND commitment in the context of a standard-setting procedure, seeking injunctive relief constitutes an abuse under Article 102 TFEU. It is notable that the Commission considers that any manoeuvre by a SEP holder that could lead an implementer to accept a higher royalty rate, or otherwise less advantageous license terms, is abusive.[75] In *Samsung*, the Commission explains how a SEP holder can avoid breaching competition law: it must first invite implementers to a negotiation process, with judicial or arbitral determination if negotiations fail. Only if an implementer fails to embark on such a process can the SEP holder seek injunctive relief without running afoul of competition law.

It should be noted that *Motorola* and *Samsung* are not the last word on this: in *Huawei/ZTE*, the CJEU has been asked to choose between these two interpretations of Article 102 TFEU: the more SEP-holder-friendly approach of the German Supreme Court, or the more implementer-friendly approach of the European Commission.[76]

Third, to deal with the proliferation of sales of patent portfolios and legal variations across countries when SEPs are transferred, many SSOs require that FRAND commitments be transferred along with patent ownership.[77]

III. Accumulation of SEPs ('patent thickets')

The third problem that might accrue is a patent thicket: patent holders are so many and the relevant patents are so dispersed among the many patent holders that users must first hack their way through a dense web of overlapping patents before they can implement a technology and commercialise their product.[78] Sometimes firms – especially smartphone

[75] *Motorola*, supra note 71, at paras. 322–383. In particular, the Commission was adamant that Apple could not be required to renounce any validity or infringement challenges to Motorola's SEPs.

[76] Case C-170/13, *Huawei/ZTE* [2013] OJ C 215/5. In all likelihood, the ruling will come in 2015 at the earliest.

[77] Maskus and Merrill, supra note 19, 47 and 81 ff. Different theories, such as traditional contract theory, competition remedies and notions akin to servitude are tested to found such a transfer (see pp. 88 ff).

[78] The definition of patent thicket is borrowed from C. Shapiro, 'Navigating the Patent Thicket: Cross Licenses, Patent Pools and Standard Setting' in E. Jaffe et al. (eds), *Innovation Policy and the Economy*, Vol. I (2001) 119–150. A more nuanced (and correct) definition can be found at P. Regibeau and K. Rockett, *Assessment of Potential Anticompetitive Conduct in the Field of Intellectual Property Rights and Assessment of the Interplay Between Competition Policy and IPR Protection* (2011); G. Van Overwalle, 'Of Thickets,

manufacturers – deliberately amass immense patent portfolios to remain competitive against rivals,[79] thus aggravating patent thickets enormously. The DVD technology was protected by more than 840 patents,[80] whereas the MPEG-2 technology was covered by more than 425 (essential) patents owned by some twenty-one patent holders.[81] Another textbook example relates to WiFi technology.[82]

A related problem is royalty stacking: the subsequent accumulation of royalties to be paid to a multitude of patent owners may result in royalty stacking.[83] In that context, it has been suggested to introduce and enforce a maximum licensing cap.[84] Indeed, as a matter of principle, there is an optimal amount of royalty that can be tacked onto a standard, and it can be determined according to the principles of monopoly pricing. That amount represents the maximum total royalty, and it is then a matter for SEP holders to agree on how to share it.

A practical solution to remedy patent thickets is the – voluntary – establishment of a patent pool (provided the pool caps the overall royalty at a reasonable level). A patent pool is an agreement between two or more patent owners to license one or more of their patents to one another and to license them as a package to third parties who are willing to pay the royalties that are associated with the license.[85] A key example of a modern patent pool in the ICT area is the pool related to the digital video compression standard known as MPEG-2 that holds more than 425 essential patents owned by some twenty-one patent holders. A second trendsetting patent pool is the DVD pool, which in fact exists of two

Blocks and Gaps: Designing Tools to Resolve Obstacles in the Gene Patents Landscape', in G. Van Overwalle (ed), *Gene Patents and Collaborative Licensing Models. Patent Pools, Clearing Houses, Open Source Models and Liability Regimes* (2009), 381–463.

79 Chia, supra note 69, 213. 80 See http://www.dvd6cla.com.

81 Baryn S. Futa (CEO and Manager, MPEG LA, LLC), *Statement before the United States Department of Justice Antitrust Division and the Federal Trade Commission Joint Hearings on Competition and Intellectual Property Law and Policy in the Knowledge-Based Economy*, 1997 (available at http://www.ftc.gov/opp/intellect/020417barynfuta.pdf).

82 For details, see Fitzgerald and Pappalardo, supra note 3, 479–480.

83 Blind et al., supra note 28, 24 and the references cited there.

84 See V. Torti, 'Enforcement of a Maximum Licensing Cap in Standardisation Environments' (2013) 35 *EIPR* 261.

85 J. Clark et al., Patent Pools: A Solution to the Problem of Access in Biotechnology Patents? (United States Patent and Trademark Office, December 2000; available at http://www.uspto.gov/web/offices/pac/dapp/opla/patentpool.pdf); R. Merges, 'Institutions for Intellectual Property Transactions: The Case of Patent Pools', in R. Dreyfuss, D.L. Zimmerman and H. First (eds), *Expanding the Boundaries of Intellectual Property* (2001) 123–65; US Department of Justice – Antitrust Division, *Business Review Letter from Joel I. Klein to Gerrard R. Beeney* (Letter 98-C relating to MPEG LA L.L.C., et al.), June 1997 (available at http://www.justice.gov/atr/public/busreview/215742.pdf).

pools: a first pool (the so-called DVD4C patent pool), uniting four of the ten core DVD developers, and a second pool (the DVD6C patent pool), assembling six other core members and covering more than 840 patents.[86] Although patent pools have attracted some scrutiny under competition law, by and large their benefits outweigh their potential anti-competitive effects, provided that certain basic conditions are respected.[87]

F. Towards a more nuanced view of standardisation – avenues for further research

In the previous headings, we have shown how intellectual property law and competition law try to come to grips with standardisation. Standardisation is comprehended as an external phenomenon that enters the realm of intellectual property or competition law if, and only from the moment, that it creates 'problems', such as the cumbersome exercise of intellectual property rights or anti-competitive conduct and outcomes. Accordingly, the manner in which these two areas deal with standardisation is tainted by the presence of problems – popularised under keywords such as 'patent ambush', 'patent holdup' or 'patent thicket' – and the need to address them.

Yet standardisation has gained such significance in our day and age that it can no longer be seen as a subordinate technological endeavour, left to its own devices unless and until 'problems' arise. It should be studied on its own merits, in a holistic manner. Furthermore, the study of standardisation should also put innovation at the core, considering that innovation has become an overarching policy objective – at least in developed countries – to which the various components of economic regulation must contribute.

Accordingly, in this section, we suggest a number of avenues for further research on standardisation and its relationship with intellectual property and competition law, in the light of the previous findings.

I. A more complete and nuanced view of standardisation

A first and foundational research endeavour would be to gain a more complete, and of necessity more nuanced, view of standardisation. Indeed

[86] For more information, see Overwalle, supra note 78, 305–324 and the references cited there.

[87] See, for EU competition law, the Guidelines on the application of Article 101 TFEU to technology transfer agreements [2014] OJ C 89/3 at para. 244 ff.

some of the literature, centred as it is on intellectual property or competition law, tends to reduce standardisation to an exercise in valorising IP rights or gaining market power. Some of the assumptions made in that strand of literature need to be investigated.

First of all, standardisation often does not start from a blank slate: there are some products already on the market. Conceivably, these products could each hold sway over a specific market (product market or geographic market), giving rise to a 'checkerboard' pattern of incompatible products. Standardisation then aims, at a minimum, to remove incompatibility, but it could also seek to replace the checkerboard with a standardised solution, as has traditionally been the case in the EU: see for instance the case of railway signaling or second-generation mobile communications. Alternatively, the products already on the market might be competing over the whole territory, giving rise to a 'marbled cake' pattern, as is often the case in the United States.

Second, from a dynamic perspective, standardisation also does not take place in a vacuum. There is a prior state. Of course, in some instances, no standardisation effort was ever undertaken beforehand and the industry attempts to standardise for the first time. But in most cases nowadays, some standardisation has already taken place beforehand. Any subsequent standard-setting effort therefore takes place against the background of prior standardisation. We would surmise that there is probably some path dependency: once standardisation has taken place, firms might be loath to abandon the advantages of standardisation and move back to a competition of incompatible products. Intermediaries and end users, in particular, stand to lose much more than they could gain. If there is path dependency, then the incentives of the various actors coming into the next round of standardisation will be affected. Furthermore, looking forward, the perspective of repeated interaction between the firms should also influence their behaviour in the course of standard-setting.

Third, still from a dynamic perspective, standardisation takes place against the background of technological evolution. It can either seek to bring some order in the aftermath of a technological jump (a breakthrough or radical innovation) or ahead of such jump. In the former case, technological evolution might be settling and standardisation is about enshrining a paradigm.[88] In the latter case, standardisation serves the organisation and management of the technological jump. Here as

[88] In the Kuhnian sense: T.S. Kuhn, *The Structure of Scientific Revolutions*, 2nd ed (Chicago: University of Chicago Press, 1970).

well, the incentives of participants might vary depending on the place of standardisation within the technological time cycle.

Fourth, firms may come to the standardisation exercise with full-fledged, fully developed products. If that is the case, then indeed standardisation can amount to anointing a winner. If not, firms are then engaging in standardisation to put together a coherent technological solution from their diverse R&D efforts. Their respective inputs into the standardisation process could be complements rather than substitutes. Building on the foregoing, the latter option seems likely in situations in which standardisation is already in place and firms decide their strategy on the assumption that further technological iterations will also be brought to standard-setting. Similarly, if standardisation takes place ahead of a technological jump, firms might be contributing pieces of a larger puzzle to be solved at the SSO. Equally, if firms are smaller and widespread, they might also be hoping to contribute part of the standard, as opposed to imposing a full-fledged solution.

Fifth, the firms engaged in standard-setting might be of various sizes and locations. Conceivably, for some markets, every one of the firms could serve the whole demand. Yet for other markets, participating firms cannot conceivably hope for more than a share of the demand for standardised products. Here as well, the incentives of firms may vary depending on their size and location. Furthermore, as mentioned earlier, firms might also have different business plans: not every firm is vertically integrated; some are pure patent holders or pure implementers.

Of course, not only might the incentives of participant firms differ according to the parameters listed in the various paragraphs, but the implications for public policy, including intellectual property and competition law, might vary according to the characteristics of the standardisation exercise under examination. It is conceivable that any legal or regulatory discourse on standardisation in general would actually cover different standardisation scenarios, depending on these parameters, so that a more differentiated discourse would be warranted. That would make policy analysis and implementation more complicated, yet it would improve the outcome, in welfare terms.

In particular, to gain a better understanding of standardisation, more research is needed on counterfactuals. Much attention has been lavished on the mobile communications industry but not sufficiently on comparators – sectors in which standardisation has not taken place or at least where the issues observed in mobile communications have not occurred. The video game industry would seem a prime candidate in that respect.

II. Standardisation and innovation

A second avenue for further research relates to the link between standardisation and innovation. Some thoughtful scholars have already started exploring the interplay between interoperability standards and innovation,[89] but further research is needed because the relation between the two is highly complex. As seen earlier, depending on the strand of literature and policy, standardisation is sometimes treated as if it was unabashedly good for innovation – for instance because it generates scale effects that enable firms to innovate on non-standardised parameters – or, to the opposite, as if it negatively affected innovation – for example, by spelling the end of alternative technologies to the standard. Beyond these simple effects lie more complicated ones that often escape decision-makers. Standardisation also has strong structuring effects. It channels the flow of innovation by creating interfaces and therefore enabling industry disaggregation: the layer model that underpins the Internet family of standards, for instance, was not only innovative in its own right, it also upended the industry structure and opened the door for a massive wave of innovation from the 'edge' of the industry.[90] Yet that is but one example: at other times, standardisation might lead to a loss of variety and impose design constraints,[91] thereby potentially retarding innovation.[92] The standardisation of physical connectors between ICT or media devices, for instance, has often adversely affected innovation; fortunately, the characteristics of that market are such that firms have been able to push through alternative solutions.[93]

Beyond the effect on innovation around the standard (incremental innovation), standardisation can also increase incentives for breakthrough innovation,[94] either to move away from the standard (escape-competition effect) or to anticipate the next generation, to continue

[89] See Gasser and Palfrey, supra note 21.

[90] See J. Zittrain, *The Future of the Internet and How to Stop It* (New Haven: Yale University Press, 2008).

[91] Farrell, supra note 22.

[92] For once standardisation has taken place, it can be hard to break out of that standard and move to another, even if the original is no longer the most suitable. See Farrell, supra note 22, 37.

[93] See, for instance, the interplay between the USB (in its various iterations), FireWire, Thunderbolt and Bluetooth standards – leaving aside all proprietary alternatives – for local connection between devices.

[94] See Curran, supra note 20, 988.

generating revenues once the standard is established (installed-base effect).[95]

Standardisation can equally distract the attention of policy- and decision-makers from other, more significant issues. Standardisation could also be seen as a remedy to the proliferation of patents and to the doubts about their average quality. In that context, the need for SSO members to put their IP portfolio on the standard-setting table would serve as a threshold mechanism, to ensure that only 'serious' firms are participating.[96] FRAND commitments would allow for patent quality issues to be sidestepped, given reciprocity amongst participating firms: each firm accepts the other ones as meaningful counterparts and accepts to forego patent validity challenges. Although SSOs might be humming along and delivering the benefits of standardisation, the outcome is not necessarily optimal because there is no guarantee that all SEPs are genuinely valuable and essential. The value of the genuine SEPs is diluted, and other patents unjustifiably gain SEP status. This becomes problematic outside the ambit of reciprocity, when financial transfers are contemplated, for instance when pure patent holders or pure implementers are involved.[97] These transfers are bound to be challenged, leading to some of the cases discussed in the previous heading. Accordingly, standardisation might in some cases offer a counterproductive quick fix for deeper problems of patent quality and proliferation.

III. The public and private faces of standardisation

The tension between the public and private sides of standardisation also deserves more attention. The scientific literature still stands out through the chasm between the literature studying standards from a public perspective – focusing on what we termed public-driven standardisation earlier – and the literature that takes a more private perspective – concentrating on private-driven standardisation. The reality is more complex: private actors can no longer be abstracted away from any study of standardisation, yet standardisation remains more than a mere private exercise. On a systemic level, pressing questions arise. First, there

95 See Regulation 1025/2012, supra note 2, Rec. (3).
96 This could partially explain why interested firms have been busy building up patent portfolios, either through their own R&D efforts or through acquisitions.
97 See supra.

is an increasing dissonance between the legal-governance perspective on patents (patents as an 'incentive to innovate') and the economic reality of patents (patents used as 'assets' on markets for technology).[98] Patents threaten to become instrumentalised and downgraded to mere entry tickets to the bargaining table.[99]

Second, considering the varying – if not diverging – interests of the private firms involved, the outcome of negotiations in and around SSOs cannot without more be assumed to fall in line with economic and social welfare.[100] As was seen earlier, some participants are pure patent holders, others are pure implementers and yet others are vertically integrated. There is a risk that standard-setting, which ought to be driven by technological considerations first and foremost, becomes captured by commercial considerations.

In that respect, it could be interesting to pit standard-setting through SSOs (as we know it now in all its variants) against alternatives on a continuum between private and public solutions.

At the private end, leaving aside the absence of any standardisation, one finds de facto standardisation, in which one firm decides to turn its technology into a relatively open platform, while retaining control over it. In its heyday, Microsoft was in such a position with its Windows PC operating system. Microsoft has often been vilified for keeping Windows proprietary, yet it did create application program interfaces for developers, and it did ensure that Windows ran on the widest range of hardware components. This contributed to ensuring a blossoming of applications,[101] as well as a jump in performance, coupled with a marked decline in price on the hardware side. Microsoft's strategy has been replicated, with

[98] See G. Van Overwalle and E. van Zimmeren, 'Functions and Limits of Patent Law', in E. Claes, W. Devroe and B. Keirsbilck (eds), *Facing the Limits of the Law* (2009) 415–442. Also see G. Van Overwalle, 'The Impact of Emerging Market Patent Systems on Europe: Awaiting 'The Rape Of Europa'?, in F.M. Abbott, C. M. Correa & P. Drahos (eds.), *Emerging Markets and the World Patent Order* (2013) 355.

[99] See D. Guellec and B. van Pottelsberghe de la Potterie, *The Economics of the European Patent System* (2007) 85 and references cited there.

[100] This seems somewhat worrisome in view of current European legislation which states that "standards for services are voluntary and should be market-driven, whereby the needs of the economic operators and stakeholders directly or indirectly affected by such standards *prevail*", even though they should take into account the public interest: Regulation 1025/2012, supra note 2, Rec. (11).

[101] So much so that competition enforcers found that Windows' dominance on the PC operating system market was protected by an 'application barrier to entry': there were so many applications written for Windows that no competing operating system could offer the same choice: see *US* v. *Microsoft* 253 F. 3d 34, 54–56 (2001).

some modifications, on the smartphone and tablet markets by Apple and Google with their iOS and Android operating systems, respectively.[102] Of course, although purely private, de facto standardisation can provide considerable benefits to society, there is no guarantee that it does so, nor is there any indication that these benefits are maximised.

At the other end of the spectrum, one finds open standards. The term 'open standard' has no single authoritative definition. A narrow definition of an open standard is a standard that is developed through an open, consensual process in which all identifiable stakeholders have been invited to participate (emphasis on 'open *process*').[103] Used in a wider sense, an open standard is a standard that is open for anyone to read at a minimal or no cost (emphasis on 'open *access*') or implement on FRAND terms (emphasis on 'open *use*').[104] Yet a still wider definition, which sets open standards apart from the mainstream practice of SSOs, is that the standard is free to read and implement, without royalties and without licensing (emphasis on 'open *and no-fee use*').[105] Setting the

[102] Apple did not open its iOS system on the hardware side, restricting it to Apple devices only. Furthermore, it exerts greater control on the application side than Microsoft used to do over Windows: applications must be vetted by Apple, and they must be sold through Apple's AppStore, where Apple takes a commission on sales. Android is theoretically an open source operating system, but nevertheless Google retains some control over it.

[103] See P. Caplan, 'Patents and Open Standards. A white paper prepared for the National Information Standards Organization' (2003) 14 *Information Standards Quarterly* 3; C. Koenig and A. Trias, 'Some Standards for Standardisation: A Basis for Harmonisation and Efficiency Maximalisation of EU and US Antitrust Control of the Standard-Setting Process' (2010) 32 *EIPR* 320, claiming as the first golden rule of standardisation: "open, transparent, non-discriminatory access to the Standard Setting Organisation and to all its potential membership categories"; J. West, 'Seeking an Open Infrastructure: Contrasting Open Standards, Open Source and Open Innovation' (available at http://journals.uic.edu/ojs/index.php/fm/article/view/1913).

[104] For the latter, see Koenig and Trias, supra note 103, in particular their fourth golden rule of standardisation. On the interaction between open standards in this sense and the right to access to information, see M. Schellekens, 'Open standaarden en de grondwet' [Open Standards and the Constitution], in H. Sleurink, J. Stedehouder and J. Broekema (eds), *Open Source Jaarboek 2009–2010* (2010), 207–220.

[105] Cf. R. Cover, 'Cover Pages: Patents and Open Standards' (available at http://xml.coverpages.org/patents.html); B. Perens, 'Open Standards, Principles and Practice' (available at http://perens.com/OpenStandards/Definition.html). For a more in-depth overview and discussion of the criteria employed in defining open standards, see K. J. Arquit, A. Y. Oruc and R. Wolfram, 'Antitrust, Intellectual Property, Standards and Interoperability', in Practising Law Institute, *Intellectual Property Antitrust* (1998); K. Krechmer, 'Open Standards Requirements' (2006) 4 *Int'l J. IT Standards & Standardisation Research* 43; J. H. Park, *Patents and Industry Standards* (2010); R.C. Shah and J.P. Kesan, 'An Empirical Study of Open Standards' (2007) and the references cited there (available

discussion on the different meanings aside for a while, open standards, in their various shades of openness, may play an important role on the intersection between technical standards incorporating patented technologies and the public interest. Some scholars have already started exploring the interface between open standards and competition law,[106] or looked into the relationship between open standards and open source using empirical data,[107] but fully fledged analyses, focusing on all related regulatory aspects of open standards, and their interplay with the public interest, remain scarce. Open standards may be considered not just as an alternative to avoid IP issues but also as a way to enhance consumer participation and social welfare. Stakeholders in standardisation processes are mostly private companies. The more open and the more transparent the standardisation process, the greater the chance that consensus is reflecting consumer interests and based on technical criteria (picking a technical winner) as opposed to commercial considerations. What is more, that consensus can also reflect social welfare arguments and provides a good balance between the various interests. In their widest definition, however, open standards might offer too little incentive to private firms to continue investing in research and development unless they can find other revenue streams.[108]

G. Conclusion

The wave of litigation surrounding standardisation and SEPs over the past decade has forced a convergence between strands of literature and policy concerning standardisation, intellectual property law and competition law. Although standardisation literature has traditionally taken a positive

at http://papers.ssrn.com/sol3/papers.cfm?abstract_id=2103646); A. Updegrove, '"Openness" and the Pursuit of Knowledge', in S. Coughlan (ed.) *Thoughts on Open Innovation. Essays on Open Innovation from Leading Thinkers in the Field* (2013) 15, 30.

106 Fitzgerald and Pappalardo, supra note 3; M. Glader, 'Open Standards: Public Policy Aspects and Competition Law Requirements' (2010) 6 *Eur Comp J* 611.

107 See Shah and Kesan, supra note 105.

108 As is the case, for instance, with firms active in the development and maintenance of Linux. They derive revenues from ancillary services, while the Linux kernel itself remains open source and royalty-free. On the relationship between 'open standards' and 'open source', see Krechmer, supra note 103; M. Välimäki and V. Oksanen, 'Patents on Compatibility Standards and Open Source. Do Patent Law Exceptions and Royalty Free Requirements Make Sense?' (2005) 2:3 *SCRIPT-ed* 397; J. West, 'Seeking an Open Infrastructure: Contrasting Open Standards, Open Source and Open Innovation' (2007) 12:6 *First Monday*.

view of its subject matter, intellectual property and competition law literature and policy took a more reserved, if not outright critical, stance toward standardisation once they became concerned with a number of issues arising out of the work of SSOs.

There is a risk that these issues – discussed under buzzwords such as 'patent ambush', 'patent holdup' or 'patent thicket' – will become the focal point for academic and policy debate on standardisation, obscuring the more complex and nuanced reality of standardisation. Accordingly, this contribution proposes a more holistic research agenda. As a first step, standardisation must be apprehended in its totality, including the parameters such as the prior state of the market, path dependency, complementarity or substitutability of contributions, which might affect the scientific and policy analysis. Once that empirical and analytical work has been carried out, broader issues concerning the relationship between standardisation and innovation, and the balance between the private and public aspects of standardisation can be tackled.

16

Standard-setting organisations

Current policy issues and empirical evidence

NEIL GANDAL AND PIERRE RÉGIBEAU

A. Introduction

Given the dramatic growth of the Internet and information technology industries in general as well as the importance of interconnection in networks, the economics of compatibility and standardisation has become mainstream economics. Most Internet and information technology products and services exhibit network effects. In such industries, questions of compatibility and standardisation are important. Because of the network effects that are inherent in such industries, successful diffusion of these products is often contingent on the emergence of a single standard. For example, Postrel (1990) blames (in part) the failure of quadraphonic sound in the 1970s on competing standards. Standardisation can occur through two main mechanisms: the de facto market dominance of a specific technology or the explicit coordination of product designs around commonly agreed technological dimensions.

This second mechanism is by far the most prevalent and operates through standard-setting organisations (SSOs). One website lists more than 1000 SSOs. Fully 260 SSOs are classified as dealing with 'interoperability' standards. Additionally, a large proportion of SSOs classified as dealing with the Internet, 'wireless and mobile', 'software' and 'Multimedia' would likely address compatibility issues. Hence, it seems therefore reasonable to assume that compatibility standards are the main activity of the large majority of SSOs on the list. The importance of SSOs is also illustrated by an estimate that IBM spent \$500 million on standards related activities in 2005 and Hewlett Packard and Sun Microsystems each belonged to more than 150 SSOs in 2003.

Despite the increasing importance of and huge growth in SSOs and the flurry of research on the economics of SSOs in the last decade, no comprehensive review of the literature exists. Such an undertaking is beyond the scope of this chapter. Here, we address a more modest goal: a selective review focusing on current policy issues regarding SSOs and empirical evidence on the performance of SSOs. A key part of the analysis here focuses on what light the empirical work on SSOs can shed on SSO performance. We should also point out that we focus our attention on 'traditional' SSOs in which participants retain the right to charge royalties for licensing their patents reading on the agreed-on standard. We do not therefore consider 'open' standard organisations.

We first provide a short review of the early empirical work on network effects (Section B I) and empirical work quantifying network benefits (Section B II). In Section C, we provide a brief primer on SSOs.

In Section D, we examine policy concerns. The policy concerns fall into two broad categories: (i) concerns about the operating performance of SSOs – that is, about how good SSOs are at 'doing their job' – and (ii) concerns regarding the impact of SSOs on the markets for products, technologies and standards. We first propose several criteria that might be used to distinguish more successful SSOs from less impressive ones. We then examine (in Section E) what light the existing empirical literature on SSOs can shed on how SSOs are meeting or not meeting these criteria. In Section F, we examine where the existing empirical work has to say about key competition policy concerns.

If standardisation is desirable and SSOs often are the only realistic route to a standard, then whether SSOs perform well or not compared to alternative arrangements may be a secondary consideration. In this view, what matters is how differences in the internal rules of SSOs and in public policy towards SSOs affect the performance of these organisations. We address this issue in Section G. Section H offers brief conclusions and suggests topics for future empirical research.

B. Benefits of standardisation: Empirical evidence

SSOs are associations that include potential rivals and make decisions that affect the conditions of competition in the relevant downstream industries. As such, as we discuss further in Section F, SSOs are immediately suspect of possibly facilitating anti-competitive behaviour. The reason why SSO activities are nevertheless encouraged is that the anti-competitive risks – minimised by an adequate competition policy – seem worth taking in light of the potential benefits from standardisation. It does therefore make sense to begin with a review of what we know about the size of such benefits.

I. Review of early empirical work on network effects

A growing empirical literature has developed to examine technological adoption of products with network effects. The primary issue addressed by the early literature is whether network effects exist; this work typically employed reduced form models. Greenstein (1993), Gandal (1994, 1995), and Saloner and Shepard (1995) provide early evidence that the value of the 'hardware' good depends on the variety of compatible complementary software.

Software for the IBM 1400 mainframe could not run on succeeding generations of IBM mainframes, and software for the IBM 360 could run on succeeding models. Greenstein (1993) finds that a firm with an IBM 1400 was no more likely than any other firm to purchase an IBM mainframe when making a future purchase. On the other hand, a firm with an IBM 360 was more likely to purchase an IBM mainframe than a firm that did not own an IBM 360. This result can be interpreted as a demand for (backward) compatible software.

Saloner and Shepard (1995) test for network effects in the ATM industry by examining whether banks with a larger expected number of ATM locations will adopt the ATM technology sooner. Because expected network size is not an observable variable, they use the number of branches as a proxy. Their results suggest that banks with more branches will adopt earlier, which is consistent with virtual network effects.

Gandal (1994) estimates hedonic (quality-adjusted) price equations for spreadsheets to examine whether spreadsheet programs that were compatible with Lotus – the de facto standard – command a premium. The results – that consumers place a positive value on compatibility – suggest (i) direct network effects because people want to share files and (ii) indirect network effects because compatible software enables the of transfer data among a variety of software programs.[1]

[1] Gandal (1995) extends the analysis to Database Management Software (DMS) and multiple standards and finds that only the Lotus file compatibility standard is significant in explaining price variations, suggesting that indirect network effects are important in the DMS market.

Empirical evidence for the existence of network effects can also be obtained indirectly. For example, Liu, Kemerer, and Slaughter (2012) argue that if network effects are indeed important, then the existence of a technology that helps mitigate those effects should have a noticeable impact on observed behaviour. In particular, the existence of a good-quality conversion technology should favour the emergence of a number of competing standards and decrease the price premium enjoyed by the sponsors of the leading technological standard. The authors test this hypothesis on data from the flash memory card market. This market is characterised by the existence of a number of different standards that were made significantly more compatible by the emergence of a digital conversion technology. The authors find that the emergence of this good quality 'adapter' did indeed lead to a substantial decrease in the premium earned by the leading formats, implying that, in the absence of such convertibility, network effects must have been significant.

II. Quantifying network benefits

The early literature just discussed was important for empirically establishing the presence of network effects across many industries. Recent work in the literature has employed structural estimation that enables researchers to better quantify network benefits and conduct counterfactuals.

Gandal, Kende and Rob (2000) develop a dynamic structural model of consumer adoption and software entry and use the model to estimate the feedback from hardware to software and vice versa in the CD industry. The authors show that a 5 percent reduction in price would have had the same effect as a 10 percent increase in CD variety in terms of increasing sales of CD players. This illustrates the importance of network effects (manifested in increases in software variety) relative to changes in the price of the hardware.

Two subsequent papers employ the Gandal et al. (2000) methodology in the video games industry. Clements and Ohashi (2005) use a logit model to test for indirect network effects in the US video game market. They find that that a 1 percent increase in game titles is equivalent to a 2.3 percent price cut in the market. Using the same market for a slightly later time period, Prieger and Hu (2006) find that a 1 percent increase in software variety is equivalent to a 0.4 percent price reduction.

Comparing these results, Gandal et al. (2000) and Prieger and Hu (2006) have similar estimates: a 2.0 to 2.5 percent increase in software variety is equivalent to a 1 percent fall in the price of the hardware. Clements and Ohashi find a larger variety effect, roughly a 0.5 percent increase in variety has the same effect as a 1 percent fall in price. Even if Gandal et al. (2000) and Prieger and Hu's (2006) lower estimates are the general rule, the importance of network effects (as manifested in the importance of changes in the software variety relative to changes in the hardware price) is strong.

Rysman (2004) developed a structural model to examine the importance of network effects in the market for Yellow Pages. The model includes a consumer adoption equation, advertiser demand for space, and a firm's profit maximising behaviour. He finds that consumers value advertising and advertiser's value consumer adoption, suggesting virtual network effects. He measures the importance of the network effect by calculating the surplus lost by the market's failure to internalise the network effect. This calculation is then compared with the (classical) deadweight loss from imperfect competition. He finds that the ratio of network deadweight loss to classical deadweight loss is 1.26, which suggests that the network effect is quite large, even in an industry with significant market power.

Finally, there are a few results that measure the magnitude of the effect of compatibility (vs. incompatibility) on the speed of technology adoption: Gandal et al. (2000) show that if it had been possible to make CD players compatible with LPs, compatibility could have accelerated the adoption process by about 1.5 years, reducing the time of adoption by more than 20 percent.

Dranove and Gandal (2003) empirically test for network effects and preannouncement effects in the DVD market by measuring the effect of potential (incompatible) competition on a network undergoing growth. They find that there are network effects and the DIVX pre-announcement reduced DVD sales by approximately 20 percent. In the end, DVD was adopted as the de facto standard. Nevertheless, the non-trivial effect of the preannouncement by such a weak entrant (see Dranove and Gandal (2003) for details) suggests that network effects were strong in this market as well.

C. A primer on SSOs

Standardisation can occur through two main mechanisms: the de facto market dominance of a specific technology or the explicit coordination of product designs around commonly agreed technological dimensions.

This second mechanism is by far the most prevalent and operates through SSOs. Not surprisingly the prevalence of SSOs varies both across sectors and over time.

In terms of sectors, we would naturally expect SSOs to arise more often where the benefits of compatibility are higher. This includes so-called complex industries, in which products need to aggregate a variety of technologies to be of any use to consumers. We would also expect efforts to achieve compatibility to be more intense when the usefulness of the products involved relies on some form of communication between users or calls for the use of complementary goods.

A rough way of getting at the importance of SSOs across technologies and sectors is to simply count the number of existing organisations. Even such a simple count runs into two main difficulties. First, there is no such thing as an official, exhaustive list of SSOs. Second, SSOs not only deal with the type of 'compatibility' standards that we are interested in, they also create quality/safety standards.

The tables that follow rely on the list of SSOs provided at http://www.consortiuminfo.org/links/. As of October 14, 2013, this list included 885 SSOs. Although we are not able to distinguish between SSOs involved in the setting of compatibility standards as opposed to those involved in quality or safety standards, we note that there are 260 SSOs classified as dealing with 'interoperability' standards. Moreover, one would think that a large proportion of SSOs dealing with the internet (141), 'wireless and mobile' (135), 'software' (174) and 'multimedia' (87) would also mostly address compatibility issues; it seems therefore reasonable to assume that compatibility standards are the main activity of the majority of SSOs on the list.[2]

The following tables show the number of SSOs by technical category. In the last column of the tables, we have tried to match the technology classification with the classification of technologies obtained by von Graevenitz et al. (2013). These authors use the pattern of patent citations to distinguish between 'complex' technologies, where many components controlled by different parties go into designing a successful product and 'discrete' technologies where products are more uni-dimensional and the

[2] Table 16.1 has more than 885 SSOs because some SSOs appear in more than one category. Table 16.2 has less than 885 SSOs because not all SSOs can be assigned to specific industries. This is not important because the goal is simply to give a sense that most of the technologies or industries in the tables are classified as complex and that compatibility issues are much likely to arise in such environments.

Table 16.1. *SSOs per technical category*

Technical category	Number of SSOs	Complexity
Audio/Video/Multimedia	87	Complex
Cloud Computing	16	NA
Credit/Debit/Smart Cards	24	NA
Electronic Media	25	Complex
Hardware	97	Complex
Imaging	30	Complex
Internet	141	Complex
Interoperability	260	NA
Languages/Protocols	27	NA
Network and Network Centric Computing	101	Complex
Operating Systems	62	Complex
Security and Cyber-security	91	Complex
Semiconductors	31	Complex
Software	174	NA
Web Services	56	Complex
Wireless and Mobile	135	Complex

Table 16.2. *SSOs per industry*

Industry	Number of SSOs	Level of complexity
Aeronautics	8	NA
Automotive	23	Complex
Bio IT and Life Sciences	19	Discrete
Clean Tech and Renewable Energy	40	Complex
Construction	14	Complex
Consumer Electronics and Content	27	Complex
Defence	10	NA
Digital and Distance Learning	22	NA
Electronics	100	Complex
Health and Medical	52	Medical Technology: Complex
Manufacturing	35	NA
Power and Smart Grid	30	Complex
Real Estate	5	Discrete
Telecom	89	Complex
Multi-Industries	84	NA

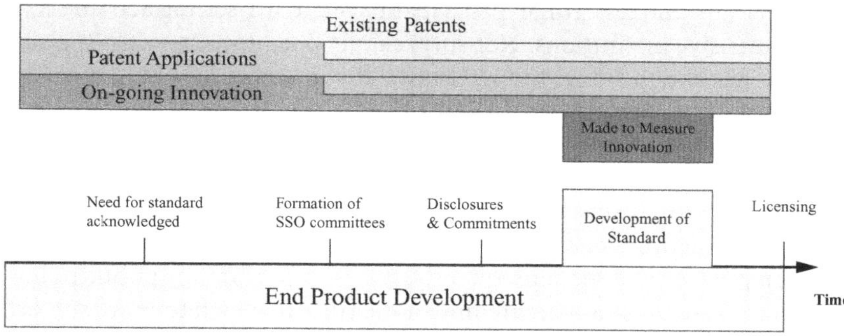

Figure 16.1. Standard-setting by SSOs – timeline

intellectual property rights (IPRs) relating to a given product tends to be held by one or very few firms. As expected, the vast majority of the technologies or industries in our tables are classified as complex, confirming our expectation that compatibility issues are much more likely to arise in such environments. Notice, however, that the categories marked NA (not available) either have few SSOs or are categories that one would intuitively think of as 'complex' anyway. (Unfortunately, we were not able to relate every one of our technologies or industries to von Graevenitz et al. [2013].)

As for any decision-making institution, the performance of SSOs ultimately depends on its membership and on how the various interests of its members get aggregated into a final decision. The following figure gives us a typical timeline for SSO activities.

The standard-setting process begins when the need for a technological standard is widely recognised in the industry. At that point, technologies that will prove relevant to that standard already exist, be it as granted patents, patent applications or even ongoing research programs. The next important stage is the formation of the relevant committees or work groups, that is, the determination of SSO *participation*. This participation has two crucial dimensions. The first one concerns the *type* of agents allowed to participate. Broadly speaking, SSO members come from three groups: IPR holders, potential users of the standards and consumers. Clearly, these groups can overlap. In particular, IPR holders are often also potential users of the standard. For example, Samsung, Motorola Mobile, Nokia and Ericsson hold significant shares of the patents declared 'essential' for the 3G standard and are all also significant manufacturers of smartphones. The second important dimension is the distinction between more and less 'equal' participants. Standard-setting activities are effort-intensive institutions. Not surprisingly, the more expertise a given member has and the more human capital that member is willing to invest into the process, the greater the influence on the design of the standard. Accordingly, large intellectual property holders and large potential users will generally have significantly more weight within the SSO than smaller firms or than final consumers. Although SSO membership is often open to all, it is sometimes restricted. On the other hand, membership is always voluntary. As we will see subsequently, this is important for public policy toward SSOs because any tightening of the rules to which they are subject is likely to increase the number of agents – and especially IPR holders – who choose to remain on the sidelines.

Although membership has its privileges, it also comes with some obligations. The two most important obligations are disclosure and pricing/(F)RAND commitments, where (F)RAND = (fair), reasonable and non-discriminatory terms, to anyone wishing to practice the standard. Disclosure requires members to reveal all existing IPRs, IPR applications and even ongoing research that might be relevant to the standard to be developed. Because the precise technological characteristics of the standard are, per force, not known at this stage, this duty of disclosure must be understood quite broadly: any proprietary knowledge that *might* be relevant to the standard should be revealed. Moreover, the duty of disclosure remains throughout the process: a new member or an existing member whose relevant portfolio of IPRs and research projects changes are both required to notify the SSO. In this sense, disclosure is a continuous process.

SSOs can also ask participating IPR holders to indicate – or even commit to – a royalty ceiling as well as a 'worst' set of other licensing conditions. Although this falls significantly short of an *ex ante* commitment to IPR licensing terms, it at least provides some information as to the likely relative 'price' of using alternative sources of technology. However, the proportion of SSOs that impose such a requirement appears to be relatively small. Much more common is a demand that participating IPR holders commit to license any price of intellectual property that actually 'reads' on the actual standard on so-called (F)RAND terms to anyone wishing to practice the standard.

Once those preconditions are set, the actual work of building a mutually acceptable standard begins in earnest. As pointed out earlier, this process is intensive in human capital – and very costly. Farrell and Simcoe (2012) cite estimates that a medium to large size firm will spend about $50,000 a year to just to participate in a single SSO in one year. This includes salary, travel and membership fees, but not any costs for contributing to the technology. In other words, it does not include the salaries of those who develop firm strategy regarding SSOs, those who are involved in technical work on behalf of the standard, or those who work on the 'legal' aspects. Chiao, Lerner, and Tirole (2007) cite an estimate that IBM spent $500 million on standards-related activities in 2005.

Moreover, in many cases, developing a standard involves much more than sifting through available technologies; it also requires the development of technology from scratch to bridge the gaps between what is currently available. One should therefore think of the set of technologies that might be relevant to the standard as being in a perpetual state of flux as the proposed standard itself evolves, patent applications are granted or rejected, ongoing research bears or does not bear fruit and standard-specific research becomes necessary.

Most SSOs claim that the standards that actually emerge from their activities are reached 'through consensus' between participants. This is of course hard to take as truth because conflicts are bound to arise between participants. Still, because most SSOs do not have explicit 'voting rules', we actually know little as to how such conflicts are resolved.

Once the standard – or at least the first generation of the standard – is determined, intellectual property holders can declare the intellectual property that they own and that they feel to be 'essential' to the standard. This process of declaration is often misunderstood. First, although SSO members are obliged to declare as essential all patents or other IPRs for which they will seek royalties from those who practice the standard, intellectual property holders who chose to sit on the sidelines and not take part in the SSO's deliberations have no such obligation. Second, in the vast majority of cases, there is no third-party verification of the truly 'essential' characteristics of the IPRs declared as such. Quite naturally, then, because failing to declare a truly essential patent would be a breach of the commitments made to the SSO, there is a natural tendency for participating IPR holders to over-declare essential IPRs.

The standard-setting process does not actually stop when an agreed-on standard is obtained. For the standard to be successful, it must be made accessible to users on reasonable terms, and it must be updated as the needs of the industry change and the set of available technologies widens. Importantly, it is not possible for users to purchase 'access to the standard' from a single agent for two main reasons. First, few SSOs mandate or even recommend that all standard essential IPRs held by members be grouped into a corresponding patent pool, where access to all intellectual assets can be obtained for a single fee (or variable royalty). So potential users of the standards must usually obtain separate licensing agreements from all SSO members holding relevant patents. Second, even if all SSO members agreed to include their relevant IPRs into a standard-specific pool, some standard essential IPRs are also bound to be held by agents who did not take part in the standard-setting process, making separate negotiations with these agents unavoidable.

Standards are not static. For example, the 3G standard has many versions. This need to update the standard raises a number of issues. Under what conditions are new members admitted to the table? Is there an efficient mechanism to ensure that existing members cannot push for continued use of their technology even though a superior technology might now be available? How does the advent of a new version of the standard affect the terms of existing licensing contracts? These are all aspects of SSOs about which we do not have any systematic, reliable information.

To get a more complete picture of the standard-setting process, it is important to realise that SSOs do not exist in splendid isolation. In particular, the past ten years have witnessed a sharp increase in standards consortia.[3] A standard consortium is an organisation that is both less formal and less inclusive than traditional SSOs. Although some consortia go on to develop their own standards – in which case we would, for the purpose of this chapter, treat them as SSOs – most of them intervene ahead of participation in more formal SSOs and involve coordination of research strategies significantly more 'upstream' than SSOs. They can therefore be seen as organisations that help reduce the duplication of R&D efforts and smooth conflict of interests ahead of formal SSO proceedings. Of course such 'coordination' is not necessarily innocuous. In this respect, it is interesting to note that Baron and Pohlman (2013) find that consortia tend to arise more frequently when the relevant technologies are more fragmented but that they tend to be made of companies with technologies that are close competitors.

[3] See Baron and Pohlman (2013).

D. Policy concerns

Policy concerns regarding SSOs can be divided into two broad families: concerns about the operating performance of SSOs – that is, about how good SSOs are at 'doing their job' – and concerns regarding the impact of SSOs in the markets for products, technologies and standards. To assess the performance of SSOs, we must first define their task – the criteria that might be used to distinguish successful SSOs from less impressive ones. We propose the following list:

- SSOs should coordinate on the best (i.e. most technologically efficient) standard, unless there are compelling cost reasons (e.g. licensing terms commitments) that justify the use of inferior substitute technologies (see section E I).
- SSOs should develop their standards in a timely manner (section E II).
- The resulting standards should be accessible (and affordable) for efficient competitors in the relevant downstream product markets. SSOs should not distort competition in either the market for technologies or the market for standards (section E III).
- The standard should be periodically updated to include more efficient technologies that have become available and SSOs should not have a chilling effect on innovation (section E IV).

The first two criteria pertain to the operating performance of SSOs, whereas the last two mostly relate to competition policy concerns. The links between these criteria and the existing empirical literature are illustrated in Figure 16.2.

E. Operating performance of SSOs

The first two criteria have been analysed theoretically by Farrell and Saloner (1988), who compare the performance of an SSO to market-based or 'de facto' standardisation. In the market-based mechanisms, markets participants can commit to a given technology in the hope that others will follow. Whether effective coordination on a standard is achieved therefore depends on how strong the 'bandwagon' effect. Not surprisingly, then, markets do not always lead to standardisation as equilibria with incompatible technology choices can emerge. On the other hand, when markets lead to de facto standardisation, they tend to do this rather quickly

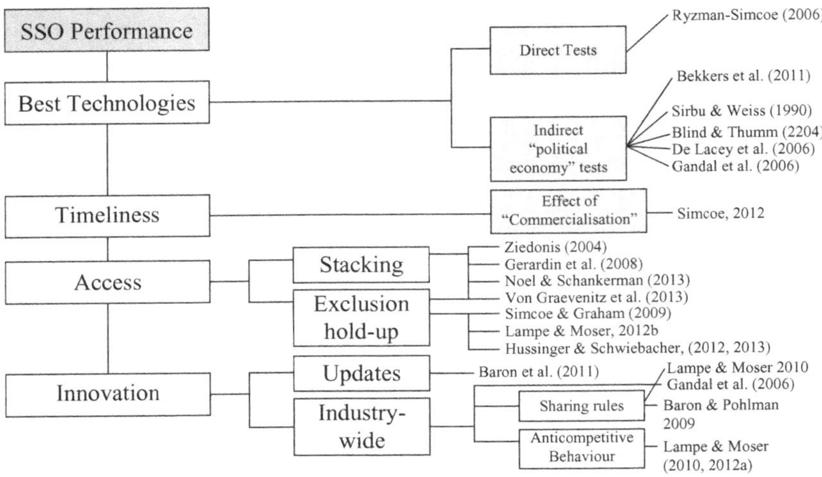

Figure 16.2. Links between performance criteria and the literature

(precisely because standardisation only arises when bandwagon effects are large). By contrast, standardisation committees work around the principle of 'unanimity', which means that participants cannot make unilateral commitments to technologies and must instead reach agreement with others. Because SSO members each have their own vested interests, agreement requires some form of negotiation. Farrell and Saloner model this negotiation process as a war of attrition. In such a game, actual delays arise in equilibrium as each player tries to convince the others that its vested interests are strong. Still, in the end, agreement is reached. Overall then, SSOs are better than the market mechanism at ensuring standardisation but they take more time to get there. Still Farrell and Saloner show that expected welfare is higher under the SSO process, although the best performance is obtained when SSO participants can also 'play the market game' by making unilateral commitments in the relevant markets. Intuitively, the threat that the market-based process might converge on a standard that they do not like spurs SSO members to settle their differences more quickly.

Although it is difficult, if not impossible, to directly examine the prediction of the Farrell-Saloner (1988) analysis empirically, there are a few empirical chapters that shed some light on the role played by standard-setting organisations.

I. Choice of technology

We are only aware of one 'direct' test of the quality of the technologies that SSOs assemble into a standard. A traditional – if imperfect – manner of measuring the importance of patented technologies is to use the number of backward citations that those patents receive. There is an obvious difficulty, however. To assess whether the SSOs end up relying on good technologies, one would want a measure of the intrinsic value of these technologies, untainted by the fact that they were selected as part of the standard. To address this issue, Rysman and Simcoe (2008) examine patent citations over time, distinguishing between the citations received before the start of the standard-setting process and those obtained afterward. They conduct their study for a sample of US patents that were disclosed during the standard-setting process at four SSOs. The citation pattern for these patents is then compared to the pattern for a control group of patents that were not used or disclosed within the SSO. The authors find that the baseline citation rate for SSO patents is about double that of non-SSO patents. Thus there is a selection effect because SSOs indeed chose technologies that are intrinsically important. They also find that disclosure itself leads to a 20 to 40 percent increase in citation rates. Hence, SSOs also 'help establish' the technologies that they endorse.

Although this is the best empirical evidence that we currently have on the equality of SSO's choices, the approach has two main drawbacks. First, technologies that are disclosed to an SSO are not necessarily actually used by an SSO. In this sense, the Rysman and Simcoe result tells us that SSOs attract agents who hold important patents, but it does not necessarily tell us that the SSO does a good job of selecting the best patents from this set. Second, there is a difference between the quality of the patents that read on the standard and the quality of the standard itself because that quality also depends on how well the technologies used for different parts of the standard actually fit together.

In the absence of more direct evidence, one can also try to infer the likely quality of the standards chosen by SSOs based on the *composition* of the relevant SSO committees. The general idea is that the better the active participants in the SSOs reflect the composition of the industry (including both users and IPR holders) and the less the internal SSO process is biased in favour of some types of participants, the more likely it is that the standard will be chosen mostly on grounds of technological merit and cost.

The earliest econometric study of standardisation committees was conducted by Weiss and Sirbu (1990). In their chapter, they considered eleven cases from the computer communications hardware sector. They focused on settings in which the standard was reached in a standards committee, two technologies were competing for inclusion into the standard, and only one of these technologies was eventually chosen by the committee. Their results suggest that the size of the firms in the coalition supporting a technology and the efforts they made through written contributions were significant determinants of which technology was chosen. In addition, when the sponsors of a technology weighed market factors more highly than technical factors, they technology was more likely to be adopted in the standards decision studied.

The lessons from the Weiss and Sirbu study are not quite clear. On the one hand, their results indicate that SSOs suffer from the same drawbacks as most political/voting process: because each party that favours a given technology has an incentive to free-ride on the efforts of those who share the same preference, highly concentrated coalitions are more likely to prevail. This would suggest that SSOs do not necessarily choose the best technological standards. On the other hand, the implications of the importance of the efforts exerted by the sponsors of the each technology are ambiguous. It might just be another consequence of the 'free rider' argument just outlined, but it might also reflect the fact that sponsors of better technologies have greater incentives to push for them, which would imply that SSOs would indeed tend to converge on the correct standard. There are two mechanisms that would imply that greater effort is spent on promoting the better technology. First, a better standard would be more likely to be widely adopted and the products designed around it would be more likely to generate large sales. This translates into both greater benefits for users and higher potential revenues for IPR holders. Second, SSO participants with interests on the user side of the market would naturally prefer the better of the two rival technologies. In this sense, the Weiss and Sirbu study is incomplete because it does not systematically analyse *who* was pushing for what. What was the composition of the SSOs in terms of users, strict IPR owners or vertically integrated entities? How did the composition of the winning coalitions compared with the composition of the SSO itself? Which type of actor exerted the most effort? Without answers to these questions, it seems hard to assess the implications of the Weiss and Sirbu analysis for the accuracy of SSO's technological choices.

Gandal and Genesove (2006) empirically examine the interaction between intellectual property and participation in standardisation committee meetings in the modem industry. They combine patent data on modems (modem in the title) and SSO 'meeting' data from the Telecommunications Industry Association (TIA,) the SSO responsible for developing voluntary (consensus) standards in the analogue modem market in the United States. Although many firms obtained 'modem' patents and many firms participated in standardisation meetings, only a small subset of forty-five firms both obtained patents and participated in the standardisation meetings. These firms accounted for a significant percentage of the patents received and the total number of meeting attendees. Their main finding is that although participation in standards meetings predicts future intellectual property (both unweighted and citation-weighted patents), the reverse is not true: patents and citations are not good predictors of future meeting attendance. The authors propose several explanations that are consistent with this phenomenon: (i) Firms with pending, but not yet granted, patents attend the committee to have the standard incorporate their intellectual property. (ii) The information garnered at these meetings help advance firms' intellectual property portfolio. (iii) Knowledge diffusion at the meetings may lead to firms citing patents of other firms attending standard meetings.

Of these three explanations, only the third one seems innocuous in terms of the SSO's ability to converge on the best possible standard. By contrast, participation in committees to further the cause of pending patents suggests that the final standard might be biased toward the (new) IPRs of participants. Finally, the gathering of information at SSO meetings lends itself to at least two, rather different interpretations. On the one hand, one might see it as a sort of 'spill-over' that improves the technological knowledge of participants and hence improve their future innovative performance. On the other hand, one might fear that participants use the knowledge gathered at SSO meeting to either file new patent applications covering unprotected aspects that seem likely to become part of the standard or – more insidiously – to modify current patent applications to ensure that the wording of their claims actually read on the standard under development. In this respect, it is interesting that Berger, Blind, and Thumm (2012) find that standard essential patents (SEPs) typically have more claims and amendments than non-essential patents.

Using a small sample of European companies, Blind and Thumm (Research Policy, 2004) find that the greater the patent intensity of the firms, the less likely they are to join an SSO standardisation process. This is clearly a concern because it indicates that SSOs might be unlikely to draw on some of the state-of-the-art technologies. This concern is reinforced by the work of Bekkers, Bongard and Nuvolari (2011). They consider the case of W-CDMA and examine the factors that influence firms to claim essential patents. They find that the involvement in the standardisation process is a stronger factor that the technical merit of the patent in determining the likelihood that a firm will declare a patent to be essential.

II. Timeliness

The main difficulty in assessing the timeliness of SSO decisions is to find some form of counterfactual: timely compared to what? Because what matters is the time spent *to design a standard*, the general performance of SSOs could only be compared to that of a different process that also achieves standardisation. This essentially leaves de facto market standardisation (or possibly open standard organisations) as possible benchmarks. However, the paucity of such benchmarks (there are few de facto standards) and the fact that different standard-setting mechanisms seem likely to emerge endogenously in industries with very different characteristics make such an exercise essentially hopeless.

However, it is still possible to get some insights on how the timeliness of SSO decisions might be affected by changes in their environment or changes in their internal rules. This is the approach used by Simcoe (2012), who looks at the time taken to develop standards for SSOs in charge of producing many of the standards used to run the Internet. Simcoe's hypothesis is that because these SSOs rely on consensus – or at least on super-majority rules, a sharpening of vested interests across members would be likely to lead to longer delays. This hypothesis is derived from a random bargaining model. In such models some negotiation delay is required to ensure that SSO participant converge on the best possible standard. However, when each party has vested interests in a particular technology, then bargaining delays become excessive. In this sense, the predictions of the random bargaining model used by Simcoe have very much the same flavour as those obtained by Farrell and Saloner (1988).

To test the prediction of the model empirically, Simcoe needs some measure of vested interests. One approach would be to compute measures of the distribution of vested interests for each SSOs, presumably relying on patent-ownership data. However, such an exercise would not be particularly reliable as such measures would be heavily influenced by one's interpretation of the technological strengths of each party and would also be vulnerable to strategic patenting behaviour. Simcoe avoids such a drawback by relying on the 'natural experiment' created by the increasing commercialisation of the Internet. When the Internet was not a source of significant wealth, the vested interests of SSO participants must also have been rather muted. As the commercial value of the Internet increased, one would naturally expect divergent preferences to become more prominent. This empirical strategy is implemented by looking at internet-related SSOs over the period 1993–2003. For each SSO, an index of commercial significance is developed on the basis of the proportion of private sector participants (the 'suit-to-beard' ratio). A special feature of the IETF rules makes it possible to adopt a difference in difference approach: the IETF publishes both standards and 'non-standard' ideas that are developed within the same type of committees (and indeed often the same committees) as standards. Because such non-standard ideas should not raise conflict of interests linked to the commercial nature of the Internet, they can be used to estimate the time that it should take to reach a decision on standards in the absence of distributional conflict stemming from vested interests. Simcoe finds that a 1 percent increase in private sector participation leads to an additional delay of almost eight days. Overall, he estimates that the total increase in private sector participation observed over the period contributed an additional eight months to the average SSO process. There is therefore reliable evidence that distributional concerns contribute significantly to the cost of standardisation through SSOs. As we discuss in Section E, this of course raises the question of how such distributive conflicts might be minimised.

III. Access on reasonable terms

There are two main facets to the issue of accessibility on reasonable terms. The first aspect relates to the FRAND commitments that the vast majority of SSOs require from participating IPR holders. As we will see, FRAND commitments are designed to address the adverse consequences of the *ex post* market power wielded by the owners of IPRs that read on the selected standard. As such, this issue is better discussed in the section on competition policy concerns.

The second aspect of accessibility – and hence pricing – is the so-called royalty-stacking issue. This is familiar territory. Once a standard has been defined, IPRs that read on this standard are complements from the point of view of the users who want to implement the standard. As has been well known since the seminal work of Cournot (1838), independent pricing of complements leads to a higher total price for the 'bundle' than if all prices had been chosen by a monopolist. Notice that this 'stacking' problem has little to do with market power: while stacking only arises if components are priced above variable costs, it can lead to large differences between the sum of independently set prices and the sum of monopoly prices even if the market power of the bundle is not strong.

The issues of hold-up and royalty stacking have been conflated in much of the recent literature on the antitrust treatment of SEPs.[4] This is unfortunate because royalty stacking is not fundamentally a competition policy issue. Competition policy deals with market power. Although royalty stacking can only arise if the owner of a given component has *some* market power, very little market power can lead to very high 'stacks' if the number of complementary components is large. Hence royalty stacking is essentially a problem arising from the sub-optimality of independent price setting decisions in the presence of strong complementarity links, which is exactly the opposite of traditional competition policy concerns about coordination of decisions across independent entities. There is another fundamental difference between the hold-up and royalty-stacking issues. Whereas an *ex ante* commitment to royalty terms and condition would solve the first issue, it would not resolve royalty stacking because the complementary goods pricing effect is the same *ex ante* as well as *ex post* (when the standard has been chosen).

From an empirical point of view, the obvious questions about stacking are 'is there stacking?' and 'how large are the consequences of stacking on the total price of accessing a bundle of technologies?' Given how much ink has recently been spilled over the stacking problem, it is remarkable that the existing empirical literature is so scarce.

Because the theoretical basis for the royalty-stacking problem is the inefficient pricing of complements by independent patent owners, it seems reasonable as a first step to study the prevalence of such 'patent thickets.' Two necessary conditions must be satisfied for the existence of such

[4] See, for example, the literature review in Gerardin, Layne-Farrar and Padilla (2008), where the authors note that early disclosure of patenting terms within SSOs has been proposed as a solution to the royalty-stacking problem. Although such disclosure would allow for a more informed choice of the economically efficient standard, it does nothing to address the issue of complementarity.

thickets: a given product (e.g. a smartphone) must involve the use of a large number of patent rights and the ownership of the relevant patents must be sufficiently dispersed.

Von Graevenitz et al. (2013) use a specificity of the European Patent Office (EPO) patent application process to build a patent-specific measure of 'thickets.' At the EPO, patent examiners rate each patent cited in a given patent application as potentially blocking on its own, potentially blocking in conjunction with others or not potentially blocking. The authors can then determine whether triples of patents held by different agents 'block' each other in the sense just defined. Relying on the incidence of such mutually blocking triplets, the authors obtain what they refer to as a measure of 'technology complexity' for the patent portfolio of each firms. These measures can then be used to rank technology areas from the most to the least 'thicket prone'. They find significant thicket presence in nine out of thirty technology areas. Notice, however, that the von Graevenitz et al. measure mostly addresses the first of the two necessary conditions for the existence of 'thickets', i.e. the need to clear a number of IPRs to produce a product. By focusing on triplets, it does not fully consider the extent of the dispersion of the ownership of such rights.

For this reason, the authors also consider a measure of 'fragmentation' first introduced by Ziedonis (2004) who measures the existence of thickets by constructing an index of the dispersion of patent-ownership for a number of firms. For each firm, the index is based on the dispersion of backward citations found in the firm's own patent portfolio, capturing the idea that the owners of the cited patents are likely to be companies from which the firm might need a license. Ziedonis herself only uses this index to study the semiconductor industry. In principle, however, the proportion of firms facing significant thickets could be used as an alternative manner of identifying 'thicket-prone' sectors of activities. Interestingly, however, von Graewenitz et al. (2013) find that the correlation between their measure of 'complexity' and the Ziedonis measure of 'fragmentation' is extremely weak. Because both of the dimensions captured by these two indices ought to be present for a thicket problem to truly arise, their results can actually be seen as casting doubt as to the prevalence of fully fledged 'thickets.'

Of course the mere existence of patent thickets is not in itself of great interest to the analysis of SSOs because, by their very operation, SSOs do inevitably create thickets of patents around the standards that they develop. On the other hand, the effects of patent thicket on firm behaviour are relevant, even if the thickets involved do not arise as part of a standard-setting process.

Even the presence of thickets does not automatically make royalty stacking a significant issue as mechanisms such as cross-licensing, patent pools and reputation might suffice to minimise the problem. It is therefore important also to have credible empirical evidence on the *effect* of thickets. Unfortunately, estimating the direct effect of thickets on the total royalty paid by downstream developers runs into a number of fearful obstacles. Not the least of these obstacles is the fact that there is little systematic information on the royalties paid by licensee, not only because such data are not collected but because licensing terms are often covered by confidentiality clauses. Moreover, even if reliable royalty data were available *and* one could identify the set of licenses required to produce a given product, simply regressing total royalties onto the number of patents and their ownership concentration (ideally with an interaction term) would only provide useful estimates if one could also control for the different 'value' of the downstream product and if one believed that the effect of other factors such as hold-up were truly independent of the pure 'complementarity effect'. Accordingly, the empirical literature has looked for indirect ways of gauging the effect of thickets on licensing behaviour.

The first approach consists at asking whether – and how – the presence of patent thickets has affected patenting behaviour. In that vein, Ziedonis (2004) tests the hypothesis that the rate of patenting increases when thickets are present. The link between thickets and patenting behaviour rests on the idea is that when transaction costs are high and ownership of required patents is dispersed, firms that want to use the technology are likely to face adverse licensing terms unless they have a significant patent portfolio of their own. This creates an incentive to invest more in licensing *for defensive* motives. Ziedonis further reasons that the desire to be well armed for *ex post* negotiation should be higher in industries where users are intensive in (sector-specific) capital, because this intensity makes them more vulnerable to *ex post* licensing bargaining in the first place. She finds that the effect of increased patent dispersion on patenting is five times higher for capital-intensive firms, confirming the initial hypothesis.

Although the Ziedonis results do suggest that dispersion of ownership matters, it might not necessarily tell us much about the existence of 'stacking', for two main reasons. First, patenting occurs for both defensive reasons and more traditional 'innovation' reasons. It is reasonable to assume that thickets increase the incentives for defensive patenting, but one should also factor in the fact that the presence of thickets – precisely because they lead to suboptimal pricing that decreases the welfare of both licensors and licensees – should also decrease incentives to patent for 'innovative' reasons. Whether one believes that Ziedonis' approach adequately controls for this depends on whether one thinks that this 'offensive' effect is also correlated with the capital-intensity of the sector. Second, and more important, it is important to be precise about the economic reasoning that underlies the tested hypothesis. When developing her hypothesis, Ziedonis argues that concentration of patent ownership is more likely to lend itself to *ex ante* licensing, while dispersion would tend to lead to *ex post* negotiations, that is, negotiations that arise when the potential licensor has already sunk a significant part of its technology-specific investment. The reason for such a difference is not dispersion of ownership itself but the fact that because of transaction costs, dispersion implies that it is harder for potential users to identify *ex ante* the set of patents that they would likely infringe.[5] Once licensing occurs *ex post*, it follows immediately that potential licensors with greater amounts of technology-specific sunk capital are more vulnerable and hence would want to protect themselves more by developing their own defensive portfolio. Putting the two parts of the argument together ensures that there will be a correlation between the dispersion of ownership of patents likely to 'read' on a given firm's products and that firm's own patent investment. However, it should by now be clear that this correlation has absolutely nothing to do with the complementary pricing issue that is at the basis of royalty stacking. At its core, the Ziedonis work is a joint test of the existence of the type of transaction costs just described and the importance of hold-up effects. From the point of view of stacking, whether negotiations occur *ex ante* or *ex post* should essentially not matter.

Noel and Schankerman (2013) also examine the effect of two types of strategic patenting behaviour in the computer software industry. The first strategic dimension is portfolio size, which, as in Ziedonis (2004), is seen as a proxy for the strength of a firm's bargaining position. In this respect, the authors find that having a large patent portfolio (compared with rivals) has a positive effect on the firm's market value, its patenting and its R&D. However, they find little evidence that portfolio expanding strategies are a sort of prisoner's dilemma where all firms try to improve their bargaining position at the eventual expense of the profitability of all

5 Transaction costs linked to actual negotiations is not a factor because it would affect *ex ante* negotiations as well as *ex post* negotiations.

involved: a proportional increase in all portfolios is still found to have a positive impact on the value of the firms. The second aspect investigated is the fragmentation of patent rights, which reduces market value but increases R&D and patenting.

So again, although the Noel and Schankerman paper does not provide us with any evidence as to the size of the 'complementary pricing' effects, which is the essence of royalty stacking, it shows that the bargaining issues (including hold-up) that arise with *ex post* licensing and the fragmentation of ownership do affect the behaviour and performance of innovative firms. Interestingly, however, their evidence also suggests that the net effect of such factors on R&D activity and the value of innovative firms is not necessarily negative.

As discussed in Gerardin et al. (2007), the price complement mechanism has received a lot of attention in both the biotech and the telecommunication industries. Although the discussion of patent thickets in telecoms is too recent to have generated significant empirical research, the debate on the importance of the 'complements' effects has been raging for a while in biotech. The growing consensus is that there is little evidence that royalty stacking (or the piling up of transaction costs) has been significant enough to cause noticeable harm. Along the same lines, because the complement pricing problem also leads to lower profits for licensees, it should be associated with lower levels of R&D and patenting, patterns that are simply not found.

IV. Updates

The impact of SSOs on the rate of innovation in a given sector of activity involves the SSO's impact on technological markets at large and is therefore a competition policy concern. By contrast, the ability of an SSO to keep updating its own standard to incorporate new developments and meet the changing needs of the sector is an important aspect of its operating performance. Unfortunately, we are not aware of any systematic empirical research investigating this aspect of SSO behaviour. Using data on 3500 ICT standards, Baron, Blind and Pohlmann (2011) find that the number of SEPs that (claim to) read on an standard has a negative effect on drastic innovation in standards, that is, there is a bias toward preserving and improving the current standard rather than adopting a new standard.

Clearly, there is room for further research. Of particular interest would be the link between the pattern of ownership of SEPs and the rate at which standards are updated. For example, does greater concentration hinder updating (because more parties have to agree to the changes), or does it favour it (because each party loses less from adding some additional patent holders to the mix)?

V. *Patent pools*

If one believes that independent pricing of complementary goods is indeed a sizeable issue, then a natural solution to the stacking problem would be to let the owners of standard essential IPRs set their licensing terms *jointly*. This could be achieved by mandating that SEP holders all join a standard-specific patent pool, where users of the standard could clear all relevant IPRs for a single fee determined jointly by pool members. Alternatively, voluntary participation in standard-specific patent pools could at least partially alleviate the royalty-stacking problem. In this sense, patent pools and their properties can be seen as an important part of the evaluation of SSOs.

The effectiveness of patent pools as a way of resolving the royalty-stacking problem hinges on three factors: pool participation, the effect of patent pool on innovation and the antitrust issues that patent pools themselves might raise. We deal with innovation in Section F. We begin with antitrust issues.

The most obvious concern related with patent pools is that they could be used to actually raise prices (royalty). Lerner and Tirole (2004) have shown that if we focus on patent pools that would form spontaneously – that is, without policy intervention – then a clause that ensures that pool members can still license their own IP separately outside of the pool suffices to ensure that the patent pools that do emerge would be socially desirable. This is true whether those pools include patents that are strict complements or whether their membership is more diverse. This seems to imply that patent pools would be an ideal solution to the SSO-related stacking problem, but this would be misleading for two reasons. First, it is not correct to think of the set of patents declared essential to a standard as strict complements. In practice, many of these patents are actually not essential to the standard. Moreover, for legal certainty reasons, it makes sense to also license some substitutes: it is not always clear whether a feature of a standard is only covered by patent A or whether patent B might also have a claim on it. Second, as soon as we depart from the case of perfect complements, the fact that the patent pools analysed by Lerner and Tirole *emerge spontaneously* is crucial to their finding. One cannot therefore assume that *mandated* pools would have the same desirable properties.

It is also important to realise that Lerner and Tirole do not directly address the issue of patent participation. Their patent pools are industry-wide. Indeed, Brenner (2009) has shown that if the extent of patent pool participation is endogenous, the 'independent licensing clause' emphasised by Lerner and Tirole is no longer sufficient to ensure the socially optimal formation of patent pool unless pool membership is determined by unanimity rule, a feature that Competition Authorities might find problematic (see Section F II). The fact that, in practice, patent pools include only about 50 percent of industry actors and that even SSO-related pools usually do not include all SEPs suggests that the issue of participation is crucial and deserves to be investigated further.

F. Competition policy concerns

To understand the various competition policy concerns related to SSOs, one must first distinguish between three types of markets that might be affected by the activities of SSOs. The first and most obvious of these markets are the markets for the products that would make use of the standard. There are two main concerns relating to behaviour in the technology market. The first concern is that SEP holders that are vertically integrated might seek to exclude rivals from the relevant downstream markets by either refusing to license the technologies required to practice the standard or by imposing discriminatory terms.[6] The second concern is the fear the SEP holders would exploit the additional market power gained through the inclusion of their IPRs into the standard to charge excessive prices. Two additional issues arise because of the vertical relationship between the technology markets and the markets for standards. First, competition authorities might suspect that SSO membership rules might be devised to exclude some holders of technology from the standard-setting process. This not only would lead to the choice of inferior standard but it might eventually reduce the innovation incentives of firms that find themselves systematically locked out. Second, when rivalry between standards is a realistic scenario, SSOs might be tempted to forbid their members from making their technology available to rival standard-setting bodies. Again,

[6] Notice that, contrary to what is often argued, 'non-practicing entities' that hold SEP are not an issue in this respect because they have no incentive to foreclose downstream market competitors. See Schamlensee (2009) for a thorough discussion.

the resulting lack of 'standard competition' would not bode well for either the quality or the timeliness of the chosen standards.

I. Exclusion

As discussed, one of the main sources of antitrust concerns with respect to SSOs (and patent pools) is the fear that these organisations might be used to exclude rivals. Some types of exclusions are easily guarded against. For example, Competition Authorities usually require that patent holders who take part in the activities of one SSO should still be able to make their patents available to a rival SSO.[7] This should be sufficient protection against the loss of inter-standard competition. Other potentially exclusive tactics are trickier to handle. Let us first consider the question of participation. Can SSO membership be restricted by a group of firms to exclude others? The simple answer is that, in that absence of appropriate competition policy, it definitely could – and indeed definitely would. Lampe and Moser (2012a) provide convincing evidence on this issue. Although their study concerns patent pools rather than SSOs, it is not hard to see why it is also relevant for SSOs, which are after all patent pools aimed at coordinating technologies (i.e. at forming a technological bundle) rather than aimed at coordinating sales of technologies (i.e. forming a 'commercial' bundle). Moreover, patent pools often arise in the shadow of standard-setting processes.

The main attraction of the Lampe and Moser analysis is that the authors look at a period over which competition policy was deliberately 'relaxed' (or even suspended) in order not to impede economic recovery in the post-depression era. Without any 'fear of the gendarme', patent pool members were found to practice exclusion with some delight. Out of the twenty patent pools reviewed by the authors, eleven did not license a single outside firm, three pools licensed up to three companies not belonging to the pool and only six pools licensed extensively outside of the pool itself. This pattern was not only explained by a reluctance to license to downstream rivals, it was also linked to 'inter-standard' competition, as the authors conclude that 'archival evidence suggests that the pools studied . . . used licensing as a means to limit competition with substitute technologies'.

[7] See the European Commission's *Horizontal Guidelines*.

II. Access: hold-up

In economic terms, there is a potential hold-up problem if at least one of two parties to an agreement must make relationship-specific investments at a time where the terms of the agreement are not set fully and irrevocably. In other words, hold-up problems only emerge if two necessary conditions are fulfilled: the relationship involves specific investments and there is room for opportunism.

Thanks to a few high-profile antitrust cases in the smartphone industry, SSOs have recently found themselves in the middle of a hold-up storm. Unfortunately, the discussion of hold-up in this context has been far from limpid. Let us therefore restate a few simple points. First, as stated earlier, hold-up requires that two conditions be met. The first condition is the presence of investments for which value is higher within the relationship between the relevant parties than outside of this relationship. In other words, one needs (partially) sunk investment. The second necessary condition is that there is room for opportunism by the party that did not make the investment. This arises if fool-proof complete contracts cannot be either written or enforced. We can then ask where and when in the standard-setting process these two conditions are met. To do this, we must first identify the relevant 'bilateral relationship'. Clearly, in this case, this is the relationship between SEP holders and potential users of the standard. This relationship involves the licensing of SEPs to the users. There are three types of investments that are linked to this relationship: the R&D investments made by the SEP holder before the start of the SSO process, the product development investments of the potential users and the effort and additional standard-specific R&D that goes into the SSO process itself.

The next step is to identify which of these investments is in fact 'relationship-specific', as required for hold-up effects to arise. This of course depends on when the terms of the licensing agreement are irrevocably set. As we have seen, although potential SEP holders who participate in SSOs often have to agree to license on 'FRAND' terms, the actual licensing terms are usually determined only after the standard is agreed-on and they are usually determined through bilateral negotiations between each SEP holder and each standard user. Even though some SSOs require the disclosure of 'worst' terms before the standard is chosen, it seems fair to say that such commitment still leave a lot to be determined once the standard is actually set. So let us proceed under the assumption that meaningful bilateral negotiations only arise after the standard is agreed on.

At that stage, the R&D costs incurred to obtain the discoveries covered by SEPs has clearly been sunk. This is true of both SEPs that existed before the start of the SSO process and of those that were developed in to 'fill in the gaps' as the standard was being developed. In terms of specificity, however, there is a potential difference between these two types of SEPs. SEPs corresponding to fill-the-gap technologies can safely be assumed to have little value outside of the chosen standard. In that sense, the corresponding investment is both sunk and relationship-specific, setting the stage for a potential hold-up *by the potential users*. Pre-existing patents that become SEPs might, on the other hand, have commercially profitable applications outside of the SSO context. To that extent, then, it would not be appropriate to consider the full value of the sunk investment as relevant to the SSO-related hold-up problem. Still, it should also be remembered that some types of technologies – such as those covering communication protocols – have by their very essence little value if they cannot be used to communicate, which usually requires some form of standard. For such technologies, then, almost the entirety of R&D costs should be seen as relationship-specific sunk costs. It should therefore be clear that, although hold-up of users by SEP-holders has gotten most of the attention, the conditions for significant hold-up of SEP holders by users are also fulfilled. As Williamson (1983) pointed out, bilateral hold-up can be a solution to the hold-up problem itself as it amounts to an 'exchange of hostages'. Hence, even if we have the right conditions for hold-up on both side of the relationship, it is not at all a priori clear that hold-up would have a significant effect on the equilibrium terms of licensing transactions.

Other costs that have been sunk – and are undeniably relationship specific – are all other costs of participating in the standard-setting decision. Indeed, if there were no such costs, then hold-up of licensees would not be an issue at all as any excessive royalty demand would lead to the (costless) redesign of the standard.

On the side of the potential licensee, the level of relationship-specific sunk investment will depend on the industry. In fast-moving industries, where first mover advantages are crucial, one should not expect product developers to wait until a standard has been finalised before starting to work on their products. Indeed, in these industries, standard-setting efforts are often a response to new needs that become apparent because potential users are working on new products. In this case, a substantial proportion of product development costs might well qualify as relevant specific sunk costs in the post-standard bilateral negotiations, exposing the users to hold-up on the part of SEP owners.

It is remarkable that, given the undeniable existence of a potential hold-up issue on both sides of the SEP holder–SEP user relationship, competition authorities have focussed their attention uniquely on the plight of the licensee. One incarnation of this concern is the series of ongoing cases regarding the use of 'injunctions' by SEP holders. The case against widely allowing SEP owners to use injunctions against potential licensees is based on the following steps. First, SEP holders benefit from the 'unearned' increased market power that having their technology included into a standard confers. There is no denying this claim as it is at the core of the hold-up issue: *ex post* market power is higher than *ex ante* market power. To avoid this issue, SSO participants usually commit to licensing on FRAND terms. FRAND terms are then usually seen as the terms that would have prevailed if negotiations had occurred *ex ante* rather than *ex post*. The Competition Authorities' fear is then that injunctions – which undeniably increase the relative *ex post* bargaining power of SEP owners compared with potential licenses – can only be used to extract terms that are 'not FRAND'. Furthermore, it is often pointed out that the problem of hold-up by SEP owners is especially worrisome because it arises from the coordination of efforts between technological rivals to design a standard.

Although broadly accepted, this approach has serious flaws. First, in the absence of any objective measure of 'FRAND' terms or even any attempt to determine what *ex ante* terms would have been, FRAND is simply a red herring. From an economic point of view, we find ourselves in a situation of bilateral hold-up. Given this, the question of whether allowing for broad injunctions is or is not desirable simply hinges on the relative strength of the two hold-up issues. If the hold-up of licensees is more severe than the hold-up of SEP owners, then injunctions are not desirable because they further increase the bargaining power of SEP owners. On the contrary, if the hold-up of patent holders is more severe than the hold-up of users, then allowing for injunctions might in fact be a way of reducing the adverse consequences of the bilateral hold-up problem. It is therefore unfortunate that the whole debate on injunction has proceeded without any evidence as to the strength of the hold-up issues on either side of the licensing relationship. It is even more unfortunate that we are not aware of any systematic evidence on this issue. Of course, the correct policy on injunctions also depends on how much of a change in bargaining power is produced by allowing the SEP owner to use injunctions fairly readily. Injunctions are a rather blunt tool that can shift the balance of power in litigation quite significantly. Even if, on balance, the hold-up issue experienced by SEP holders were more serious than the hold-up issue faced by potential licensee, it might be that a relatively free use of injunctions would switch bargaining power too drastically in favour of the SEP owners. This possibility seems most realistic in industries where the market moves fast: by being absent from the market even for a relatively short period of time, a licensee under an injunction could lose not only large volumes of sales (think Christmas season) but could also be left behind in the race to establish its product. By contrast, the absence of an injunction might not have such a quick and destructive impact on SEP holders, at least if they can be assured that an infringing standard users will eventually be made to pay appropriate royalties.

The traditional argument against injunctions is also flawed in a more subtle manner. It just is not correct to claim that the hold-up by SEP owners is worse (or more worthy of antitrust interest) than the hold-up by licensees 'because the power to hold-up was obtained through concerted action'. To assess the competition law aspects of standards achieved through SSOs, the correct counterfactual cannot be a world without a standard. It must be a world in which a standard would have been achieved without coordination between rivals. This could occur through de facto standardisation. Alternatively, for argument's sake, one can also think of a counterfactual in which a standard would be obtained by randomly choosing a (compatible) technology for each of the standard's components. In either case, the resulting hold-up issue is just as severe as if the standard had been chosen 'jointly by rival'. It is therefore erroneous to link the issue of hold-up, and hence the issue of FRAND, to the cooperative nature of the standard-setting process. Of course, this cooperative nature is relevant for other issues such as the incentives to exclude rival technologies from the design or the incentives to foreclose alternative standards, but it is simply irrelevant to the main issue of hold-up and FRAND commitments.

Although there is no systematic evidence regarding the relative extent of hold-up on the two sides of the licensing relationship and we also do not have any evidence as to the overall effect of this two-sided hold-up problem, there is some related empirical work.

Simcoe and Graham (2009) find that SEPs have higher litigation rates than comparable non-SEP patents. This is especially true for SEPs held by small firms. The authors interpret this pattern as resulting from the fact that small firms are less likely to be 'repeat players' in the standard-setting process so that they face less of a reputational cost from 'violating' FRAND commitments and trying to use their enhanced market power to extract high royalties. Alternatively, they suggest that litigation might be more frequent with small firms because their limited portfolio makes traditional 'cross-licensing' settlements harder to achieve. Taken at face value, these conclusions do imply that there is indeed a significant problem of hold-up by SEP holders. If there was not, then the first explanation proposed by the authors would not make sense. Although this is the implication of one of the interpretations proposed by the authors, it is not clear that this conclusion necessarily follows from the results obtained by the authors. There are two issues. First, litigation rates are computed as litigation per patent. However, SEPs are likely to be in higher demand than non-SEPs with the same characteristics – precisely because they are part of a standard. This means that, as SEPs will be involved in many more licensing negotiations, there are also more opportunities for disagreement and hence litigation. Ideally, then, the litigation rates should be computed with respect to the total number of bilateral licensing negotiations. As information on such negotiations is not generally available, this is cannot be done, but then it is not clear what the higher rate of litigation of SEPs actually indicates.

Turning to the differential behaviour of small SEP owners, there are other explanations for their behaviour than the fact that they might feel less constraint not to exploit their 'hold-up' power. For example, small SEP holders might be more vulnerable to litigation than larger firms with more experience in defending their IP rights.

Another type of evidence is provided by Hussinger and Schwiebacher (2012, 2013). In these papers, the authors examine the benefits to firms from disclosing their IP to SSOs. Using data for the 1986 to 2005 period, they find that disclosure of patents to standard-setting organisations are correlated with firms' valuations. Thus they conclude that the loss from FRAND commitment (and hence the inability to license exclusively) is outweighed by product market benefits from a standardised technology. Although interesting, this result lends itself to many possible interpretations. The first one is indeed that the commitment made to have one's patented technology included in a standard are adhered to and that such commitments are worth incurring in light of the potential benefits from reading on the chosen standard. However, one could also see this as evidence that, despite the commitment made, SEP owners are able to extract higher royalties from potential licensees. In this sense, this could be seen as indirect evidence that the hold-up by SEP owners might be more severe than the hold-up by potential licensees. Yet another possibility is that the effect on firm's valuation has actually nothing to do with the specific licensing process of SEPs and hence with hold-up. Inclusion in a standard might just be seen as external validation/certification of the quality of the firm's R&D efforts.

III. Innovation

Competition policy also worries about the impact of firms' behaviour – and hence the impact of SSOs – on innovation. In what follows, we take standardisation as a fait accompli, that is, we do not ask whether standardisation itself leads to more or less innovation than a world without standards. Rather we ask whether the way that SSO operate is likely to be harmful to the innovation incentives of both patent holders and potential standard users. In particular, what is the likely effect of royalty stacking, hold-ups and other features associated with SSO on the overall rate of innovation?

In terms of theory, standard-setting organisations can affect innovation in a large variety of ways. For example, SSOs that favour the 'political strength' of firms with large portfolios at the expense of simply choosing the best combination of available technologies would clearly encourage firms to pursue 'numbers' instead of quality. In this sense, any empirical evidence that suggests strong 'political economy' bias within SSOs should also alert us to the possibility of a corresponding dynamic inefficiency.

Another, more subtle mechanism, is that dominant SSOs turn R&D investment into 'winner-takes-all' races because, especially for standard-oriented patents such as telecom patents, only patents that are retained within the standard are likely to produce significant income streams. As is well known, such winner-takes-all environments encourage excessive duplication of effort and might also discourage the pursuit of an optimal variety of research routes.

Most obviously, of course, SSOs might affect the balance between innovation in the market for standard-related innovations and the market for innovations related to the downstream products that apply the standard. The greater the hold-up by patent holders compared with the hold-up by licensees, the more SSOs would change R&D incentives in favour of standard-related technologies. The same would be true if SEPs turned out to involve a truly higher rate of litigation than other similar patents.

Finally, SSOs can have a more direct effect on innovation if they provide an environment that promotes knowledge sharing. Such sharing has the traditional opposite effects of potentially decreasing *ex ante* innovation while increasing the social value of any given amount of innovation. In this respect, Gandal, Gantman and Genesove's (2006) results suggest the possibility of knowledge diffusion or spill-overs at standardisation meetings, which would have dynamic benefits. The meeting data examined by the authors for the TIA TR-30 committee (which was responsible for setting analogue standards in data transmission systems and equipment) showed that the committee and its subcommittees met regularly, with approximately five to six meetings per year. Further, many of the participants attended consistently over a period of time. Hence, the committees created an environment for knowledge spill-overs, similar to those at academic conferences.

There is little direct empirical evidence on the non-spill-over channels through which SSOs can affect innovation. There are, however, a few useful papers on the link between patent pools and innovations. Because SSOs have some common features with patent pools and, moreover, patent pools can be used to minimise the royalty-stacking issue to which SSOs might give rise, it is interesting to briefly review this small literature. Two papers by Lampe and Moser (2010, 2012a) are of particular interest.

Lampe and Moser (2010) look at the effect of the nineteenth-century Sewing Machine Patent Pool. Using differences in differences, they first examine the effect of the formation of the pool on the patenting behaviour of both participants and firms that remained outside the pool. They find a significantly negative effect of the pool on the number of patents obtained by both members and non-members. It is not just that patenting decreased when the pool was formed. It also picked up as soon as the pool was dissolved. Of course there are many reasons why patenting activity might not adequately reflect innovation efforts. Note, however, that, for pool members at least, the traditional hypothesis is that pool participation leads to more patenting for strategic reasons. This is because pools often split the spoils based – partially at least – on the number of patents held by each member. If the strategic patenting incentive truly is to patent more once one joins a pool, then the Lampe and Moser result is only strengthened: the fact that less patenting was observed is even more likely to indicate that innovation actually slowed down. Nevertheless, the authors also construct a measure of 'real' innovation that is independent of patent counts: the evolution of the number of stitches per minute that the machines were able to perform. The analysis of this indicator confirms their patent-based findings; the formation slowed down the pace of progress, which picked up again once the pool was dissolved. Although the authors do not explain why innovation by pool members has decreased, they trace the decrease in the innovation of non-members to the increased litigation following the formation of the pool (pool members also pool resources for litigation).

In a similar vein, Lampe and Moser (2012a) examine patent pools in twenty industries over the period of the 'New Deal' when antitrust enforcement was deliberately weak. Using a difference in difference approach, they find that on average the formation of a pool decreased patenting by 16 percent.

Overall then, there is rather little empirical evidence about the effect of SSOs on innovation. There seems to be reasons to believe that SSOs encourage knowledge diffusion. On the other hand, there is fairly strong evidence that (unregulated) patent pools actually decrease the rate of innovation. However, although this points to a cost of using pools to address the royalty-stacking issue stemming from SSOs, it is not clear that the patent pool results can be directly translated to the SSO themselves. In particular, SSOs do not get involved in litigation and, absent an accompanying pool, SEP owners enforce their patents separately. The 'litigation' mechanism suggested by Lampe and Moser cannot therefore apply to SSOs themselves.

G. Internal SSO rules

So far we have reviewed the – relatively scarce – empirical literature that sheds some light on various aspects of SSOs performance. However, one could argue that such assessments do not help us answer the relevant question. If standardisation is desirable and SSOs often are the only realistic route to a standard, then whether SSOs perform well on average is largely irrelevant: whatever their drawbacks are, this is the cost that society has to bear to reap the benefits of standardisation. In this view, what should be of interest is how differences in the internal rules of SSOs and in public policy toward SSOs affect the performance of these organisations. We have already discussed in Section F how competition policy might be used to minimise the anticompetitive effects of SSOs. In this section, we turn to the SSOs' own internal rules.

As we saw in Section B, SSOs are political/economic institutions in which influence within the SSO might matter as much as technical merit when it comes to have one's patents included in the agreed-on standard.

A natural target for improving the performance of SSOs is therefore the rules that determine how power within the SSO is divided. In this spirit, Farrell and Simcoe (2012) theoretically examine which type of consensus rule is optimal for SSOs. They note that firms often have vested interests in promoting their own technology. This makes 'absolute' consensus difficult to achieve and can significantly slow down the standardisation process. Hence, they examine alternative (weaker) notions of consensus. They find that allowing neutral players to arbitrate and break deadlocks can improve on outcomes.

Another crucial determinant of SSO performance is participation. It is only if participation is broad and includes the firms with the best technology that there is any hope the standard will be as good as technically (and economically) possible. In this respect, Layne-Farrar and Lerner (2011) find that pools that adopt proportional sharing rules based on the number of essential patents in the standard are less likely to attract potential participants. In particular, firms with high value patent portfolios are less likely to join pools with numeric proportionality.

Chiao, Lerner and Tirole (2007) empirically examine the rules of SSOs. They find that a provision requiring royalty-free licensing is negatively correlated with a provision requiring disclosure of intellectual property. This is as expected: if licensing does not involve payment, then there is no room for either patent ambush or hold-up. In a similar flavour, they find that RAND licensing requirements are positively correlated with a provision requiring such disclosure. Unfortunately, this type of analysis does not tell us much about how the actual performance of SSOs might depend on the type of internal rules that they adopt. A huge empirical effort in this direction is needed if we are to further our empirical understanding of SSOs.

H. Conclusion

The analysis in Sections E, F, and G shows that although a recent flurry of empirical work on SSOs has been conducted in the past decade or so, we still lack empirical evidence on key issues. Our chapter has identified holes in the empirical literature and suggests what might be fruitful topics for future research.

SSOs raise two broad types of policy issues. The first relates to the actual performance of SSOs: do they select high-quality standards in a timely manner, make them available at reasonable terms and update them efficiently?

Direct evidence on technology choice suggests that SSOs do indeed rely on 'good-quality' innovations. Unfortunately, such direct evidence is scarce and must therefore be completed by studies of the internal politics of the SSO, under the hypothesis that the best technology is more likely to be chosen if all interests are adequately represented and can actually influence the selection process. Unfortunately, available studies often fail to document the nature of the internal SSO debates (i.e. who supported what) and are therefore of limited use. The 'internal politics' of SSOs also proves determinant in terms of the speed of the proceedings: the more participants care about obtaining a greater 'share' of the surplus created by standardisation, the slower the process appears to be.

The issue of accessibility can be broken into two aspects: the definition of FRAND terms and the problem arising from the 'stacking' of royalties by independent intellectual property owners. Given that the very concept of FRAND is still mostly undefined, there is no significant empirical work on the first aspects. By contrast we now have a large literature that documents the existence of 'patent thickets' and show that they affect the behaviour of industry participants. However, we are not aware of a single reliable empirical studies that would give us an idea of 'how big' the problems arising from patent thickets and royalty stacking are. It would therefore seem prudent to reign in the debate about how royalty stacking should be handles until we have some evidence that there actually is a significant problem to be dealt with.

The updating of standards is a further area in which new empirical work is urgently needed. Currently, we have little information about the procedures that SSOs follow when introducing new technology into an existing standard. We essentially do not know anything about how different rules affect the effectiveness of such updates.

The second set of policy concerns relate to the effect of SSOs on the relevant product markets as well as on the markets for technologies and standards. The main issues are exclusion, hold-up and innovation. Recent empirical work strongly suggests that – unchecked – SSOs can actually be used as an exclusionary tool. Competition law scrutiny in this respect is therefore fully warranted.[8] Although the situations of hold-up that can arise from the standardisation process are well understood,

[8] See also Larouche and van Overwalle's contribution in this volume.

competition policy has mostly focused on the potential hold-up of standard *users*, essentially ignoring the potential hold-up of standard *contributors* who have already sunk their R&D investments. No meaningful policy debate on hold-up, FRAND and injunctions can be had without empirical evidence on the relative importance of these two types of hold-up, complemented by evidence on the effects of the two types of hold-up (e.g. how good are private parties at finding solutions to each type of hold-up?). Unfortunately, no such evidence is currently available. Finally, in Europe at least, competition policy is also supposed to foster innovation. Although empirical evidence suggests that SSOs might increase innovation by increasing the cross-fertilisation of knowledge, there are also a number of recent empirical studies that suggest that patent pools have a net effect on the rate of innovation. Given the similarities between SSOs and patent pools, this should sound a cautionary note. Further empirical work on the link between SSOs and innovation is clearly needed.

References

Baron, J., K. Blind and T. Pohlman, 2011, 'Essential patents and standard dynamics', presented at the 7th International Standardization and Innovation in Information Technology (SIIT) Conference.

Baron, J., and T. Pohlman, 2013, 'Who cooperates in standards consortia – rivals or complementors?', *Mimeo, Northwestern University Law School.*

Bekkers, R., R. Bongard and A. Nuvolari, 2011, 'An empirical study of the determinants of essential patent claims in compatibility standards', *Research Policy*, 40(7): 1001–1015.

Berger, F., K. Blind and N. Thumm, 2012, 'Filing behaviour regarding essential patents in industry standards', *Research Policy*, 41(1): 216–225.

Blind, K., and N. Thumm, 2004, 'Interrelation between patenting and standardization strategies: Empirical evidence and policy implications', *Research Policy*, 33(10): 1583–1598.

Brenner, S. 2009, 'An optimal mechanism of patent pool formation', *Economic Theory*, 40(3): 373–388.

Chiao, B., J. Lerner and J. Tirole, 2007, 'The rules of standard-setting organisations: An empirical analysis', *Rand Journal of Economics*, 38(4): 905–930.

Clements, M., and H. Ohashi, 2005, 'Indirect network effects and the product cycle: Video games in the US, 1994–2002', *The Journal of Industrial Economics*, 53(4): 515–542.

Cournot, A., 1838, *Researches on the Mathematical Principles of the Theory of Wealth.*

De Lacey, B. J., K. Herman, D. Kiron and J. Lerner, 2006, 'Strategic behaviour in standard-setting organizations', Mimeo, University of Texas.

Dranove, D., and N. Gandal (with David Dranove), 2003, 'The DVD vs. DIVX Standard War: Empirical Evidence of Network Effects and Preannouncement Effects', *Journal of Economics and Management Strategy*, 12: 363–386.

European Commission. *Horizontal Guidelines.* Available at: http://ec.europa.eu/competition/antitrust/legislation/horizontal.html

Farrell, J., and G. Saloner, 1988, 'Coordination through committees and markets', *Rand Journal of Economics*, 19(2): 235–252.

Farrell, J., and T. Simcoe, 2012, 'Choosing the rules for consensus standardisation', *Rand Journal of Economics*, 43(2): 235–252.

Gandal, N., 1994, 'Hedonic price indexes for spreadsheets and an empirical test for network externalities', *The RAND Journal of Economics*, 25: 160–170.

Gandal, N., 1995, 'Compatibility standards and complementary network externalities in the PC software market', *Review of Economics and Statistics*, 77: 599–608.

Gandal, N., N. Gantman and D. Genesove, 2006, 'Intellectual property and standardization committee participation in the U.S. modem industry', in S. Greenstein and V. Stango eds, *Standards and Public Policy*, Cambridge University Press, pp. 208–230.

Gandal, N., M. Kende and R. Rob, 2000, 'The dynamics of technological adoption in hardware/software systems: The case of compact disc players', *RAND Journal of Economics*, 31: 43–61.

Gerardin, D., A. Layne-Farrar and J. Padilla, 2008, 'The Complements Problem with Standard Setting: Assessing the Evidence on Royalty Stacking', *Boston University Journal of Science and Technology Law*, 14(2): 144.

Gerardin, D., and M. Rato, 2007, 'Can standard setting lead to exploitative abuse: A dissonant view on patent hold-up, royalty stacking and the meaning of FRAND', *European Competition Journal* 101. Available at SSRN: http://ssrn.com/abstract=946792 or http://dx.doi.org/10.2139/ssrn.946792

Von Graevenitz, G., S. Wagner and D. Harhoff, 2013, 'Incidence and growth of patent thickets: The impact of technological opportunities and complexity', *Journal of Industrial Economics*, LXI(3): 521–563.

Greenstein, S., 1993, 'Did installed base give an incumbent any (measurable) advantages in federal computer procurement?' *RAND Journal of Economics*, 24: 19–39.

Hussinger, K., and F. Schwiebacher, 2012, 'The market value of standard-setting activities for technology providers', Paper presented at the DRUID Society 2012 meeting. Available at: http://druid8.sit.aau.dk/acc_papers/u7gikmsxtyrih83fa3q0yyjr1fi8.pdf

Hussinger, K., and F. Schwiebacher, 2013, 'The value of disclosing IPR to open standard-setting organizations', ZEW Discussion Paper 13–060.

Köhler, F., K. Blind and N. Thumm, 2013, 'An analysis of the filing behaviour of patents in standards', Institut d'Economie Industrielle (IDEI). Available at: http://idei.fr/doc/conf/sic/seppapers2013/thummweb.pdf

Lampe, R., and P. Moser, 2010, 'Do patent pools encourage innovation? Evidence from the 19th century sewing machine industry', *Journal of Economic History*, 70(4): 898–920.

Lampe, R., and P. Moser, 2012a, 'Do patent pools encourage innovation? Evidence from 20 US industries under the New Deal'. NBER Working Paper 18316.

Lampe, R., and P. Moser, 2012b, 'Patent pools: Licensing strategies in the absence of regulation', in S.J. Jalal, B.S. Silverman and M.A. Cusumano (eds), *History of Strategy* (Advances in Strategic Management, Vol. 29), Emerald Group Publishing Ltd, pp. 69–86.

Layne-Farrar, A., and J. Lerner, 2011, 'To join or not to join: Examining patent pool participation and rent-sharing rules', *International Journal of Industrial Organization*, 29(2): 294–303.

Lerner, J., and J. Tirole, 2004, 'Efficient Patent Pools', *American Economic Review*, 94(3): 691–711.

Liu, C., C. Kemerer and S. Slaughter, 2012, 'Standards competition in the presence of digital conversion technology: An empirical analysis of the flash memory card market', *MIS Quarterly*, 36(3): 921–942.

Noel, M., and M. Schankerman, 2013, 'Strategic patenting and software innovation', *Journal of Industrial Economics*, LXI(3): 481–520.

Postrel, S., 1990, 'Competing networks and proprietary standards: The case of quadraphonic sounds', *Journal of Industrial Economics*, 39: 169–185.

Prieger, J., and W. Hu, 2006, 'An empirical analysis of indirect network effects in the home video game market'. NET Institute Working Paper No. 06-25. Available at SSRN: http://papers.ssrn.com/sol3/papers.cfm?abstract_id=941223.

Rysman, M., 2004, 'Competition between networks: A study of the market for Yellow Pages', *Review of Economic Studies*, 71: 483–512.

Rysman, M., and T. Simcoe, 2008, 'Patents and the performance of voluntary standard setting organizations', *Management Science*, 54(11): 1920–1934.

Saloner, G., and A. Shepard, 1995, 'Adoption of technologies with network externalities: An empirical examination of the adoption of automated teller machines', *RAND Journal of Economics*, 26: 479–501.

Schmalensee, R., 2009, 'Standard-setting, innovation specialist and competition policy', *Journal of Industrial Economics*, 57(3): 526–552.

Simcoe, T. S., and S. J. H. Graham, 2009, 'Competing on standards: Entrepreneurship, intellectual property and platform technologies', *Journal of Economics and Management Strategy*, 18(3): 775–816.

Simcoe, T. S., 2012, 'Consensus governance for shared technology platforms', *American Economic Review*, 101(1): 305–336.

Weiss, M. B. H., and M. Sirbu, 1990, 'Technology choice in voluntary standards committee: An empirical analysis', *Innovation and New Technology*, 1(1–2): 111–133.

Williamson, O., 1983, 'Credible commitments: Using hostages to support exchange', *American Economic Review*, 73(4): 519–540.

Ziedonis, R., 2004, 'Don't fence me in: Fragmented markets for technology and the patent acquisition strategies of firms', *Management Science*, 50(6): 804–820.

17

Technical standards and the draft General Data Protection Regulation

ELENI KOSTA AND KEES STUURMAN

A. Introduction

Privacy issues relate to a wide range of aspects of technical operations, including availability and integrity of services and data, confidentiality, data classification mechanisms, vendor lock-in, certification, audit and testing, to name a few examples. The application of the European legal framework on the protection of personal data in online environments, in particular the Data Protection Directive (1995/46/EC), is a highly debated issue, not only in relation to the trans-border transfers of data originating from a European Union (EU) Member State but also to the practical application of the principles contained in the Directive in data processing operations. The current statutory framework for data processing includes a number of main principles that are directly linked to the architecture of systems and applications without any explicit reference to technical standards. However, the Directive is currently under review, and the revised European data protection framework seems to have a stronger focus on the role and significance of technical standards. The European Commission proposed the replacement of the Directive with a Regulation, which not only contains updates to existing principles and provisions of the Directive but also introduces novelties with regard to the processing of personal data. One such novelty is the explicit reference to the importance of technical standardisation initiatives in relation to data protection. De Hert and Papakonstantinou commented that 'despite the fact that [technical standards] focus on the effectiveness of processes rather than an adequate level of (human rights) protection, [they] are of relevance to

The authors thank Dr John Borking for his valuable comments. Any mistakes remain the authors' alone.

the international data privacy field'.[1] The protection of personal data has traditionally been seen as part of the protection of information in computer systems. However, the proposed General Data Protection Regulation (GDPR) puts special emphasis on the need for the development of specific standards for a number of data protection issues. The enactment of the GDPR may for instance introduce a 'right to data portability' (i.e. the right to transfer data from one electronic processing system to and into another, without being prevented from doing so by the controller). As a precondition and to further improve access of individuals to their personal data, the GDPR provides data subjects with the right to obtain from the controller a copy of those data in a commonly used electronic and structured format and establishes the data protection by design and by default principle. Such novelties will put even more emphasis on the key importance of technical standards in relation to privacy compliance. Therefore, the focus of this chapter is on the role of technical standards on data processing operations in view of the ongoing review of the European data protection framework.

In particular, this chapter explores the complex relation between data protection and technical standards from a legal perspective. It first describes the relationship between technical standards and the European data protection legal framework. The chapter also reflects on the development of privacy standards in complex technological environments, such as cloud computing and radio-frequency identification (RFID) applications. It finally examines the role of technical standards in the GDPR and analyses the specific provisions of the GDPR that relate to standardisation.

B. The (complex) relation between data protection regulation and technical standards

I. The interplay between standards and legislation

Traditionally, EU rules have not always facilitated the use of standards developed by (often non-European) fora and consortia. Various EU policy initiatives are facilitating the necessary change. The so-called New Approach, which was introduced in the field of product legislation in 1985, and European Standardisation has led to the adoption of a large number of European Directives and Regulations that regulate issues relating to the adoption of standards in various areas that are of interest

[1] P De Hert and V Papakonstantinou, 'Three scenarios for international governance of data privacy: Towards an international data privacy organization, preferably a UN agency' (2013) 9(2) *Journal of Law and Policy*, 271, at 295.

for the European internal market, ranging from ecodesign and energy labelling, to packaging waste, toys safety and RFIDs.[2]

The Technical Standards and Regulations Directive 98/34[3] defined a standard in its Article 1(6) as

> a technical specification[4] approved by a recognised standardisation body for repeated or continuous application, with which compliance is not compulsory and which is one of the following:
>
> - international standard: a standard adopted by an international standardisation organisation and made available to the public;
> - European standard: a standard adopted by a European standardisation body and made available to the public; and
> - national standard: a standard adopted by a national standardisation body and made available to the public.

The European Regulation 1025/2012 on European standardisation[5] defines a standard in a way closer to the functioning of standards in practice[6] as

[2] A list of most of the European Directive and Regulations can be found at: http://www.newapproach.org/Directives/DirectiveList.asp. For a more thorough understanding of the new approach to European standardization, see European Commission, Guide to the implementation of directives based on the New Approach and the Global Approach, 2000, available online at http://www.gnb-cpd.eu/filelibrary/New_Approach_Guide_-_99_Edition_EN_(Final_Version).pdf, accessed on 22 July 2014.

[3] Directive 98/34 of the European Parliament and of the Council of 22 June 1998 laying down a procedure for the provision of information in the field of technical standards and regulations and of rules on Information Society services [1998] OJ L 204, p. 37, as amended.

[4] A technical specification is defined in Article 1(3) of Directive 98/34 as 'a specification contained in a document which lays down the characteristics required of a product such as levels of quality, performance, safety or dimensions, including the requirements applicable to the product as regards the name under which the product is sold, terminology, symbols, testing and test methods, packaging, marking or labelling and conformity assessment procedures.

The term 'technical specification' also covers production methods and processes used in respect of agricultural products as referred to Article 38(1) of the Treaty, products intended for human and animal consumption, and medicinal products as defined in Article 1 of Directive 65/65/EEC (2), as well as production methods and processes relating to other products, where these have an effect on their characteristics'.

[5] Regulation (EU) No 1025/2012 of the European Parliament and of the Council of 25 October 2012 on European standardisation, amending Council Directives 89/686/EEC and 93/15/EEC and Directives 94/9/EC, 94/25/EC, 95/16/EC, 97/23/EC, 98/34/EC, 2004/22/EC, 2007/23/EC, 2009/23/EC and 2009/105/EC of the European Parliament and of the Council and repealing Council Decision 87/95/EEC and Decision No 1673/2006/EC of the European Parliament and of the Council [2012] OJ L 316, p. 12.

[6] Paul de Hert, Vagelis Papakonstantinou, Irene Kamara, The new cloud computing ISO/IEC 27018 standard through the lens of the EU legislation on data protection, Brussels

a technical specification, adopted by a recognised standardisation body, for repeated or continuous application, with which compliance is not compulsory, and which is one of the following: (a) 'international standard' means a standard adopted by an international standardisation body; (b) 'European standard' means a standard adopted by a European standardisation organisation; (c) 'harmonised standard' means a European standard adopted on the basis of a request made by the Commission for the application of Union harmonisation legislation; (d) 'national standard' means a standard adopted by a national standardisation body (Art. 2(1) Regulation 1025/2012).

The adoption of this regulation will facilitate the necessary changes, especially in the context of public procurement because it aims at the establishment of 'rules with regard to the cooperation between European standardisation organisations, national standardisation bodies, Member States and the Commission, the establishment of European standards and European standardisation deliverables for products and for services in support of Union legislation and policies, the identification of ICT technical specifications eligible for referencing, the financing of European standardisation and stakeholder participation in European standardisation'.[7] The Regulation allows contracting authorities to formulate the technical specifications in procurement documents by reference to information and communications technology (ICT) specifications produced by at least certain fora and consortia. The relevant bodies will be identified by the Commission after consultation of the European Multi-Stakeholder Platform on ICT standardisation.[8] Specific demands are put on (i) the nature of the specifications involved (specifically in relation to European standards/specifications), (ii) the source of the specification (requirements regarding the standard setting body) and (iii) the principles that should have been complied with in the standard-setting procedure. In formulating these policy objectives, there has been a strong focus on the need and benefits of achieving interoperability. It remains to be seen to what extent this development will be relevant for the use of standards in regulating data protection.

Privacy Hub Working Paper Vol. 1 No 2, November 2014, p. 6, available at http://ssrn.com/abstract=2542125.

[7] Article 1, Regulation (EU) 1095/2012.

[8] Communication from the Commission to the European Parliament, the Council, the European Economic and Social Committee and the Committee of the Regions, Against lock-in: building open ICT systems by making better use of standards in public procurement. COM(2013) 455 final, p. 8.

Moreover, the European Commission published in 2013 a proposal for a directive on network and information security.[9] The development of technical standards for network and information security (NIS) plays a prominent role in the Directive and will undoubtedly influence the standardisation of security requirements.

II. *European data protection and technical standards*

1. Importance of technical standards for the regulation of data protection

Research published in 2001 by the Deutsches Institute für Normung (DIN; German Institute for Standardisation) already aimed at identifying the economic effects of standardisation for businesses and society as a whole, showing not only that standardisation leads to significant cost reduction but also that standardisation reduces the economic risk of R&D activities. The study concluded that standards as a form of technology transfer have a positive influence on innovation (and that a lack of standards creates significant hurdles to innovation) and that relevant standards for society lead to withdrawal of standards and practices that have outlived their technical, scientific or societal significance.[10]

Technical standards are of significant importance for the regulation of data protection, for the wide deployment and scalability of privacy-enhancing technologies in Privacy by Design, in a number of data processing purposes and in a number of steps of data processing operations. Privacy standards would promote built-in privacy protection and would wipe out products and services that lead to privacy intrusion. They would stimulate and open markets for new products and services with built-in privacy protection as well.[11]

First, standards can be used to define certain data protection concepts; the relatively recent difficulties in defining the concept of 'Do Not Track',

[9] Proposal for a Directive of the European Parliament and of the Council concerning measures to ensure a high common level of network and information security across the Union, 07.02.2013, COM/2013/048 final – 2013/0027 (COD).

[10] DIN German Institute for Standardization, 'Economic benefits of standardization, Summary of results, Final report and practical examples' (2000), available at http://www.din.de/sixcms_upload/media/2896/economic_benefits_standardization.pdf, accessed 23 July 2015.

[11] J Borking, 'Privacy Standards for Trust', *Jusletter* 3 Oktober 2005, Bern.

an HTTP mechanism for expressing the user's preference regarding tracking, have clearly illustrated the need for and the importance of standards for the definition of data protection concepts.[12]

Second, the implementation of certain standards can be imposed as regulatory obligation to ensure that specific requirements of the prevailing data protection regulation (e.g. regarding security) are met. Winn refers to specific national legislation imposing regulatory obligations to implement information security management standards: the US Department of Health and Human Services issued in 2003 the Health Insurance Portability and Accountability Act (HIPAA) Security Rule imposing obligations on health care providers to adopt management process standards, and the UK Gambling Commission required in a 2007 document, which was replaced in 2009,[13] remote and software gambling operators to implement security procedures based on the ISO/IEC 270xx standard series, which is an information security risk-management standard.[14]

Third, standards may prevent or protect anonymity of users, like the ISO 15408:1999 Common Criteria Technology Security Evaluation.

Fourth, standards can be used to structure the communication between data controller and data subject; early initiatives on the development on a technical standard that would enable the communication between website operators and their potential visitors about the information that is collected about the visitors and its future use were carried out already since 1997 by the World Wide Web Consortium (W3C) for the development of the Platform for Privacy Preferences (P3P) specifications.[15]

Fifth, standards are of vital importance for auditing purposes related to data protection, and finally standards may be used in public tenders

[12] A-M McDonald and J-M. Peha, 'Track gap: Policy implications of user expectations for the 'Do Not Track' Internet privacy feature', Telecommunications Policy Research Conference, 23–25 September 2011, at 4, available at http://papers.ssrn.com/sol3/papers.cfm?abstract_id=1993133: 'Researchers and a Mozilla engineer submitted a DNT (Do Not Track, C.S.) proposal to the IETF Internet standards body. Microsoft submitted a different proposal to the W3C standards body', accessed 23 July 2015.

[13] UK Gambling Commission, 'Remote gambling and software technical standards' (2009), available at http://www.gamblingcommission.gov.uk/PDF/Remote%20gambling%20and%20software%20technical%20standards%20-%20August%202009.pdf, accessed 23 July 2015.

[14] J-K Winn, 'Technical standards as data protection regulation', in S. Gutwirth et al. (eds), *Reinventing data protection?*, (2009), 191, at 203; Gambling Commission, supra note 13.

[15] Available at http://www.w3.org/P3P/. The W3C Working Group issued on 3 November 2006 a note on The Platform for Privacy references 1.1 (P3P1.1) Specification, giving it a provisionally final state: http://www.w3.org/TR/P3P11/, accessed 23 July 2015.

to ensure that data protection–compliant products and services are being acquired by public authorities.

2. The Data Protection Directive and (the lack of direct reference to) technical standards

The Data Protection Directive does not make any specific reference to technical standards. This does not, however, imply that the legislator does not attribute any importance to standards in the context of the Directive. Especially with regard to the security obligations that are imposed on the data controller (and to an extent to the data processor), the wording of Article 17 and Recital 46 of the Directive on the security of data processing implicitly recognises the importance of standards. More concretely, Recital 46 states: 'the processing of personal data requires that appropriate technical . . . measures be taken, . . . at the time of the design of the processing system . . . , particularly in order to maintain security and thereby to prevent any unauthorised processing'.

Nevertheless, the lack of any specific and clear reference to technical standards in the Data Protection Directive gave rise to a long-lasting debate about the necessity of data protection standards on the one hand and the potential role of the European Commission in the development of such standards on the other. The many initiatives have been and are being undertaken for the development of standards, as illustrated in the industry driven ISTPA (International Security, Trust and Privacy Alliance) framework project, which demonstrated that in the absence of standards, vendors are actually proceeding to develop their own version of standards, mostly at a lower data protection level than provided for in the Data Protection Directive.[16]

C. Standards and the draft GDPR

I. Introduction to the draft GDPR

In January 2012, the European Commission presented its proposals for the reform of the European data protection legal framework (hereafter 'Commission Proposal'), proposing the replacement of the Data Protection

[16] ISTPA, 'The ISTPA Privacy Framework' (2001), available at http://emoglen.law.columbia.edu/LIS/archive/privacy-legis/ISTPA-FrameworkWhitePaper013101.pdf, accessed 23 July 2015. The website www.istpa.org was not available.

Directive with a General Data Protection Regulation (GDPR),[17] which was the outcome of consultations and debates lasting three intense years.[18] The proposed GDPR puts special emphasis on the need for the development of specific standards for a number of data protection issues. Following the 2012 Commission Proposal, the European Parliament (hereafter 'EP') – after more than two years of discussions and lobbying[19] – adopted in plenary a legislative resolution on 12 March 2014 (hereafter 'EP first reading').[20] More than a year later, the Council of the European Union (hereafter 'Council') agreed on its first reading position on the GDPR on 15 June 2015 (hereafter 'Council's first reading').[21] The discussions in this section are mainly based on the aforementioned documents produced by the Commission, the EP and the Council.

The draft GDPR identifies the strengthening of individual control of data subjects over their personal data as a cornerstone of the data protection reform, while emphasising the enforcement of the substantive data protection principles, transparency and accountability. The Commission Proposal contains explicit references to the use of technical standards in relation to certification, the principle of data protection by design and by default, data security and data portability. The EP first reading made

[17] European Commission, Proposal for a Regulation of the European Parliament and of the Council on the protection of individuals with regard to the processing of personal data and on the free movement of such data (General Data Protection Regulation) COM (2012) 11 final – 2012/0011 (COD), 25.01.2012, available at http://ec.europa.eu/justice/data-protection/document/review2012/com_2012_11_en.pdf, accessed 23 July 2015.

[18] The legal framework on data protection proposed by the European Commission, consists on the aforementioned draft Data Protection Regulation, as well as a proposal for a Directive on data protection in relation to police authorities and criminal justice, which repealed the Framework Decision on data protection in the third pillar (Council of the European Union, Framework Decision 2008/977/JHA of 27 November 2008 on the protection of personal data processed in the framework of police and judicial cooperation in criminal matters [2008] OJ L350/60).

[19] A number of Opinions were published by various Committees of the European Parliament, such as the Legal Affairs Committee, the Employment and Social Affairs Committee, the Industry, Research and Energy Committee, the Internal Market and Consumer Protection Committee, which tabled almost 4.000 amendments. On 21 October 2013, the Committee on Civil Liberties, Justice and Home Affairs ('LIBE') of the European Parliament adopted amendments to the Commission's proposal.

[20] First reading of the European Parliament, available at http://www.europarl.europa.eu/oeil/popups/ficheprocedure.do?lang=en&reference=2012/0011%28COD%29, accessed 23 July 2015.

[21] European Council, preparation of a general approach, 9565/15, 11.6.2015 (Full Text adopted at JHA Council Meeting on 15.6.2015), available at http://data.consilium.europa.eu/doc/document/ST-9565-2015-INIT/en/pdf

further references to the use of standards with regard to the right to object, the processing of data concerning health and the processing of personal data for historical, statistical and scientific research purposes. The Council first reading, shows that the Council wishes to further differentiate from positions expressed by the European Commission and the Parliament on specific issues relating to technical standards, as illustrated later.

II. Certification

The Commission Proposal introduced the use of data protection certification mechanisms, seals and marks that would allow data subjects to assess the level of protection for the processing of personal data that is provided by data controllers and data processors (Art. 39 Commission Proposal). The establishment of such mechanisms could contribute to the demonstration of compliance of controllers and processors and function as accountability mechanisms. The European Commission would have the possibility to lay down technical standards for data protection certification mechanisms, seals and marks (Art. 39(3) Commission Proposal). The Commission Proposal specified in Article 87 the Committee procedure that is necessary for conferring implementing powers on the Commission, such as the one described in Article 39(3) for the laying down of technical standards for certification, seals and marks.[22] The Committee established via Regulation 182/2011[23] shall assist the Commission with the exercise of implementing powers.[24] More specifically, the examination procedure, which is specified in Article 5 of Regulation 182/2011, will be followed.[25]

The EP first reading deleted the possibility of the Commission to lay down technical standards for data protection certification mechanisms, seals and marks. It rather entrusted the European Data Protection Board[26] with the power to certify on its own initiative that 'a data protection-enhancing technical standard is compliant with [the GDPR]' (Art. 39(1i)

22 See also Recital 130 Commission Proposal.

23 Regulation (EU) No 182/2011 of the European Parliament and of the Council of 16 February 2011 laying down the rules and general principles concerning mechanisms for control by Member States of the Commission's exercise of implementing powers, OJ L55/13, 28.02.2011.

24 Art. 87(1) Commission Proposal.

25 Art. 39(3) and Rec. 131 Commission Proposal.

26 The draft GDPR proposes the replacement of the Article 29 Working Party with the European Data Protection Board. The latter is established though Article 64 of the draft GDPR and 'shall be composed of the head of one supervisory authority of each Member State and of the European Data Protection Supervisor' (Art. 64(1) Commission Proposal).

EP first reading). The EP amendment did not specify any kind of involvement of the European Commission in the development and implementation of technical standards with regard to data protection certification mechanisms, seals and marks, facilitating in this way the development of industry-driven standards from any standardisation body. The EP also introduced a standardised data protection mark, the so-called European Data Protection Seal (Art. 39(1e) EP first reading). The European Data Protection Seal would be granted by the Data Protection Authorities to data controllers and data processors, who are audited and have been certified for processing personal data in compliance with the GDPR (Art. 39(1e) EP first reading). The establishment of standardised marks should be encouraged (Rec. 77 EP first reading) because they would allow data subjects to 'quickly, reliably and verifiably assess the level of data protection of relevant products and services' (Rec. 77 EP first reading). The Council in its first reading upheld the authority of the Commission to lay down technical standards for data protection certification mechanisms, seals and marks via implementing acts (Art. 39a(8), Council's first reading). The Council also abolished the European Data Protection Seal that was proposed by the EP and broadened the scope of data protection certification mechanisms, seals and marks that would demonstrate compliance with the GDPR of processing operations carried our by controllers and processors (Art. 39(1), Council's first reading). Moreover the Council introduced more detailed provisions on the competent certification bodies and the procedures to be followed for certification and for the accreditation of the certification bodies (Arts. 39 and 39a, Council's first reading).

III. The principles of data protection by design and by default

Article 23 of the Commission Proposal introduced the principles of data protection by design and by default. The data protection by design principle introduces an obligation for data controllers to 'implement appropriate technical and organisational measures and procedures' that will ensure that the processing of personal data will meet the requirements of the GDPR and will safeguard the rights of the data subjects. Such measures and procedures shall be implemented both at the design phase as well as during the carrying out of the processing, and they shall take into account the state of the art and the cost of implementation (Art. 23(1) Commission Proposal). Note should be made to the fact that the GDPR does not make use of the word 'design' in the provisions setting out the substantive requirements of the principle.[27] The EP extended the obligations borne under the data protection by design principle also to data processors, it removed any reference to the cost of implementation and added as relevant factors that the controllers and the processors need to take into account for the implementation of the principle the current technical knowledge, international best practices and the risks represented by the data processing. The EP clarified that data protection by design shall apply to 'the entire lifecycle management of personal data from collection to processing to deletion' (Art. 23(5) EP first reading), linking in this way the data protection by design principle to *the risks represented by the data processing* and subsequently to the security requirements.[28]

The principle of data protection by default introduces an obligation for data controllers to implement, by default, mechanisms that will ensure that only the necessary data for each specific purpose are processed and that the principles of data minimisation and data conservation are respected both during the collection and storage of the data (Art. 23(2) Commission Proposal).

Similar to the possibility that was given to the European Commission to lay down technical standards for data protection certification mechanisms, seals and marks, Article 23(4) of the Commission Proposal provides the European Commission with the power to adopt implementing acts to lay down technical standards from the requirements relating to the data protection by design and by default principles. The translation of the data protection by design requirements into practical measures through technical standards might be a challenging issue because the development of too detailed or too specific standards may result in undermining the objective of technology neutrality.[29] However, the EP removed any reference to technical standards with regard to the data protection by design and by default requirements, probably identifying the challenges that are entailed with regard to the adoption of technical standards, especially as an initiative driven by the European Commission. The Council's

[27] K W Hon, E Kosta, Ch Millard, D Stefanatou, 'Cloud accountability: The likely impact of the proposed EU data protection regulation' (2014), available at http://ssrn.com/abstract=2405971, accessed 23 July 2015.

[28] Ibid.

[29] Ibid; EJ Koops and R Leenes, 'Privacy regulation cannot be hardcoded. A critical comment on the 'privacy by design' provision in data protection law' (2013) *International Review of Law, Computers & Technology* 3; EJ Koops, 'Should ICT-regulation be technology-neutral?' in EJ Koops and others (eds), *Starting points in ICT regulation. Deconstructing prevalent policy one-liners* (TMC Asser Press 2006) 77ff.

first reading complies with the approach of the EP and does not contain any reference to technical standards in relation to the principles of data protection by design and by default.

Van Rest et al. point out that the GDPR makes extensive references to the concept of 'Privacy by Design' but fails to specify what exactly is meant under the term, highlighting difficulties in the unravelling of the terms taking into account economic incentives, legacy systems and the lack of adoption of trust of end-users and consumers in privacy by design. They claim that in addition to introducing the principle of privacy by design into the GDPR, designing Privacy by Design should be an ongoing, transparent dialogue between representatives of data subjects and data holders/system designers. It is a *conditio sine qua non* for system designers that the terminology used in the legislation and the privacy requirements become clearer.[30] Standardisation can assist to this end because it will give guidance for the implementation and construction process, as well as when testing the resulting system for compliance.

IV. Security requirements

The Commission Proposal contained a provision on the security of processing of personal data, similar to Article 17 of the Data Protection Directive, establishing an obligation not only for data controllers, but also for data processors, to 'implement appropriate technical and organisational measures to ensure a level of security appropriate to the risks represented by the processing and the nature of the personal data to be protected, having regard to the state of the art and the costs of their implementation' (Art. 30(1) Commission Proposal). Security threats for which such measures shall be taken are 'accidental or unlawful destruction or accidental loss and to prevent any unlawful forms of processing, in particular any unauthorised disclosure, dissemination or access, or alteration of personal data' (Art. 30(2) Commission Proposal). The EP linked the security requirements with the results of data protection impact assessments (Art. 30(1) EP first reading) and specified what shall be included in a security policy (Art. 30(1a) EP first reading) and what shall be achieved as a minimum through the adoption of security measures (Art. 30(2) EP first reading). The EP added a clarification that objectives of security requirements are to 'ensure the ongoing confidentiality, integrity, availability and

[30] J Van Rest, 'Designing Privacy-by-Design', in B. Preneel and D. Ikonomou, *First Privacy Technologies and Policy*, Heidelberg 2014, LNCS 8319, 55–72.

resilience of systems and services processing personal data' (Art. 1a(b) EP first reading). Hon et al. have recommended that the GDPR contain specific reference to examples of security measures: 'express reference . . . to encryption and backups as examples of security measures to protect confidentiality (for the former), and integrity and availability (for the latter) would help to raise awareness regarding these measures. Requirements regarding resilience, if retained, should be defined clearly, but again in general terms'.[31]

The Commission Proposal empowered the European Commission to adopt delegated acts for the aforementioned technical and organisational measures (Art. 30(3) Commission Proposal), as well as implementing acts in order to specify the security requirements 'in particular (a) prevent any unauthorised access to personal data; (b) prevent any unauthorised disclosure, reading, copying, modification, erasure or removal of personal data; (c) ensure the verification of the lawfulness of processing operations' (Art 30(4) Commission Proposal). The European Parliament deleted the powers of the Commission to adopt delegated and implementing acts in relation to security. Although Article 30 of the Commission Proposal did not make any reference to technical standards, Recital 66 stipulated that 'When establishing technical standards and organisational measures to ensure security of processing, the Commission should promote technological neutrality, interoperability and innovation, and, where appropriate, cooperate with third countries'. The wording of Recital 66 reveals the intention of the Commission to be involved in the development and establishment of technical standards with regard to security requirements, something that seems to be supported by the EP because Recital 66 was not amended during the first reading of the Parliament. Although the Parliament removed the Commission's power to lay down technical standards relating to the requirements for the implementation of the data protection by design and by default principles, such technical standards can be adopted as part of security requirements.[32] In the Council's first reading the reference to standards has been removed from the text of Recital 66.

[31] Hon et al., supra note 27.

[32] It should be noted that with regard to security risk analysis and information security management, the state of the art has been defined for several years by the ISO/IEC 270xx standard series. A Habtamu and J Borking, 'Risk analysis methods and practices, privacy risk analysis methodology', Oslo 2012, available at https://www.researchgate.net/publication/259471777_Risk_Analysis_Methods_and_Practices_Privacy_Risk_Analysis_Methodology, accessed 23 July 2015.

V. *Data portability*

The Commission Proposal recognises a right of the data subject to data portability. Under 'data portability', the Commission recognised the right of the data subject to transfer data from one electronic processing system to and into another, without being prevented from doing so by the controller and to receive a copy of their data by service providers in a format that they can further use in other online services. The Commission proposal provides for a general right to obtain a copy of processed data 'in an electronic and structured format which is commonly used and allows for further use by the data subject' (data portability) applicable to all cases 'where personal data are processed by electronic means and in a structured and commonly used format', irrespective of the legal ground based on which the data processing took place (Art. 18(1) Commission Proposal). What is interesting from a technical point of view is that Article 18(2) of the Commission Proposal introduced a *stricto sensu* right to data portability, a separate right – applicable only when the processing is based on the consent of the data subject or on a contract – entitling data subjects to transmit their personal data from an automated processing system (such as a social networking service[33]) into another one in an electronic format which is commonly used. The European Data Protection Supervisor suggested that the right of the data subject to data portability should be extended to all cases of processing of personal data irrespective of the processing ground.[34] The actual impact of the provision has been criticised because it is applicable only when the processing is based on data subject's consent or on contract.[35]

The EP deleted Article 18 and treated it as part of the right to access and to *obtain* data: 'where the data subject has provided the personal data where the personal data are processed by electronic means, the data subject shall have the right to obtain from the controller a copy of the provided personal data in an electronic and interoperable format which is commonly used and allows for further use by the data subject without hindrance from the controller from whom the personal data

[33] Recital 55 Commission Proposal. In the Council Proposal the reference to social networks no longer returns in the text of Recital 55.

[34] European Data Protection Supervisor, Opinion on the data protection reform package (2012), para. 151.

[35] C Cuijpers, N Purtova and E Kosta, 'Data protection reform and the Internet: the draft Data Protection Regulation', in A Savin and J Trzaskowski (eds), *Research Handbook on EU Internet Law* (Edward Elgar 2014), at 558–559.

are withdrawn' (Art. 15(2a) EP first reading). The EP did not make any reference to technical standards. Nevertheless, technical standards would undoubtedly need to be developed to support data portability, even if only the general right to obtain a copy of processed data survives in the final GDPR, as suggested by the EP. During the debates in the Council one of the heated issues of discussion was whether the European Commission would be recognised with the authority to set technical standards relating to data portability. The Council in its final first reading deleted the relevant provision of Article 18(3) and does not make any reference to technical standards in relation to the right to data portability.

VI. Right to object

The right to object is a fundamental right of the data subjects, who can object in specific cases, at any time to the processing of personal data, 'unless the controller demonstrates compelling legitimate grounds for the processing which override the interests or fundamental rights and freedoms of the data subject' (Art. 19(1) Commission Proposal). The data subjects have a right to object to the processing of their personal data when the processing is necessary for the protection of their vital interests (Art. 6(1)(d) Commission Proposal), for the performance of a task carried out in the public interest or in the exercise of official authority vested in the controller (Art. 6(1)(e) Commission Proposal) and for the purposes of the legitimate interests of the data controller (Art. 6(1)(f) Commission Proposal). The EP further specified the right of the data subject to object to processing of personal data that is based on legitimate interests of the data controller, clarifying that 'the data subject shall have at any time and without any further justification, the right to object free of charge in general or for any particular purpose to the processing of their personal data' (Art. 6(2) EP first reading). Although the Commission Proposal does not make any reference to technical standards with regard to the right to object, the EP introduced a special provision about the right to object in the context of the use of information society services, where the right to object 'may be exercised by automated means using a technical standard which allows the data subject to clearly express his or her wishes' (Art. 19(2b) EP first reading). The exercise of the right to object in the context of the use of information society services shall not affect the provisions of Directive 2002/58/EC on the protection of personal data in the electronic communications sector. The development of a technical standard for the exercise of a right to object would harmonise the application of this right and would empower the data subjects in the exercise of their right in the information society. However, the Council in its first reading removed any reference to technical standards in relation to the right to object.

VII. The European Parliament on the processing of data concerning health and the processing of personal data for historical, statistical and scientific research purposes

The European Commission introduced specific provisions for the processing of health data (Art. 81 Commission Proposal). Health data, which are necessary for historical, statistical or scientific research purposes, such as 'patient registries set up for improving diagnoses and differentiating between similar types of diseases and preparing studies for therapies' shall be processed in accordance with the specific provisions of Article 83 for the processing of personal data for historical, statistical or scientific research purposes (Art. 81(2) Commission Proposal). More specifically, Article 83 of the Commission Proposal required that personal data can be used for historical, statistical or scientific research only under two conditions: (a) the use of the personal data is absolutely necessary for the historical, statistical or scientific research purposes, which cannot be fulfilled by processing non-personal data and (b) the identification data ('data enabling the attribution of information to a data subject') are kept separately from the other information as long as this is possible for the fulfilment of the purposes (Art. 83(1) Commission Proposal). The EP further specified this latter condition, requiring that the data are kept 'under the highest technical standards, and all necessary measures are taken to prevent unwarranted re-identification of the data subjects' (Art. 81(2a) EP first reading). The EP inserted an additional requirement that the use of health data for historical, statistical or scientific research purposes, which can be only permitted if the data subject consents to the processing (Art. 81(2) EP first reading). However, the EP allowed Member States to introduce exceptions from the consent requirement for research, when such research serves a high public interest, if that research cannot possibly be carried out otherwise. The EP required that 'the data shall be anonymised, or if that is not possible for the research purposes, pseudonymised under the highest technical standards, and all necessary measures shall be taken to prevent unwarranted re-identification of the data subjects' (Art. 81(2a) EP first reading). The choice of the EP to refer to technical standards in relation to the storage of data and in relation to pseudonymisation and anonymisation procedures[36] is in line with its reference to technical standards with regard to security. The EP has mainly associated the development and implementation of technical standards to the area of security, which is very much in line with the current industry initiatives. The Council in its first reading did not support the approach of the EP. The Council deleted Article 81 and removed from Article 83 and reference to technical standards.

D. Towards the introduction of technical standards in the European data protection legislation

I. The development of data protection standards in Europe

In the broader context of the New Approach to technical standards, the European Commission explored back in 1999 whether there was support by the industry for the Commission to contribute to the development of data protection standards.[37] Because the industry did not warmly support such an initiative, the Commission via CEN/ISSS[38] established the Initiative for Privacy Standardisation in Europe (IPSE), which aimed at investigating whether standardisation could facilitate the implementation of the 1995 Data Protection Directive and, in case of positive outcome, at providing recommendations to that direction.[39]

In the context of this initiative, Jos Dumortier and Caroline Goemans prepared a discussion paper on data privacy and standardisation for the CEN/ISSS Open Seminar on data protection that took place in Brussels in March 2000.[40] After an analysis of the different legal frameworks on data protection in the EU and in the USA, which were found to raise barriers to the international transfers of personal data, and an assessment of the potential role of standardisation in the area of personal data

[36] Article 29 Working Party, Opinion 05/2014 on Anonymisation Techniques, 0829/14/EN WP216, Brussels, 2014.

[37] Winn, supra note 14, 198.

[38] CEN/ISSS (Information Society Standardisation System) was created in mid-1997 by CEN (European Committee for Standardisation) as the focus for its ICT standards activities, according to http://ec.europa.eu/idabc/en/document/6990.html, accessed 23 July 2015. More information available at http://www.cen.eu/isss.

[39] J Dumortier and C Goemans, 'Online data privacy and standardization: towards a more effective protection?', in J. Dumortier et al. (eds.), *A decade of research at the crossroads of law and ICT*, Larcier, Brussels (2001), 53, at 69.

[40] J Dumortier and C Goemans, 'Data privacy and standardization', Discussion paper prepared for the CEN/ISSS Open Seminar on data protection, Brussels, 23/24 March 2000, available at http://www.law.kuleuven.be/icri/en/docs/publications/90cen-paper2f90.pdf, accessed 23 July 2015.

protection, the discussion paper proposed three standardisation activities: (1) the development of a general data protection standard that would set out operational steps that would assist an organisation in complying with the data protection legislation (mainly the Data Protection Directive), (2) the launching of a number of sector-specific initiatives (e.g. eCommerce, human resources management), (3) the development of data protection technical tools, e.g. tools that would facilitate and enhance the controller-processor relationship or for reporting privacy policy and privacy performance to supervisory authorities.[41]

CEN/ISSS published its final report on 13 February 2002 with seven core recommendations on the following issues: identification of a common European set of voluntary best practices for data protection, avoidance to initiate management standards, development of a generic set of contract clauses on security requirements, stock-taking of data protection auditing practices, further exploration of the role of web seals as a basis for considering further standardisation work in this area, development of a coherent approach to the assessment of the impact of technologies on data protection and finally compilation and delivery of educational and guidance material on privacy-related standardisation issues.[42] The report did not result in concrete recommendations for the adoption of technical standards for data protection. Nevertheless, the Article 29 Data Protection Working Party welcomed the CEN/ISSS initiative urging for its continuation and favouring the development of specific recommendations.[43]

Following the publication of the aforementioned report, CEN/ISS continued its efforts on the issues of data protection and standardisation. The CEN/ISSS Data Protection and Privacy Workshop (CEN/ISSS/WS/DPP)[44] announced in 2008 a new work plan with three further areas of work: 'One project would identify a common European set of voluntary "Best Practices" for data protection management; a second would develop privacy audit tools to permit managers to perform self-assessments of their compliance with data protection laws; and a third would establish a "voluntary technology dialogue system" to maintain open lines of

[41] Dumortier and Goemans, supra note 39, 46.

[42] CEN/ISSS, Initiative on Privacy Standardization in Europe, IPSE, Final Report, 13 February 2002, available at http://danskprivacynet.files.wordpress.com/2008/06/ipse_finalreport1.pdf, accessed 23 July 2015.

[43] Article 29 Data protection Working Party, Opinion 1/2002 on the CEN/ISSS Report on Privacy Standardisation in Europe, WP57/30.05.2002.

[44] Available at http://www.cen.eu/cenorm/businessdomains/businessdomains/isss/activity/wsdpp.asp, accessed 23 July 2015.

communication between regulators and private enterprises developing new technologies that may affect data protection rights'.[45] The CEN/ISSS Data Protection and Privacy Workshop produced a number of recommendations, before it was dissolved by the European Commission in 2012.[46]

The importance of technical standards in the area of data protection and privacy has also been highlighted in the Data Protection and Privacy Commissioners Conference. In 2004 the Data Protection and Privacy Commissioners Conference adopted a resolution supporting the development of an effective and unanimously accepted international privacy technology standard[47] followed by a resolution in 2007 on the development of international standards.[48] The Conference expressed its support to the development of effective and universally accepted international privacy standards, focusing especially on the role of ISO in the development of privacy standards. The conference further urged Data Protection and Privacy Commissioners to be 'more actively involved in the ISO standards development process via their respective national standards development organisations'[49] and to promote the participation in such process of other stakeholders, such as academics, non-governmental organisations and research centres.

Recognising this development, the European Commission stated in its 2011 policy document 'A strategic vision for European standards: Moving forward to enhance and accelerate the sustainable growth of the European economy by 2020' that

45 Available at http://www.privacyconference2007.gc.ca/workbooks/terra_incognita_workbook5_e.html, accessed 23 July 2015.

46 CWA 15499–1:2006 Personal Data Protection Audit Framework (EU Directive EC 95/46) – Part I: Baseline Framework; CWA 15499–2:2006 Personal Data Protection Audit Framework (EU Directive EC 95/46) – Part II: Checklists, questionnaires and templates for users of the framework; CWA 15262:2005 Inventory of data protection auditing practices; CWA 15263:2005 Analysis of privacy protection technologies, privacy-enhancing technologies (PET), privacy management systems (PMS) and identity management systems (IMS), the drivers thereof and the need for standardization; CWA 15292:2005 Standard form contract to assist compliance with obligations imposed by article 17 of the Data Protection Directive 95/46/EC (and implementation guide); CWA 16111:2010 Voluntary Technology Dialogue Framework (VTDF); CWA 16112:2010 Self-assessment framework for managers; CWA 16113: 2010 Personal Data Protection Good Practices.

47 Borking, supra note 11, p. 7.

48 29th International Data Protection and Privacy Commissioners' Conference, Resolution on Development of International Standards, 26–28 September 2007, Montreal, Canada, available at http://privacyconference2012.org/wps/wcm/connect/535413004ae86ad89a449aa0fea628d8/2007_M4.pdf?MOD=AJPERES, accessed 23 July 2015.

49 29th International Data Protection and Privacy Commissioners' Conference, supra note 48.

> A European standardisation system supporting the Europe 2020 Strategy for smart, sustainable and inclusive growth will need to meet the following strategic objectives: (1) Standards need to be quickly available – especially but not only – to assure the interoperability between services and applications in the field of information and communications technologies so that Europe can reap the full benefits of ICT. The most relevant ICT standards developed by the ESOs or by global ICT Fora and Consortia, provided that these standards comply with a set of quality criteria, should play a more prominent role in fulfilling public policy objectives and societal needs. It should become possible to use these standards in public procurement, or to facilitate policy making and legislation.[50]

II. Standards, data protection and electronic communications

Technical standards are being gradually more introduced in practice in the context of the processing of personal data in the electronic communications sector. Directive 2002/58/EC (hereafter ePrivacy Directive)[51] regulates specific issues of relating to the protection of personal data and privacy in the electronic communications sector and was amended in 2009.[52] One of the novelties introduced to the ePrivacy Directive was the notification obligation for personal data breaches, which was established in Article 4 of the ePrivacy Directive.

In April 2012 the European Network and Information Security Agency (ENISA) produced recommendations on the technical implementation of the data breach notification, where the importance of technical standards is highlighted.[53] Although the scope of the notification obligation

[50] 'A strategic vision for European standards: Moving forward to enhance and accelerate the sustainable growth of the European economy by 2020', Communication from the Commission to the European Parliament, the Council and the European Economic and Social Committee, COM(2011) 311 final, p. 4.

[51] Directive 2002/58/EC of 12 July 2002 concerning the processing of personal data and the protection of privacy in the electronic communications sector (Directive on privacy and electronic communications) [2002] OJ L201/37.

[52] The 'Citizens' Rights Directive' amended the ePrivacy and the Universal Service Directives: European Parliament and the Council of the European Union, Directive 2009/136/EC amending Directive 2002/22/EC on universal service and users' rights relating to electronic communications networks and services, Directive 2002/58/EC concerning the processing of personal data and the protection of privacy in the electronic communications sector and Regulation (EC) No 2006/2004 on cooperation between national authorities responsible for the enforcement of consumer protection laws ('Citizens' Rights Directive') [2009] OJ L337/11.

[53] ENISA, Recommendations for technical implementation of Art.4, 04.05.2012, available at http://www.enisa.europa.eu/activities/identity-and-trust/risks-and-data-breaches/dbn/art4_tech, accessed 23 July 2015.

for personal data breaches in Europe is currently limited to the electronic communications sector, in light of the changes proposed in the draft General Data Protection Regulation to widen the notification scope, ENISA intends its recommendations to be of relevance for all data breach contexts. The Recommendations propose a systematic approach to personal data breach management, inter alia, by recommending specific steps that the data controller needs to take, with regard to, among others, security requirements, as well as the detection and assessment of the personal data breaches, especially, the methodology to assess the impact and severity of detected personal data breaches. The data breach management approach that is followed in the ENISA recommendations is based on the existing security incident response management procedures, such as ISO standard 27035:2011–09(E), which involves five basic steps: (i) plan and prepare, (ii) detect and assess, (iii) notify and respond, (iv) collect evidence and carry out forensic analysis and (v) review and improve.

III. Cloud computing

Cloud computing has been defined as 'a model for enabling ubiquitous, convenient, on-demand network access to a shared pool of configurable computing resources (e.g. networks, servers, storage, applications, and services) that can be rapidly provisioned and released with minimal management effort or service provider interaction'.[54] Cloud computing is a complex environment, where the adoption of technical standards for data protection certification mechanisms, seals and marks in accordance with the draft GDPR will allow cloud users to assess the level of data protection provided by cloud providers, even when they are not established in the EU. Indeed, the Article 29 Data Protection Working Party has found that 'adoption of privacy-specific standards and certifications is central to the establishment of a trustworthy relationship between cloud providers, controllers and data subjects'.[55] ENISA has compiled a list of different certification schemes, which could be relevant for potential cloud computing customers, the Cloud Certification Schemes List (CCSL).[56] The

[54] P Mell and T Grance, 'The NIST definition of cloud computing – Recommendations of the National Institute of Standards and Technology', September 2011, available at http://csrc.nist.gov/publications/nistpubs/800-145/SP800-145.pdf, accessed 23 July 2015.

[55] Article 29 Data Protection Working Party, Opinion 05/2012 on Cloud Computing, WP196 (01.07.2012), at 22.

[56] Available at https://resilience.enisa.europa.eu/cloud-computing-certification, accessed 23 July 2015.

CCSL provides information about the underlying standards for existing certification schemes.

The analysis of the provisions of the draft GDPR where technical standards are explicitly mentioned has already illustrated the fact that technical standards will play a gradually stronger role in the data protection arena. At the same time, cloud computing is tilting the traditional approaches to data protection, and therefore the use of technical standards in this context will aim at ensuring the data protection rights of the individuals. In 2012 the International Working Group on Data Protection in Telecommunications (IWGDPT) recommended that cloud computing must not lead to a lowering of data protection standards compared with conventional data processing and that further efforts should be put into standardisation (as well as third-party certification, Privacy by Design technologies etc.) for the achievement of a desired level of trust in cloud computing.[57]

Several Standards Developing Organisations (SDOs) have recently started to study cloud computing in relation to the development of information privacy protection standards.[58] In the early days of operation of ETSI's Cloud Standards Coordination initiative,[59] more than forty sources for cloud-related standards were identified, while currently there are talks for more than one hundred such standardisation groups.[60] Organisations involved include formal standards bodies (ETSI, ISO, ITU, BSI, NIST etc.) to fora and consortia, including the Cloud Security Alliance Cloud Controls Matrix (CSA CCM), the Open Certification Framework (OCF), the Payment Card Industry Security Standards Council, IETF, EuroCloud, ODCA, OASIS and the OpenID Foundation.[61] Several of

[57] International Working Group on Data Protection in Telecommunications (IWGDPT), Working Paper on cloud computing – Privacy and data protection issues 'Sopot Memorandum', 24.04.2012, available at http://www.datenschutz-berlin.de/attachments/873/Sopot_Memorandum_Cloud_Computing.pdf?1335513083, accessed 23 July 2015.

[58] S Guilloteau and V Mauree, 'Privacy in cloud computing, ITU-T Technology Watch Report', (*ITU*, March 2012) 14–16, available at http://www.itu.int/dms_pub/itu-t/oth/23/01/T23010000160001PDFE.pdf, accessed 23 July 2015.

[59] Available at http://www.etsi.org/index.php/news-events/news/600-2012-11-cloud-standards-coordination-launch, accessed 23 July 2015.

[60] EP, Cloud Computing, May 2012, at 71, available at http://www.europarl.europa.eu/document/activities/cont/201205/20120531ATT46111/20120531ATT46111EN.pdf, accessed 23 July 2015.

[61] Output Breakout Session #1 Security and Data protection, ETSI Cloud Coordination Standards, Cannes 4–5 December 2012), available at http://docbox.etsi.org/Workshop/2012/201212_CSC/, accessed 23 July 2015. In the final CSC report the work of more than twenty-five cloud-relevant standardizations organizations was considered.

these standardisation groups are dealing with issues of data protection in cloud computing.

The ISO and IEC developed in July 2014 the ISO/IEC 27108 standard 'Information Technology – Security Techniques – Code of Practice for Protection of Personally Identifiable Information (PII) in Public Clouds Acting as PII Processors'. The standard is a practical tool for cloud service providers that function as processors and follows the European terminology, which distinguishes, between controllers and processors. Therefore the scope of the specific standard is limited to specific types of cloud service providers.[62]

However, the involvement of a large number of standardisation bodies to the development of data protection standards in cloud computing environments does not necessarily ensure success in their development and market adoption. The International Telecommunications Unit concluded in its report on data protection in cloud computing: 'Despite growing concern about data protection in cloud computing, there is still a lack of standards. To foster adoption of cloud services, standards will set requirements for the assessment and selection of solutions that meet the expected level of security and data protection. Work on standards is now in progress. Over the last few years, several bodies have been involved in efforts to develop an information data protection standard, and the main SDOs have now begun to study cloud computing'.[63]

One of the main action points of the EU Action plan 'Unleashing the Potential of Cloud Computing in Europe' is 'Cutting through the jungle of technical standards so that cloud users get interoperability, data portability and reversibility'.[64] The necessary standards were to be identified by 2013. Although data protection is not being referred to explicitly in the scope of this action point, the reference to data portability makes clear that this action plan is of obvious importance for this field as well.

To implement this EU action point and to foster cloud computing in the EU, ETSI's Cloud Standards Coordination initiative (CSC) was

[62] For an analysis of the ISO/IEC 27108 standard in relation to the protection of personal data, see Paul de Hert, Vagelis Papakonstantinou, Irene Kamara, supra note 6, p. 6.

[63] Data protection in cloud computing, ITU-T Technology Watch Report, March 2012, at 14.

[64] European Commission, Communication: Unleashing the Potential of Cloud Computing in Europe, Brussels, 27.9.2012, COM(2012) 529 final, available at http://media.loeb.com/eu_cloud.pdf, accessed 23 July 2015.

established upon the request of the European Commission. The aim of the group was to map existing standards relevant for cloud computing. The CSC has been set up open for all stakeholders, ETSI members as well as non-ETSI members, including fora, consortia, trade and users associations. CSC focused on IOP (interoperability, data portability and reversibility), SEC (security and data protection) and SLA (service level agreements) and all documents were shared on a public website.[65] On the basis of their analysis of the development of standards related to cloud computing, Gleeson and Walden conclude that there does not seem to be so much of a standards problem in terms of interoperability, data or application portability. There seems, however, to be a demand for informational and evaluative standards reassuring users about data security and data protection in the cloud.[66] The adoption of the GDPR will actually provide the adequate legal framework for the development of technical standards in the field of cloud computing.

IV. RFID applications

The European Commission has already undertaken the necessary steps for a standard for a Privacy Impact Assessment of RFID applications that can be used as an example for Privacy Impact Assessments of other applications or processes.

In response to the growing deployment of RFID systems in Europe, the European Commission published in 2007 a Communication that proposed 'follow-up steps to overcome barriers to wide take-up to benefit society and the economy while incorporating appropriate privacy, health and environmental safeguards',[67] and one year later, it addressed a Mandate M/436 to the European standards organisations CEN, the European Committee for Electrotechnical Standardisation (CENELEC) and ETSI on data protection, privacy and information aspects of RFID.[68] The work that was executed in two phases resulted in the development

[65] The Final Report of the CSC was delivered in November 2013 (see http://www.etsi.org/news-events/news/734-2013-12-press-release-report-on-cloud-computing-standards).

[66] N Gleeson and I Walden, 'It's a jungle out there'?: Cloud computing, standards and the law', in *European Journal of Law and Information Technology* (2014), 5(2): 21–22.

[67] European Commission, Radio Frequency Identification (RFID) in Europe: steps towards a policy framework {SEC(2007) 312} /* COM/2007/0096 final */, available at http://eur-lex.europa.eu/legal-content/EN/ALL/?uri=CELEX:52007DC0096, accessed 23 July 2015.

[68] C Tételin, RFID and Privacy Impact Assessment (PIA), URSI-France, Journées scientifiques 25/26 mars 2014, 7–14, at available at http://webistem.com/ursi-f2014/output_directory/website/data/articles/000011.pdf, accessed 23 July 2015.

of the European standard EN 16571:2014 within the framework of CEN. This European Standard contains the specification and description of the RFID privacy impact assessment (PIA) process and was published on 25 June 2014.[69]

E. Concluding remarks

This chapter has discussed the role of standards in the European data protection framework in view of the proposed provisions of the GDPR that refer to the development of technical standards. The European Commission in its 2012 proposal vested itself with the power to adopt a large number of delegated and implementing acts that would allow it to set out, among others, technical standards in several areas. Many of the references to delegated and implementing acts were removed from the text of the GDPR later on in the legislative process. The focus of the criticism was that the option of having delegated and implemented acts to be set by the Commission would leave data controllers and data processors uncertain of their obligations and could seriously impede long-term business planning since the rules could be changed at any time. The procedure for the adoption of the GDPR is still ongoing, and thus it is still unclear how the final text of the Regulation will ultimately read, pending further consultations (the so-called trilogue process) between the European Parliament, the Council and the European Commission.

Nevertheless, the analysis of the initial Commission Proposal, the EP first reading and the Council first reading showed a tendency of the EP to explicitly strengthen the reliance on technical data protection standards mainly in the area of security: *stricto sensu* security requirements, use of technical standards to prevent unwarranted re-identification of the data subjects, or in relation to pseudonymisation and anonymisation of personal data. Technical standards are also discussed in relation to personal data marks and the European Data Protection Seal in particular, while technical standards for the exercise of the data subjects' right to object would empower them in the exercise of their right in the information society. The Council in its first reading removed almost all of the direct references to technical standards that were introduced by the EP. However, the lack of explicit reference to technical standards in relation to specific

[69] NEN-EN 16571:2014 en, Information technology – RFID privacy impact assessment process, available at http://www.nen.nl/NEN-Shop-2/Standard/NENEN-165712014-en.htm, accessed 23 July 2015.

data protection issues does not mean that the development of such standards should be avoided, but rather that the initiative for them should be left to the industry. The ENISA recommendations on the technical implementation of personal data breach notification are a prominent example in which technical standards are extremely important in data protection. More concretely the ENISA recommendations are based on existing security incident response management procedures, such as ISO standard 27035:2011–09(E). At the same time, technical standards relating to data protection and privacy have been developed in relation to cloud computing and RFID applications. Moreover, the International Organisation for Standardisation has developed a series of relevant standards that are best practice guidelines in the area of information security,[70] for which there is actually a strong demand in practice and relate to the safeguarding of privacy and data protection.[71] All these initiatives are paving the way to more technical standards in relation to privacy and data protection.

Standards development will have to be fostered and coordinated to obtain the high quality standards needed for a proper framework enhancing compliance with data protection regulation. This is not only of importance for the industry and citizens/consumers who want to make full use of the potential advantages of complex technological environments but also for the industry itself, as data protection has become a sensitive issue for the core business of the information technology sector.

[70] ENISA, 'Security certification practice in the EU – Information Security Management Systems – A case study', 21 November 2013, available at http://www.enisa.europa.eu/activities/identity-and-trust/library/deliverables/security-certification-practice-in-the-eu-information-security-management-systems-a-case-study, accessed 23 July 2015. In August 2014 ENISA published an overview of standards relevant for cloud computing security, available at https://resilience.enisa.europa.eu/cloud-security-and-resilience/Cloudstandards.pdf, accessed 23 July 2015.

[71] According to a 2012 information security breaches survey, 41 percent of the customers of large organisations and 13 percent of small organisations asked companies to comply with a recognised international standards, such as ISO 27001: PriceWaterhouseCoopers, in collaboration with BIS, 'Information security breaches survey – Technical report 2012', available at http://www.pwc.co.uk/en_UK/uk/assets/pdf/olpapp/uk-information-security-breaches-survey-technical-report.pdf, accessed 23 July 2015.

Conclusion: a future research agenda

PANAGIOTIS DELIMATSIS

International standardisation is key for economic growth and innovation. International standard-setting bodies (SSBs) constitute a forum for competition of ideas in which one of them (or a combination thereof) will become, through standardisation, the reference point for production processes and methods globally. Technical merit and non-economic considerations battle for relevance, whereas the relevant institutions try to ensure their smooth functioning within an ever-increasing number of interested parties/stakeholders; as such, clashes of conflicting values can easily lead to a stalemate. Hearsay about lack of representativeness and inclusiveness is not uncommon, and anecdotal evidence about power politics and strategic behaviour that permeate standard-setting processes exists and makes headlines from time to time, as the recent reform of the intellectual property policy of the Institute of Electrical and Electronics Engineers (IEEE) amply demonstrates.[1] SSBs are faced with a daunting task in this regard; they walk the tightrope between identifying technical superiority (best-available technology) and ensuring openness, transparency and effective participation for those countries that find certain SSBs to be largely intractable.

This volume has demonstrated that the standard-setters and stakeholders involved in international standardisation are of mixed nature and, like self-regulators, have a conflict of interest inherent in their functions: they are there to serve the interests of their constituents but also the national interest.[2] In many cases, domestic industries are so deeply convinced of

[1] See R. Lloyd, 'The big patent debate in which Google (for once) is on the sidelines', 23 April 2015, available at: http://www.iam-media.com/Blog/Detail.aspx?g=e3ed24e1-e4a0-4572-864b-5583f4087523 (visited 30 June 2015).

[2] See also the US Trade Agreements Act of 1979, which provides that 'the representation of US interests before any private international standards organization shall be carried out by the organization member.' The latter is defined as 'the private person who holds membership in a private international standards organization.' See 19 U.S.C. 2543, quoted in H. Schepel, 'The Empire's Drains: Sources of Legal Recognition of Private Standardization Under the

the superiority of their standards that they believe they promote the public interest – along with their own – when they strive for the application of standards, which actually reflect characteristics of their own standards. At this level, industry representatives and SSBs participating in the international standard-setting process become the missionaries of a strikingly paradoxical message of globalisation: as we move toward closer integration and interdependence at a global level,[3] societies may prove to be less prepared to abandon long-established practices and important values that have shaped their lives for decades or even centuries. When they discuss and negotiate in a globalised context, they tend to defend them fiercely or to attempt to impose their domestic preferences and values. The ongoing negotiations between the EU and the United States relating to coordination of technical regulations and standard-setting on the two sides of the Atlantic are telling. This probably explains why creating globalised standards or achieving harmonisation of technical regulations is a long-term project and, in some areas, a utopian ideal.[4] Thus, the formula '1+1+1', that is, 'one standard, one test and one conformity assessment procedure accepted everywhere', is not realistic.

This volume is brimful with insights and suggestions for future research relating to various aspects of standardisation. We attempted to classify them in four categories: those relating to standard-setting in general; those relating to standardisation and trade; those regarding standard-setting and development; and, finally, those relating to standardisation and innovation.

A. Standardisation in general

This volume improves our understanding of the importance of standard-setting – be it public-, private- or hybrid-driven, occurring at the supranational, regional or international level – and the processes used, but it also advocates for more comprehensive and holistic research efforts. A more systematic approach toward standardisation research is warranted in view of the importance of technical standard-setting, as de Vries finds

TBT Agreement' in C. Joerges and E.-U. Petersmann (eds), *Constitutionalism, Multilevel Trade Governance and International Economic Law* (Hart, 2011), 397, at 404.

3 D. Lazer, 'Regulatory Interdependence and International Governance', 8:3 *Journal of European Public Policy* (2001) 474.

4 A. Sykes, 'The (Limited) Role of Regulatory Harmonization in International Goods and Services Markets', 1 *Journal of International Economic Law* (1999) 49.

(Chapter 1). This may require viewing standardisation as a distinct academic discipline. This approach would allow for testing certain hypotheses and assumptions that are – sometimes light-heartedly – made with respect to the mechanics, dynamics, actors and values of the standard-setting ecology.

Standardisation research should also expand to analyse new phenomena such as the proliferation of value chains transcending national borders or the creation of standards in the service sector. In the latter area in particular, much 'learning by doing' takes place. More interestingly, due to the importance of public interest, the State may be invited or compelled by the circumstances and the topic (e.g. cybersecurity or health care) to contribute effectively in the preparation of such standards.

Furthermore, standardisation and sustainability remains a largely underexplored area when it comes to areas other than the environment. The broader effects on society and social structures, including consumer protection, are issues that research on standardisation remains to deal with. As Blind suggests, broader questions as to the appropriate methodologies for measuring the impact of standardisation arise, notably when linked to the area of quality infrastructure. Blind rightly emphasises that accreditation and metrology are two areas in which our knowledge is limited, and the same can be said with respect to the interaction of the various components of quality infrastructure.

Additionally, more careful analysis of the different steps that lead to the adoption of standards will need to be undertaken. For instance, the adequate protection of the political dimension in a standard-setting process is a topic that has been neglected in the existing literature. Various contributions in this volume are adamant about this.[5] To mention only one of the manifold possibilities for further research, internal review mechanisms appear to be underdeveloped in various SSBs beyond the national level. For a national SSB, typically an assessment of such mechanisms can be made by the courts. However, at the international level, such an assessment would be virtually impossible.[6] Absent such a possibility, it is for the members of a given SSB to ensure the establishment or maintenance of such a procedural and substantive guarantee within a given institution.

[5] See the chapters by Lindahl, Mattli and Sendon, Arcuri, Delimatsis and Schepel, among others.

[6] Delimatsis (Chapter 5), however, argued that the World Trade Organization can function as an arbiter of the legality of international standardisation processes, including with regard to openness and inclusiveness.

More generally, the politics of standardisation have grown in prominence in recent years. The virtual non-involvement of the State in standard-setting is also a meaningful strategy because regulatees are more interested in buying private rules rather than public rules, leaving the State with no incentive to invest in regulating efficiently.[7] Several contributions to this volume showed that the erosion of State consent in international law and the transfer of power to technocrats does not come without consequences.

The ensuing phenomena, notably those relating to (inter-)institutional dynamics, are in dire need of being mapped and addressed. International standard-setting is emblematic of such challenges. Anecdotal evidence suggests that international standard-setters often behave as a de facto inaccessible club, excluding voices that increasingly seek to be heard in such fora.[8] There are examples demonstrating the political imbalances that create distortions and elitist standards. However, demands for more openness and representativeness that come from emerging economies and other developing countries suggest that the current formation of rule-making in this arena reflects anachronistic international geopolitics and can no longer hold. Power is now more equally distributed among the nations of the world. Such redistribution has to be reflected within international organisations as well, notably those that may have substantial effects on the economic emancipation and welfare of a given country.

At the same time, procedural fairness is not necessarily a panacea; it should not be viewed without regard to the addressees and the structure of the supply chain at issue. In sectors such as information and communications technology, standards are developed independently from any public intervention and to date have reached such a high level of sophistication that it would be increasingly difficult for new players from developing countries with low levels of R&D (public or private) investment to catch up and participate effectively despite the unprecedented diffusion of new technologies globally. New ideas would have to be developed that could ensure knowledge transfer and/or knowledge sharing in fair terms. This would allow for continuous but also socially beneficial and inclusive innovation. As one would expect, we have only scratched the surface regarding the identification of appropriate distributional mechanisms that would

[7] See F. Parisi and V. Fon, *The Economics of Lawmaking* (Oxford University Press, 2009).

[8] See, for instance, the criticism against Global G.A.P.: N. Hachez and J. Wouters, 'A Glimpse at the Democratic Legitimacy of Private Standards. Assessing the Public Accountability of GlobalG.A.P.', 14 *Journal of International Economic Law* 14, 677.

broaden the circle of beneficiaries of creativity and innovation without hampering incentives to innovate.

More generally, whereas the contributions to this volume have shed light on standard-setting processes used in various fora and discussed specific cases of standards development, they also pointed to the need for more empirical research on standard-setting. It appears that, even in hybrid (i.e. public-private) bodies, the distribution of power and influence is uneven. This may have different repercussions and systemic implications depending on whether the standards created remain voluntary or whether they are part of a mandate given to the SSBs by the State (as is the case with the mandates given by the European Commission to the European Committee for Standardisation). The latter aim to become part of future technical regulations that will no longer be voluntary but mandatory requirements to comply with to gain market access.

This may be the result of a considerate attempt by a given State to influence the direction that SSBs take and seek the exploitation of first-mover advantages. For instance, within the EU a more concerted effort is being made to link standard-setting to the broader EU-wide industrial policy.[9] Clearly, a more strategic view of such matters is in its infancy within the EU.[10] The introductory text of the recent EU Standardization Regulation corroborates this view, as it calls on the EU to 'promote bilateral approaches with third countries to coordinate standardization efforts and *promote European standards*, for instance, *when negotiating agreements* or by seconding standardization experts to third countries'[11] (emphasis added). However, the discrepancy between a hierarchical, well-integrated EU-wide standard-setting system and a less integrated industrial policy (which is still very much decided still at the member State level) may put significant hurdles to such plans.

In addition, many times the incorporation of private standards may be made *ex ante* in legislative instruments. In the case where private and public interest intermingle, such as standard-setting in health care or finance, the State is an important stakeholder that may need to be involved more

[9] See also the EU's flagship initiative on industrial policy: European Commission Communication, 'An Integrated Industrial Policy for the Globalization Era', COM(2010) 614 final.

[10] European Commission Communication, 'A strategic vision for European standards: Moving forward to enhance and accelerate the sustainable growth of the European economy by 2020', COM(2011) 311 final.

[11] Ibid, Rec. 6.

actively, in particular with respect to standards closely related to the role of the State such as cybersecurity, privacy or health care.[12] Although the forum where this process takes place plays an important role in the final shape of the rules, future research should empirically clarify the optimal timing, diversity and specificity of input stemming from the stakeholders involved, including consumers, trade unions, public authorities or companies.

B. Standardisation and trade

Many times, standards, no matter how well crafted, can impede trade. This is mainly because standards reflect preferences and values of a given populace that may – and usually do – diverge, thereby inflating compliance costs for companies.[13] Thus international standards would reduce compliance costs and address network externalities and information asymmetries.[14] As a result, international standardisation, both government- and private-sector-driven, soon became the preferred layer of regulatory action.[15] Globalisation vindicates this choice: as global supply chains become increasingly important, international standards only grow in prominence.[16]

Addressing technical barriers to trade became fashionable in recent times on the occasion of mega–preferential trade agreements (PTAs; Transatlantic Trade and Investment Partnership [TTIP] and Trans-Pacific Partnership [TPP]) that are currently being negotiated among the biggest trade powers globally. Good regulatory practice and regulatory convergence through cooperation comes to the forefront time and again. Such discussions should not be viewed in isolation; rather, they take place both *within* and *outside* the WTO. For instance, at the WTO level, discussions on good regulatory practice are held within the TBT Committee,

[12] See also American National Standards Institute (ANSI), 'United States Standards Strategy', 2010.

[13] R. Staiger and A. Sykes, 'International Trade, National Treatment and Domestic Regulation', 40 *Journal of Legal Studies* (2011) 149.

[14] See WTO, *World Trade Report 2005*.

[15] T. Büthe and W. Mattli, 'Setting International Standards – Technological Rationality or Primacy of Power?', 56 *World Politics* (2003) 1; K. Tamm Hallström, *Organizing International Standardization – ISO and the IASC in Quest of Authority* (Edward Elgar, 2004).

[16] World Economic Forum (WEF), *The Global Enabling Trade Report 2012 – Reducing Supply Chain Barriers*, 2012.

whereas international standardisation processes became a central issue in the non-agricultural market access (NAMA) negotiations.[17]

The EU and the US have experimented for years on upstream regulatory cooperation without success, and TTIP has revived hopes for more serious efforts in achieving coherence and compatibility this time around. One of the most ambitious goals of the TTIP negotiations on regulatory issues is the creation of a systematic and institutional transatlantic framework among relevant standard-setting bodies on the two sides of the Atlantic.[18] This will prove to be anything but a walkover; substantial differences exists between the two sides already as far as the definition of an international standard is concerned. In the European Union's view, various standards that the United States considers international (including those used for public policies by reference) are in fact developed by US-based standards bodies and thus not sufficiently representative of non-US interests.[19] Therefore, scholars should work on identifying best practices and bringing forward new ideas, because to date most PTAs have been modelled after the existing multilateral trading system framework, simply limiting the number of beneficiaries. In this respect, regulatory harmonisation should be closely scrutinised by the World Trade Organization (WTO) system to avoid trade diversion and the emergence of new trade barriers.

The world of voluntary standards has evolved to affect more parties globally than initially thought. The early immunity that it enjoyed at the international level due to its non-public nature has eroded after the 'multilateralisation' of the Technical Barriers to Trade (TBT) Agreement. The voluntary character of standards remained, but the advent of the WTO changed the impact of international standards forever. Regulatory convergence and reduction of non-tariff barriers would now pass through the adoption of these standards, which were regarded as authoritative expressions of technical state of the art internationally. The TBT took an orthodox view vis-à-vis international standards: standards improve

[17] WTO, Negotiating Group on Market Access, 'Market Access for Non-Agricultural Products – International Standards in Support of Trade and Economic Development: Strengthening the Contribution of the Committee Decision', Proposal from the United States, TN/MA/W/141, 29 March 2011 (referring to the TBT Committee Decision relating to the development of international standards, see below, Section B).

[18] See European Commission, 'EU-US Transatlantic Trade and Investment Partnership: Technical Barriers to Trade – Initial EU position paper', 2013.

[19] See CEN/CENELEC, 'Position Paper on EU-US Transatlantic Trade and Investment Partnership (TTIP) – Technical Barriers to Trade – Initial EU Position Paper', September 2013.

product efficiency and facilitate trade, notably when adopted at the international level, as compliance costs are reduced. Therefore, adherence to international standards becomes a necessary condition for the very attainment of the TBT objectives.

This explains why the WTO, a generally reluctant international organisation regarding the use of non-WTO material to assess the WTO compliance of a given Member, is more lenient vis-à-vis the output of ISSBs. Much of previous WTO case law, most prominently, cases such as *EC – Hormones* and *EC – Sardines*,[20] exemplify this deferential approach. In both cases, non-consensual international standards were considered as relevant benchmarks for assessing WTO compatibility of national measures. For a consensual organisation such as the WTO, in which the legacy of consensus is one of the overarching legitimating artefacts of the multilateral trading system, this is quite extraordinary.

Importantly, no grandfathering would be accepted with respect to output of ISSBs: as established by the WTO Appellate Body in *EC – Sardines*, even standards adopted *before* the creation of the WTO would be considered to assess the consistency of the regulatory instruments of a given WTO member with the TBT. Non-compliance with these mandatory benchmarks for domestic technical regulations would raise suspicions as to the good intentions of governmental intervention.

The much-needed reform in international SSBs and the International Organisation for Standardisation (ISO) was the result of developments within the TBT Committee, most notably the adoption of the TBT Committee Decision in 2000. Analysis of this Decision and its potential impact has been surprisingly scarce to date. It is the first time in the WTO that the delegation of regulatory power to ISSBs was made conditional to adherence to a series of principles, mainly of procedural due process nature. This Decision was vested with substantial legal value after the *US – Tuna II* ruling, showing the potential bite of WTO adjudication as an *ex post* legitimating device for international standardisation: output by standards organisations will from now on be critically reviewed and perhaps discarded if it fails to satisfy the principles set out in the Decision.

One of the topics for future research on the relationship between WTO law and standard-setting consists of analysing the substantive scope of

[20] Nowadays, mention is exclusively made of the European Union and no longer of the European Community, as a result of the entry of the Treaty of Lisbon into force. See Article 1 para. 3 of the European Union Treaty.

these principles but also the possibility of additional principles[21] against which international standards should be juxtaposed. More empirical research is needed to develop analytical tools and heuristic devices that would objectively assess the mechanics of international standard-setting and, if needed, make the case for institutional reform within international SSBs with a view to increasing deliberation, representativeness, openness, transparency, due process and accountability without jeopardising economic and technological progress. New methods measuring compliance with such principles would be needed to assess the performance of SSBs and feed into their regular reform mechanisms. Such methods would necessarily have both quantitative and qualitative attributes.

Future research should put more emphasis as to whether technocratic (or output) legitimacy may be sufficient to discharge ISSBs such as ISO or the FSB of the obligations that they have as global standard-setters. It is telling that, in certain cases, public benefits were by-products rather than the outcome of a genuine benevolent efforts of the standardisation initiative.[22] ISO has undertaken serious good-faith efforts to address complaints regarding its standard-setting processes, notably with respect to participation, relevance and coherence. Its observer status at the TBT Committee also helped getting the pulse of those WTO Members, developing countries for the most part, which believe that access to ISO standard-setting is still intractable.

Such challenges may not only relate to ISO but to Codex Alimentarius and other international standard-setters as well. The cases of ractopamine and ISO 26000 discussed in this volume are indicative of the problems and challenges that international SSBs, both private and public, currently face. Global dynamics alter the map of global rule-making and interdependence is more fragile than ever. For instance, what would happen if the newly emerging economies that start having significant monopsony power experiment with the creation of their own standards that diverge from ISO standards, based on arguments of effective exclusion within ISO? This could have undesirable consequences for all ISO members and everyday business and trade. Further research on the functioning of international SSBs would be necessary to shed light on best practices and policies as well as procedural deficiencies that perpetuate an unbalanced standard-setting landscape at the international level.

[21] See the European Parliament Resolution of 21 October 2010 on the future of European standardisation (2010/2051(INI)), paras. 8–9.

[22] See Marchetti's contribution in this volume (Chapter 6).

C. Standardisation and development

The costs for choosing this locus of standardisation may be minimal or substantial, depending on the level of sophistication and the available resources that the relevant firms display. Active participation in standard-setting requires significant funding that not all firms may be able to dedicate. But this is changing for the most advanced parts of the developing world. The emergence of new players in the international scene seeking to shape rule-making in international affairs more actively has played no less a significant role in increasing the impact of international standard-setting.

Previously run, in essence, by the developed world, the international standardisation community attracted the interest of an ever-increasing number of parties and thus had to become more inclusive without, however, putting into significant jeopardy its effectiveness with respect to consolidating technological advances. Strengthening procedural rights is the gateway to more inclusive and representative international standards. In turn, only higher levels of participation can reduce the knowledge gap between developed and developing countries. However, inclusiveness comes at a cost, and only empirical research can showcase the costs and benefits of such an endeavour, perhaps demonstrating the need for a more tailor-made approach and a better division of tasks, as the contribution by Bijkmakers and van Calster (Chapter 12) has shown.

The consumer would be a net winner due to this development: economic theory would suggest that an international standard reduces consumer costs, as information becomes more readily available and prices more readily comparable.[23] However, the consumer in today's world has a dual role to play: not only the rule of the *final* user who buys a product embodying an international standard, but increasingly the role of the *stakeholder* in shaping that standard. Finding innovative ways to sensitise and subsequently allow for consumers' active involvement in standard-setting remains one of the pressing challenges for future transnational decision-making.

Consumer's participation is either impeded in international SSBs[24] or loosely organised in a way that also raises concerns of representation. New virtual platforms should allow for more concentrated expression of the

[23] See WTO, World Trade Report 2012, p. 136.

[24] See Marx and Wouters' contribution in this volume (Chapter 10).

willingness of the consumers. However, when it comes to the developing world, such a change inevitably goes through domestic standard-setting processes as well.

Furthermore, one of the key lessons to draw is that technical standardisation has peculiarities that need to inform any proposal for change and reform. Indeed, convergence may in theory be desirable, and developing countries may draw useful lessons as to their domestic institutional structures, but it may not be beneficial across the board. One-sided regulatory harmonisation ('race to the top') needs to be well considered before implemented; coupling it with capacity-building efforts may be necessary, but still insufficient. The contribution by Cadot et al. (Chapter 11) sends a powerful message that harmonisation is meaningless if it creates trade opportunities while destroying previous trade opportunities. Just as with most research areas relating to development, more empirical work is needed to offer additional, more conclusive insights as to the interrelationships among initiatives, strategies and practices. This applies to promotion of CSR through investment as well.

D. Standardisation and innovation

As suggested by Lerner and Tirole,[25] SSBs perform three essential functions. The first is technical, as it relates to the discovery or engineering – the learning about and certification of the value of functionalities. The second function relates to standardisation, that is, steering market expectations towards a particular technology, as the SSB typically selects one among several options. This is the point where standard-setting and intellectual property meet; certain patents that are critical for the standard to produce its beneficial effects become standard-essential patents (SEPs). The third function of SSBs, the regulation function, results from the second and aims, inter alia, to diffuse the benefits of the standardisation function rather than allow for monopoly power of any of its members, notably SEP holders. For this reason, among other rules, international SSBs adopt rules relating to intellectual property rights – patents in particular. SEPs are typically to be found in the area of telecommunications and consumer electronics products, but in other sectors as well. SEPs become important due to the high level of interconnectedness or interoperability in these sectors.

[25] J. Lerner and J. Tirole, 'Standard-Essential Patents', 123:3 *Journal of Political Economy* (2015), 547.

Patents have traditionally been regarded as incentives to more innovation.[26] Much of the current innovative advances are in fact due to the patent system, which, through various means such as disclosure requirements when registering patents, has allowed for cumulative innovation, innovation that builds on prior generations of related technologies.[27] More recently, the increased number of bad patents and the sometimes dubious strategies of non-practising entities; the ever-increasing number of different patents incorporated in a single product; the ownership fragmentation; and allegations of patent holdup were accused of raising transaction costs, stifling innovation and ultimately draining consumer welfare.[28]

For example, Gandal and Régibeau (Chapter 16) as well as Larouche and Van Overwalle (Chapter 15) do underline the importance of more careful empirical work on standards-related environments to understand the drivers and dynamics of innovation in rapidly changing sectors. Indeed, we currently lack solid, conclusive, large-scale evidence relating to the behaviour, strategies and practices of some of the most important private-driven SSBs globally. Most of the existing work is one-sided and mono-disciplinary. This is an important gap in the standardisation-related scholarship and a promising field for interdisciplinary research.

Another important area for future research at the intersection between standardisation and innovation is the issue of dynamic standards design and how law should confront the phenomena and strategic interactions prevalent in – often competing – standard-setting fora. For instance, there seems to be ample room for research based on principles such as equity, fairness or transparency.

[26] This approach has been based on the premise that one product incorporate a single patent. See also M. Lemley and C. Shapiro, 'Patent Holdup and Royalty Stacking', 85 *Texas Law Review* (2007), 1991.

[27] See A. Galasso and M. Schankerman, 'Patents and Cumulative Innovation: Causal Evidence from the Courts', 130:1 *Quarterly Journal of Economics* (2015), 317. The US Supreme Court put this in an eloquent way in *KSR International Co.* v. *Teleflex Inc*, 550 U.S. 398 (2007):

> We build and create by bringing to the tangible and palpable reality around us new works based on instinct, simple logic, ordinary inferences, extraordinary ideas, and sometimes even genius. These advances, once part of our shared knowledge, define a new threshold from which innovation starts once more. And as progress beginning from higher levels of achievement is expected in the normal course, the results of ordinary innovation are not the subject of exclusive rights under the patent laws.

[28] See M. Frakes and M. Wasserman, 'Does the U.S. Patent and Trademark Office Grant Too Many Bad Patents?: Evidence from a Quasi-Experiment', 67 *Stanford Law Review* (2015), 613.

Finally, additional research, both analytical and empirical, is warranted in the area of technical standardisation and the protection of privacy in the digital age. Learning by doing appears to be a central feature of this rapidly growing sector, as evidenced by the pre-adoption phase of the general data protection regulation within the EU. These are not exclusively EU-specific concerns; large-scale data would need to be managed, handled with care and safety and instantly available anywhere around the world. Issues associated with 'big data' will certainly keep scholars busy in multiple disciplines, including law, economics, ethics, computer science and engineering.

INDEX

Lightning Source UK Ltd.
Milton Keynes UK
UKOW06n1801250216

269128UK00004B/44/P

9 781107 128330